T0331438

miRNAomics and Stress Management in Plants

Plants, being sessile, are negatively affected by the biotic and abiotic (environmental) stresses, reducing plant growth, productivity, and quality to a larger extent. Plants have evolved different physiological, biochemical, and molecular mechanisms to combat these stress conditions to maintain their growth, development, and productivity. Understanding the mechanisms involved in the plant response to stress conditions is the foremost step in the development of stress-tolerant plants. One of the important adaptations to stress conditions is the precise and fine regulation of gene expression in both time and space. Although gene regulation occurs at different levels through different mechanisms, the most crucial is at the level of transcription. One of the important post-transcriptional gene regulatory mechanisms used by the plants to restore and maintain cellular homeostasis during the stress conditions are microRNAs.

MicroRNAs, a group of approximately 22-nucleotide-long, non-coding RNAs, have recently been identified as a new class of regulators governing gene expression at the post-transcriptional level. MiRNAs can silence genes either by guiding the degradation of the target mRNAs or by repressing the mRNA translation. Plant miRNAs have been demonstrated to regulate many genes involved in various developmental processes, for example, auxin signaling, organ polarity/radial patterning, developmental transitions, and secondary metabolism regulation. Moreover, increasing evidence indicates the role of plant miRNA-guided gene regulation in response to biotic and abiotic stresses. High-throughput sequencing approaches have significantly elucidated the identification and functional characterization of numerous miRNAs in plants. Understanding the role and mechanism of action of miRNAs during abiotic and biotic stresses can potentially offer new approaches to improve plant growth and productivity.

This unique book covers the different aspects of plant microRNAomics including the discovery, biogenesis, role in different stress conditions, and applications of microRNAs in developing stress-tolerant plants. Chapters cover the updated knowledge in the field of plant microRNA research. The book, *miRNAomics and Stress Management in Plants*, intends to demonstrate the breadth of research and the significant advances that have been made in understanding the role of miRNAs in the plant development and stress management.

This comprehensive volume will be of value to plant physiologists, plant biochemists, geneticists, molecular biologists, agronomists, environmental researchers, and graduate and undergraduate students of plant science.

miRNAomics and Stress Management in Plants

Edited By

Peerzada Yasir Yousuf
Government Degree College Anantnag, Jammu and Kashmir, India

Peerzada Arshid Shabir
Government College for Women, M. A. Road, Srinagar
Jammu and Kashmir, India

Khalid Rehman Hakeem
Professor at King Abdul-Aziz University, Jeddah, Saudi Arabia

CRC Press
Taylor & Francis Group
Boca Raton London New York

CRC Press is an imprint of the
Taylor & Francis Group, an **informa** business

Cover illustration: © Shutterstock

First edition published 2025
by CRC Press
2385 Executive Center Drive, Suite 320, Boca Raton, FL 33431

and by CRC Press
4 Park Square, Milton Park, Abingdon, Oxon, OX14 4RN

© 2025 selection and editorial matter, Peerzada Yasir Yousuf, Peerzada Arshid Shabir, and Khalid Rehman Hakeem; individual chapters, the contributors

CRC Press is an imprint of Taylor & Francis Group, LLC

Library of Congress Cataloging-in-Publication Data
Names: Yousuf, Peerzada Yasir, editor. | Shabir, Peerzada Arshid, editor. |
   Hakeem, Khalid Rehman, editor.
Title: MiRNAomics and stress management in plants / edited by Peerzada Yasir Yousuf, Peerzada Arshid Shabir, Khalid Rehman Hakeem
Description: First edition | Boca Raton, FL : CRC Press, 2025 | Includes bibliographical references and index
Identifiers: LCCN 2024014436 (print) | LCCN 2024014437 (ebook) |
   ISBN 9781032164175 (hardback) | ISBN 9781032164182 (paperback) |
   ISBN 9781003248453 (ebook)
Subjects: LCSH: Plants—Effect of stress on—Genetic aspects. | MicroRNA. |
   Plant genetic regulation.
Classification: LCC QK981 .M57 2025 (print) | LCC QK981 (ebook) |
   DDC 581.7/88—dc23/eng/20240617
LC record available at https://lccn.loc.gov/2024014436
LC ebook record available at https://lccn.loc.gov/2024014437

ISBN: 9781032164175 (hbk)
ISBN: 9781032164182 (pbk)
ISBN: 9781003248453 (ebk)

DOI: 10.1201/9781003248453

Typeset in Times
by Apex CoVantage, LLC

Contents

Preface ix
Editor Biographies xi
List of Contributors xiii

Chapter 1 Introduction to Plant miRNAs.. 1

*Nasir Aziz Wagay, Shah Rafiq, Amanulla Khan, Abdul Hadi, Aabida
Ishrath, Peerzada Yasir Yousuf, and Peerzada Arshid Shabir*

1.1 Introduction 1
1.2 Historical Perspective 2
1.3 miRNA Biogenesis... 2
1.4 The miRNA Gene Families 3
1.5 The Origin of Plant miRNA Genes 4
1.6 Evolution of Plant miRNAs 5
1.7 miRNA Gene Clusters (The Formation of miRNA Gene Families)................ 6
1.8 Functions of miRNA.. 6
    1.8.1 Role of miRNAs in Signal Transduction 7
    1.8.2 Role of miRNAs in Gene Regulation 8
    1.8.3 miRNA Role in miRNA and siRNA Biogenesis and Function........... 9
    1.8.4 miRNA Role in Developmental Processes........................ 10
    1.8.5 Role in Stress 10
1.9 Artificial miRNAs (amiRNAs)... 11
1.10 Conclusion and Future Prospects 11
References 12

Chapter 2 Plant miRNAs: Biogenesis, Mode of Action, and Their Role 20

*Bipin Maurya, Lakee Sharma, Nidhi Rai, Vishnu Mishra, Ashish Kumar,
and Shashi Pandey Rai*

2.1 Introduction 20
2.2 Origin and Evolution of Plant miRNAs....................................... 21
2.3 Comparison Between Animal and Plant miRNAs......................... 23
2.4 miRNA Gene Transcription.. 24
2.5 Post-Transcriptional Control of miRNA Biogenesis.................... 24
2.6 Formation and Function of miRNA Effector Complex 27
2.7 Translational Repression 27
2.8 Role of miRNAs in Plant Development.. 28
2.9 Concluding Remarks 30
2.10 Future Prospects... 30
    2.10.1 Acknowledgement... 31
References 31

Chapter 3 miRNAs and Plant Development.. 36

*Shah Rafiq, Nasir Aziz Wagay, Abdul Hadi, Aabida Ishrath,
and Zahoor Ahmad Kaloo*

3.1 Introduction 36

3.2   Role of miRNAs in Plant Growth and Development.........................36
        3.2.1   Role in Vegetative Development.....................................37
        3.2.2   Role of miRNA in Shoot Meristem 37
        3.2.3   Role of miRNAs in Leaf Development 37
        3.2.4   Role of miRNAs in Vascular Development....................41
        3.2.5   miRNAs and Reproductive Development 42
        3.2.6   Organ Development 44
        3.2.7   Role in Other Processes 45
3.3   Interaction Between Plant Hormones and miRNAs....................46
3.4   Conclusion and Future Perspectives 47
References...48

Chapter 4   Dynamic Function of miRNAs in Sensing and Signaling Nutrient
             Stress in Plants 55

             Samina Mazahar, Yasheshwar, and Shahid Umar
4.1   Introduction...55
4.2   Macronutrient Sensing and Signaling in Plants as Regulated
        by miRNAs 55
        4.2.1   miRNAs Responsive to Nitrogen (N) Uptake
                and Transport 55
        4.2.2   miRNAs in Maintaining Phosphorus (P) Homeostasis...58
        4.2.3   miRNA Responsive to Sulphur....................................59
4.3   Micronutrients Sensing and Signaling as Regulated
        by miRNAs 61
4.4   Conclusion 61
References...61

Chapter 5   Salt-Stress-Responsive Plant miRNAs 65

             Sajad Hussain Shah, Shaistul Islam, Zubair Ahmad Parrey,
             and Firoz Mohammad
5.1   Introduction...65
5.2   Roles of miRNAs in Growth and Development of Plants
        Under Diverse Conditions...66
5.3   miRNAs and Mitigation of Salt Stress in Plants 67
        5.3.1   miRNAs Regulate Target Genes Associated
                With Salt Stress...69
5.4   Conclusion and Future Prospects..71
References...72

Chapter 6   Heavy Metal-Regulated miRNAs...76

             Zubair Ahmad Parrey, Shaistul Islam, Sajad Hussain Shah,
             and Firoz Mohammad
6.1   Introduction...76
6.2   Biogenesis of miRNAs..77
6.3   Role of miRNAs in Plants...77
6.4   miRNAs Mediated Phytohormones Signalling Under HM Stress.....78
        6.4.1   miRNA Target Genes Under HM Stress........................79
6.5   Conclusion 83
References...83

Chapter 7 Tolerance to Radiation Stress in Plants With Reference to microRNAs 88

*Sumira Malik, Shilpa Prasad, Rahul Kumar, Shristi Kishore,
and Nitesh Singh*
7.1   Introduction 88
7.2   Different Types of Plant miRNAs Involved in Managing
       Radiation Stress 89
7.3   Biogenesis of miRNAs 90
7.4   Mechanism of Action of Plant miRNAs in Response
       to Radiation Stress 91
7.5   Involvement of Computational Tools in This Field 94
7.6   Summary and Perspectives 94
References 94

Chapter 8 miRNAs and Plant-Pathogen Interactions 97

Bipin Maurya, Vishnu Mishra, and Shashi Pandey Rai
8.1   Introduction 97
8.2   Role of miRNAs in Plant-Pathogen Interactions 99
       8.2.1   Role of miRNAs in *Arabidopsis* During Pathogen Interaction ...100
       8.2.2   Role of miRNAs in the Solanaceae Family During Pathogen
               Interaction 101
       8.2.3   Role of miRNAs in Rice During Pathogenesis 101
       8.2.4   Involvement of miRNAs in *Pulwonia Species*
               During Pathogenesis 109
8.3   Some Other miRNAs During Plant-Pathogen Interaction........... 109
8.4   Use of In Silico Approaches in Advancing Plant miRNA Research.............111
8.5   Biotechnological Applications of Novel Technologies 112
8.6   Conclusion and Future Perspectives 112
       8.6.1   Acknowledgements 113
       8.6.2   Author Contributions 113
       8.6.3   Conflict of Interest 113
References 113

Chapter 9 Plant Response to Bacterial Infections: miRNAomics Approach 119

Sumira Malik, Shristi Kishore, Nitesh Singh, and Rahul Kumar
9.1   Introduction 119
9.2   miRNA Detection Methods 119
9.3   Bacterial Pathogens and miRNAs 119
9.4   Defensive Roles of Plant miRNAs in Response to Bacterial
       Infections.. 121
       9.4.1   Application of miRNAomics in Plant's Response
               to Bacterial Infections.. 123
9.5   Conclusions and Future Prospective 123
References 124

Chapter 10 miRNAs: A Novel TARGET for Improving Stress Tolerance
in Plants Using Transgenics 127

C. Deepika, S. R. Venkatachalam, A. Yuvaraja, and P. Arutchenthil
10.1   Introduction 127

10.2   miRNA – A Novel Transgene Candidate...127
10.3   Strategies for Developing miRNA Based Transgenic Plants.........................128
        10.3.1   miRNA as Positive Regulators on Targets 128
        10.3.2   miRNA as Negative Regulators on Targets......................................129
        10.3.3   Artificial miRNAs 129
10.4   miRNAs and Stress Response in Transgenic Plants....................................129
        10.4.1   Biotic Resistance by miRNA...130
        10.4.2   Abiotic Stress Resistance Transgenics by miRNA..........................132
10.5   Databases for Identification, Functional Characterization, and Target
        Prediction 139
10.6   Conclusion and Future Prospects..141
References...141

Chapter 11 Engineering Stress Tolerance in Plants Using miRNAomics Approach:
               Challenges and Future Perspectives 148

               Mubeen Fatima, Safdar Hussain, Sidqua Zafar, and Noureen Zahra
11.1   Introduction 148
11.2   Biosynthesis of miRNA in the Plants 149
11.3   miRNA Application for Environmental Stress Tolerance in Plants 150
11.4   Functional Analysis and Computational Tools of miRNAs 151
11.5   miRNA-Based Strategies for Plant Improvement..152
        11.5.1   Case Studies...153
11.6   Challenges and Future Prospects...155
References...156

Index...161

Preface

Plants, being sessile, are negatively affected by the biotic and abiotic (environmental) stresses, reducing plant growth, productivity, and quality to a larger extent. Plants have evolved different physiological, biochemical, and molecular mechanisms to combat these stress conditions to maintain their growth, development, and productivity. Understanding the mechanisms involved in the plant response to stress conditions is the foremost step in the development of stress-tolerant plants. One of the important adaptations to stress conditions is the precise and fine regulation of gene expression in both time and space. Although the gene regulation occurs at different levels through different mechanisms, the most crucial being at the level of transcription. One of the important post-transcriptional gene regulatory mechanisms used by the plants to restore and maintain cellular homeostasis during the stress conditions are MicroRNAs.

MicroRNAs, a group of approximately 22-nucleotide-long, non-coding RNAs, have recently been identified as a new class of regulators governing gene expression at the post-transcriptional level. MiRNAs can silence genes either by guiding degradation of the target mRNAs or by repressing the mRNA translation. Plant miRNAs have been demonstrated to regulate many genes involved in various developmental processes, for example, auxin signaling, organ polarity/radial patterning, developmental transitions and secondary metabolism regulation. Moreover, increasing evidences indicates the role of plant miRNA-guided gene regulation in response to biotic and abiotic stresses. High-throughput sequencing approaches have significantly elucidated the identification and functional characterization of numerous miRNAs in plants. Understanding the role and mechanism of action of miRNAs during abiotic and biotic stresses can potentially offer new approaches to improve plant growth and productivity.

miRNAomics and Stress management in Plants is a thoughtful, provocative, and up-to-date volume that covers the different aspects of plant microRNAs, including their discovery, biogenesis, role in different stress conditions, and their applications in developing stress-tolerant plants. Chapters cover the updated knowledge in the field of plant microRNA research.

With scant knowledge and literature available on the subject, the intended volume should be quite beneficial for the audience in understanding the role of microRNAs in tolerance of a different biotic and abiotic stresses in plants and the role they might play in the development of smart stress-tolerant plants through biotechnological interventions.

This volume should be quite useful for the readers in understanding the microRNAs, their biogenesis, mode of action, and their role in the tolerance of both biotic and abiotic stresses.

The book is an attempt at teaming up with global leading research experts working in the field of plant molecular stress physiology to compile an up-to-date and wide-ranging volume. The main purpose of this book is to present information on the most recent developments including the different modern approaches and methodologies that are being currently employed in the field of plant stress tolerance. We trust that this comprehensive volume will familiarize the readers with the detailed mechanisms of miRNAs and its regulation and their roles in different plant stresses.

Editor Biographies

Dr. Yasir Yousuf Peerzada (PhD, MSc, CSIR-NET, ARS-NET, SET, GATE)

Dr. Peerzada Yasir Yousuf earned his BSc from the University of Kashmir in 2008 and MSc and PhD degrees in Botany from Jamia Hamdard, New Delhi, in 2010 and 2016 respectively. He worked in Molecular Ecology Lab as a senior research fellow in the Department of Botany Jamia Hamdard, New Delhi. He taught environmental science as guest faculty in Jamia Hamdard and in different faculties like the Faculty of Unani Medicine, Faculty of Information Technology, etc. In his PhD program he worked on the biochemical and molecular aspects of stress tolerance in plants. Dr. Yousuf has qualified for many competitive examinations like UGC-CSIR NET (AIR 39), ARS-NET, GATE (AIR 45), and JK-SET. He completed a certification course on mitochondria from Harvard University, United States. He also pursued NPTEL online certification course on "Bioenergetics of Life Processes" from Indian Institute of Technology (IIT) Kanpur with Gold + Elite category and was among toppers (Score 91%) besides pursuing an NPTEL online certification course on "Introduction to Proteomics" from the Indian Institute of Technology (IIT) Bombay with Elite category (Score 71%). He was selected for the SERB National Post-Doctoral fellowship (N-PDF) by the Department of Science and Technology, Government of India, in 2016.

Dr. Yousuf has a research experience of 6 years and teaching experience of 11 years. Currently, he holds the position of assistant professor of botany in the Department of Higher Education Jammu and Kashmir. He has published several research articles and book chapters in globally reputed journals like *Frontiers in Plant Science*, *Genes*, *Plos One*, *Protoplasma*, *Plant Growth Regulation*, *Environmental Science and Pollution Research*, etc. Dr. Yousuf has two edited books to his credit. He has presented his work at several national and international conferences. He has delivered extension lectures in different fields of plant sciences at different institutes. He is a lifetime member of different national and international organizations besides being the recipient of some fellowships. He also serves as a member of the editorial board and reviewer of many research journals. He is actively engaged in the research fields of plant physiology and molecular ecology.

Dr. Arshid Shabir Peerzada (PhD, MSc, CSIR-NET)

Dr. Peerzada Arshid Shabir received his PhD in the field of plant reproductive ecology and genetic diversity and is presently working as an assistant professor in the Department of Higher Education, Jammu and Kashmir, India. He is actively engaged in the teaching of plant reproductive ecology, plant physiology, cell and molecular biology, and genetic engineering of plants. He has published many research papers, review articles, and book chapters in reputed international journals. He has also been a reviewer to many reputed international journals and has presented many papers in various national and international conferences. His current research interests are to study the genomic bases of phenotypic plasticity and the study of the genomic bases of adaptation and speciation. Dr. Peerzada Arshid Shabir has been conferred with junior and senior research fellowships by Counsel of Scientific and Industrial Research, New Delhi. He has ten years of research experience and four years of teaching experience.

Khalid Rehman Hakeem (PhD, FRBS)

Khalid Rehman Hakeem is a professor at King Abdul-Aziz University, Jeddah, Saudi Arabia. After completing his doctorate (botany; specialization in plant eco-physiology and molecular biology) from Jamia Hamdard, New Delhi, India, he worked as a lecturer at the University of Kashmir, Srinagar, India, for a short period. Later, he joined Universiti Putra Malaysia, Selangor, Malaysia, and worked there as a post-doctorate fellow and fellow researcher (associate [rofessor) for several years. Dr. Hakeem has more than ten years of teaching and research experience in plant eco-physiology, biotechnology and molecular biology, medicinal plant research, plant-microbe-soil interactions, as well as in environmental studies. He is the recipient of several fellowships at both national and international levels. He has served as a visiting scientist at Jinan University, Guangzhou, China. Currently, he is involved with several international research projects with different government organizations. To date, Dr. Hakeem has authored and edited more than 35 books with international publishers, including Springer Nature, Academic Press (Elsevier), and CRC Press. He also has to his credit more than 95 research publications in peer-reviewed international journals and 55 book chapters in edited volumes with international publishers. At present, Dr. Hakeem serves as an editorial board member and reviewer for several high-impact international scientific journals from Elsevier, Springer Nature, Taylor and Francis, Cambridge, and John Wiley Publishers. He is included in the advisory board of Cambridge Scholars Publishing, UK. He is also a fellow of the Plantae group of the American Society of Plant Biologists, member of the World Academy of Sciences, member of the International Society for Development and Sustainability, Japan, and member of the Asian Federation of Biotechnology, Korea. Currently, Dr. Hakeem is engaged in studying the plant processes at eco-physiological as well as molecular levels.

Contributors

P. Arutchenthil
Department of Genetics and Plant Breeding
Horticultural College and Research Institute for
    Women
Tamil Nadu Agricultural University

C. Deepika
Tapioca and Castor Research Station
Tamil Nadu Agricultural University
Salem, Tamil Nadu, India

Mubeen Fatima
Centre for Applied Molecular Biology (CAMB)
University of the Punjab
Lahore, Pakistan

Abdul Hadi
Department of Botany
University of Kashmir, Hazratbal Srinagar
Jammu and Kashmir, India

Safdar Hussain
Centre for Applied Molecular Biology (CAMB)
University of the Punjab
Lahore, Pakistan

Aabida Ishrath
Department of Botany
University of Kashmir, Hazratbal Srinagar
Jammu and Kashmir, India

Shaistul Islam
Plant Physiology and Biochemistry Section
Department of Botany
Aligarh Muslim University
Aligarh, India

Zahoor Ahmad Kaloo
Department of Botany
University of Kashmir, Hazratbal Srinagar
Jammu and Kashmir, India

Amanulla Khan
Anjuman Islam Janjira Degree College of Science
Lokmanya Tilak Road, Murud – Janjira
Raigad, Maharashtra, India

Shristi Kishore
Amity Institute of Biotechnology
Amity University Jharkhand
Ranchi, Jharkhand, India

Ashish Kumar
Biotechnology Division
Central Institute of Medicinal and Aromatic
    Plants (CIMAP)
Lucknow (Uttar Pradesh), India

Rahul Kumar
Amity Institute of Biotechnology
Amity University Jharkhand
Ranchi, Jharkhand, India

Sumira Malik
Amity Institute of Biotechnology
Amity University Jharkhand
Ranchi, Jharkhand, India

Bipin Maurya
Laboratory of Morphogenesis
Department of Botany, Institute of Science
Banaras Hindu University
Varanasi (Uttar Pradesh), India

Samina Mazahar
Dyal Singh College
University of Delhi

Vishnu Mishra
Laboratory of Morphogenesis
Department of Botany, Institute of Science
Banaras Hindu University
Varanasi (Uttar Pradesh), India
National Institute of Plant Genome Research
New Delhi, India

Firoz Mohammad
Plant Physiology and Biochemistry Section
Department of Botany
Aligarh Muslim University
Aligarh, India

Shashi Pandey Rai
Laboratory of Morphogenesis
Department of Botany, Institute of Science
Banaras Hindu University
Varanasi (Uttar Pradesh), India

Zubair Ahmad Parrey
Plant Physiology and Biochemistry Section
Department of Botany
Aligarh Muslim University
Aligarh, India

Shilpa Prasad
Amity Institute of Biotechnology
Amity University Jharkhand
Ranchi, Jharkhand, India

Shah Rafiq
Department of Botany
University of Kashmir, Hazratbal Srinagar
Jammu and Kashmir, India

Nidhi Rai
Laboratory of Morphogenesis
Department of Botany, Institute of Science
Banaras Hindu University
Varanasi (Uttar Pradesh), India

Peerzada Arshid Shabir
Department of Botany
Government College for Women, M. A. Road
   Srinagar
Jammu and Kashmir, India

Sajad Hussain Shah
Plant Physiology and Biochemistry Section
Department of Botany
Aligarh Muslim University
Aligarh, India

Lakee Sharma
Laboratory of Morphogenesis
Department of Botany, Institute of Science
Banaras Hindu University
Varanasi (Uttar Pradesh), India

Nitesh Singh
Department of Botany
Indira Gandhi National Tribal University
Amarkantak, Madhya Pradesh, India

Shahid Umar
School of Chemical and Life Science
Jamia Hamdard
New Delhi, India

S. R. Venkatachalam
Tamil Nadu Agricultural University

Nasir Aziz Wagay
Department of Botany
Government Degree College Boys Baramulla
   (Boys)
India

Yasheshwar
Acharya Narendra Dev College
University of Delhi

Peerzada Yasir Yousuf
Department of Botany
Government Degree College Anantnag
Jammu and Kashmir, India

A. Yuvaraja
Tapioca and Castor Research Station
Tamil Nadu Agricultural University
Salem, Tamil Nadu, India

Sidqua Zafar
Centre for Applied Molecular Biology (CAMB)
University of the Punjab
Lahore, Pakistan

Noureen Zahra
Centre for Excellence in Molecular Biology
   (CEMB)
University of the Punjab
Lahore, Pakistan

1 Introduction to Plant miRNAs

Nasir Aziz Wagay, Shah Rafiq, Amanulla Khan,
Abdul Hadi, Aabida Ishrath, Peerzada Yasir Yousuf,
and Peerzada Arshid Shabir

1.1 INTRODUCTION

In the past two decades, the "RNA world" has evolved into the "RNA molecular world," where RNA influences gene expression in addition to its role in enzymatic processes. The discovery of short non-coding RNAs has been one of the most important developments in this transition, which has been carried out on a variety of fronts. Plant short non-coding RNAs can be divided into two major types: siRNA or small interfering RNAs and miRNAs/miRsor microRNAs. MiRNA loci are translated to form ssRNA, which led to the formation of miRNAs; however, siRNAs are created from dsRNA (Voinnet, 2009). To regulate a wide range of biological functions, miRNAs in plants have been reported to be species-specific and highly reserved. Since a little more than a decade, scientists have been studying the role of miRNAs in plants. While certain elements of miRNAs, such as their biogenesis, have been thoroughly uncovered, others, such as their origins, are still unknown. Researchers may analyze biological processes and create exogenous treatments to alter gene expression when they have a full grasp of the role of miRNAs. miRNA research also sheds light on evolutionary processes and may potentially point to significant phenotypic evolution events, such as the origin of plants with land-adapted phenotypes (Taylor et al., 2014).

MicroRNAs (miRNAs) are non-coding RNA molecules, endogenously produced, with short sizes ranging from 20 to 25 bases (Ying et al., 2008; O'Brien et al., 2018) that are converted from larger hairpin precursors (Jones-Rhoades et al., 2006). They exert negative post-transcriptional regulation on gene expression (Zhang et al., 2012). One of the most prevalent groups of gene regulatory molecules, they control the expression of numerous protein-coding genes involved in growth and development throughout the whole life cycle of a multicellular organism (Bartel, 2004). The numerous and significant functions played by this novel class of short RNAs in plant growth and development have been demonstrated by several experimental and computational methods (Bartel, 2004).

Historically, two methods have been used to identify most plant miRNA genes: molecular cloning of short RNAs (Llave et al., 2002; Lu et al., 2006; Park et al., 2002; Rajagopalan et al., 2006) or computational prediction of miRNA genes based on preservation of sequence and their secondary structure (Adai et al., 2005; Tuskan et al., 2006). However, molecular cloning is the most effective method for finding miRNAs, bioinformatic techniques have been a helpful addition. For instance, computational methods can find homologous miRNAs in databases of genomic or cDNA sequences when new miRNAs are identified by molecular cloning, establishing the copy number and the spectrum of evolutionary conservation of the miRNAs. The correct and extensive identification of gene content in these databases depends on the precise identification of known miRNA families in the expanding collection of genomic and cDNA sequences.

In the past two decades, plant miRNA biology has advanced significantly. Initially, the function of miRNAs in plant growth was discovered via inhibition of transcription factors (TF) (Rhoades et al., 2002; Park et al., 2002; Palatnik et al., 2003). As a result, abiotic stress and pathogen contact were added to these regulatory molecules' range of potential actions (Jones-Rhoades et al., 2006; Rajagopalan et al., 2006). The production of miRNAs has been thoroughly studied, and the

DOI: 10.1201/9781003248453-1

significance of miRNA turnover is still being worked out. In the field of miRNA biology, there has been a recent surge in publications.

1.2 HISTORICAL PERSPECTIVE

Lin-4 is the first miRNA to be discovered in *Caenorhabditis elegans* by Victor Ambros and his associates in 1993. This miRNA suppressed the lin-14 protein levels of a gene involved in the developmental process. Instead of producing a protein, the lin-4 gene produces two short RNAs. One RNA has a length of around 21 nucleotides, whereas the second RNA is 60 nucleotides long. It was expected that the longer RNA, which folds into a stem-loop structure, is the precursor of the 21-mer RNA. This shorter lin-4 RNA served as the first member of the widely distributed miRNA class of tiny regulatory RNAs. The 3′-UTR of lin-14 mRNA contains partial antisense complementary sequences that the lin-4 RNA binds to suppress translation without affecting the amount of lin-14 transcript (Lee et al., 1993). In 2000, the same organism led to the discovery of let-7, another miRNA, which came about seven years after lin-4 (Reinhart et al., 2000). Lin-41, a gene necessary for changing a cell's destiny from late larval to adult, is regulated by miRNA let-7. Since almost all animal phyla included the let-7 miRNA, this demonstrated the biological significance of miRNAs.

Later, several researchers succeeded in cloning miRNAs from various organisms. Small RNAs were also shown to be present in all plants after being cloned. In 2002, researchers found the first plant miRNA in the plant *Arabidopsis thaliana*. A sizable collection of *Arabidopsis* miRNAs, most of which were 21 to 24 nucleotides in length, were cloned by Llave et al. in 2002. About 90% of the 125 sequences were found to come from intergenic regions, whereas a small number came from protein-coding genes. Some of these miRNAs had developmentally controlled accumulation. Surprisingly, a group of miRNAs was found to be present in transposons of the *Arabidopsis* genome (Mette et al., 2002). Eight among 16 miRNAs of *Arabidopsis* with differential expression patterns were found to be preserved in the genome of *Oryza sativa*. The accumulation of miRNA was stopped by the mutation in a homolog of RNase III family enzymeDicer. CARPEL FACTORY resulted in the conclusion that plants have a comparable system to animals for controlling the processing of miRNA (Park et al., 2002; Reinhart et al., 2002). Even while animal and plant miRNAs shared certain commonalities, these two groups' differences stood out significantly. This includes differences in gene structure, preference for 5′ nucleotides, mature miRNA sizes, and most importantly, the lack of genetic conservation between the two populations. It became abundantly obvious from the mechanistic investigations that these two categories varied greatly in their processes of biogenesis and functions. When seen as a whole, it seems that these two groups had separate evolutionary ancestries.

1.3 MIRNA BIOGENESIS

Typically, plant genomes include 100 to several hundred MIRNA (MIR) genes, many of which are organized into families (Nozawa et al., 2012; Budak & Akpinar, 2015). The miRNAs can be categorized as "intergenic" and "intronic" depending on where they are located in the genome. The first type is processed by the RNA Polymerase II from introns of their host transcripts, while the latter is processed between two protein-coding genes and transcribed as individual units. (Millar & Waterhouse, 2005; Budak & Akpinar, 2015). Primary miRNAs are capped at 5′ end, polyadenylated at 3′ end, and undergo a splicing mechanism similar to other products of pol II (Xie et al., 2005; Rogers & Chen, 2013). The primary miRNA structure is similar to the structure of a hairpin, formed by an upper and lower stem, terminal loop, region of miRNA, and two arms, which are recognized and further processed by Dicer-like RNase III endonucleases (DCLs). There are varying amounts of DCL proteins in various plant species, four identified in the plant *Arabidopsis thaliana*. DCL1 catalyzes the creation of the majority of miRNAs with the aid of auxiliary proteins like Serrate (SE) – a zinc-finger protein and Hyponastic Leaves 1 (HYL1) – a dsRNA binding protein (Fang & Spector, 2007; Dong et al., 2008). miRNA production can be potentially influenced by other DCLs. For instance, in *Oryza sativa* OsDCL3a produces a family of 24-nt miRNAs which regulate DNA methylation similarly to hc-siRNAs, whereas AtDCL4 produces miR822 and miR839 (Rajagopalan et al., 2006; Wu et al., 2010). Pri-miRNA stem-loops have significantly more complicated structures and much greater length variability than their 70 nt mammalian counterparts (from 60 nt to over 500 nt) (Xie et al., 2005; Bologna & Voinnet, 2014). The processing of pri-miRNAs by plants might occur either from the loop-distal site to the loop-proximal site or vice versa (Bologna et al., 2009, 2013; Mateos et al., 2010; Song et al., 2010; Werner et al., 2010). The nascent miRNA/miRNA duplex processed by DCL has two hydroxyl groups at 3′ end (3′ OH and 2′ OH) and phosphate at 5′ end on each strand. Both –OH groups are important, but short RNA methyltransferase HUA Enhancer 1 (HEN1) methylates exclusively at the 2′-OH site (Yu et al., 2005; Yang et al., 2006).

It is reported that the animal Exportin 5 (EXPO5) homolog protein Hasty (HST) exports methylated miRNA duplexes/miRNA (Park et al., 2005) where the RISC gathered was unknown for a very long period. Bologna et al. (2018) have demonstrated that RISC is primarily assembled in the nucleus and subsequently transported to the cytoplasm via EXPO1. Some miRNAs are transported in duplex forms and assembled in the cytosol. The passenger strand of the miRNA/miRNA duplex is expelled and destroyed, and the guide strand of the duplex is specifically assembled into the Argonaute (AGO) protein. There are ten AGO proteins in *Arabidopsis*, with AGO1 serving as the main effector protein for miRNAs (Zhang et al., 2015).

Through base pairing, miRNAs cause gene silencing by translation inhibition or target cleavage with the help of RISC. However, new reports imply the involvement of RISC/AGO1, which regulates their transcription process (Dolata et al., 2016; Liu et al., 2018; Yang et al., 2019). The synthesis of secondary siRNAs known as easiRNAs and/or phasiRNAs can be started by different miRNAs, including miR173, miR390, and miR845 (Fei et al., 2013; Creasey et al., 2014; Deng et al., 2018). Despite the presence of non-canonical targeting, animals only require a short base-pairing to the seed region of miRNAs (positions 2–8) to identify their targets (Helwak et al., 2013; Agarwal et al., 2015). Plants, on the other hand, use a stronger base-pairing regulation, with almost flawless matching in the 5′ end (with just one mismatch) and slack, but plentiful pairing in the 3′ region (maximum four mismatches and only minor bulges are permitted) (Schwab et al., 2005; Axtell & Meyers, 2018). Theoretically, animals have a larger number of target genes than plants. Even though target cleavage appears to be more common, it is more significant since it is necessary for the development of plants after germination.

1.4 THE MIRNA GENE FAMILIES

Depending on the number of target sites they have, animal miRNAs directly control more than one-third of all protein-coding genes (Mallory & Vaucheret, 2006; Lewis et al., 2005); however in plants, known miRNAs and their targets are significantly less in number (Jones-Rhoades et al., 2006). Accordingly, genomes of animal have a significant number of small miRNA gene families, whereas plant genomes contain fewer miRNAs than animal genomes but bigger miRNA gene families. The majority of miRNA gene families from animal species typically contain less than two members, but the average size of plant miRNA gene families is roughly 2.5 genes with an exception: the zebrafish (*Danio rerio*). The miR-430 gene family has around 100 members and is one of the larger miRNA families that belongs to it (Chen et al., 2005; Giraldez et al., 2005). Nucleotides 2–8, which appear to be crucial for target recognition (Lewis et al., 2005), are similar in MiR-430 genes. In their 3′ region, they are likewise quite homologous, but their central and terminal nucleotides are different. The miRNA gene family average size in zebrafish would remain at 2.72 after removing this family, which is comparable to the size in *Arabidopsis*.

Plant miRNAs generated from the same gene family are typically almost similar, in comparison to animal miRNA genes, which have experienced divergence even on mature miRNA sequences. The commonalities not only occur at the mature miRNA regions but also throughout the genes, indicating that the growth of plant miRNA gene families has a recent beginning and may still be in process. The model plant *A. thaliana* has at least 22 miRNA gene families, the majority of which are shared by monocots and eudicots (Jones-Rhoades & Bartel, 2004; Griffiths-Jones et al., 2006). The rice genome is more than three times larger than the *Arabidopsis* genome and has three times more miRNA genes. However, the Registry 8.1 of MicroRNA, lists 46 and 47 miRNA gene families from the rice and *Arabidopsis* genomes, respectively indicating that both monocots and eudicots have a similar number of distinct miRNA gene pools.

Even among species that are only distantly related, homologs of the majority of miRNAs have been discovered to exist. For instance, the human genome contains at least one-third of the miRNAs discovered in *C. elegans*. (Lim et al., 2003; Zhang et al., 2006), indicating that these miRNAs have retained much of their functional diversity across different animal lineages. Similar findings in plant genomes were also reported. Recent research by Zhang et al. (2006) has identified a total of 481 miRNAs from 71 distinct plant species that are members of 37 miRNA families from more than 6 million plant EST sequences. It has been shown that plant miRNAs are preserved in gymnosperms, ferns, mosses, and liverworts (Floyd & Bowman, 2004). Plant genomes (Tanzer et al., 2005; Maher et al., 2006; Guddeti et al., 2005; Lee et al., 2002) show a physical clustering of several miRNA gene families' members. Animal genomes frequently contain such clusters of non-homologous miRNA genes (Tanzer et al., 2005; Maher et al., 2006; Guddeti et al., 2005; Lee et al., 2002). As a result, the genomic organization of various species' miRNA gene families varies.

At first, miRNAs were found in rice (*Oryza sativa*) and *A. thaliana* that belonged to families that are shared by both species (Llave et al., 2002; Park et al., 2002; Jones-Rhoades & Bartel, 2004). The discovery of miRNAs from ancient families was made possible by computational prediction and sequencing techniques (Axtell & Bartel, 2005). Several miRNA families, including miR396, miR397, miR398, miR828, miR2111, and miR403, are preserved in all types of plants, between seed plants and vascular plants (Cuperus et al., 2011). MiR156 is substantially conserved between lower plants and angiosperms and is involved in the transition of plant shoot growth from the juvenile to the adult stage. In addition to miR156, miR166, and miR171 have been reported to be conserved across several plant species. However, other miRNAs, including miR415 and miR417, are not conserved and have only been identified in angiospermous plants. It's interesting to note that *Chlamydomonas reinhardtii*, a single-cell alga, lacks miRNAs and signals an early evolutionary split or divergence from higher plants. MiRNA families exhibit cross-kingdom conservation, according to Arteaga-Vázquez et al. (2006). Two miRNA identified families, miR854 and miR855, are conserved throughout animal and plant lineages. However, retrotransposons contain loci from these families.

Along with conserved miRNAs, other types of miRNAs, such as species-specific or non-conserved miRNAs, have also been discovered. Deep sequencing of *A. thaliana* small RNA populations revealed that certain miRNAs are unique to the family Brassicaceae, although some miRNAs are conserved (Lu et al., 2006; Rajagopalan et al., 2006; Fahlgren et al., 2007, 2010; Ma et al., 2010). A substantial number of species-specific miRNA genes have been identified in *Selaginella moellendorffii* and *Physcomitrella patens* (Axtell et al., 2007), rice (Heisel et al., 2008; Lu et al., 2008a; Sunkar et al., 2008; Zhu et al., 2008), and *Medicago truncatula* (Szittya et al., 2008; Lelandais-Brière et al., 2009), and *Glycine max* (Subramanian et al., 2008). Other instances that show plants have recently developed miRNA loci is the abundance of non-conserved miRNAs in plants.

The evolutionarily stable and changeable portions of the miRNA region, which are the results of extensive evolution, may be separated into two categories. While most of the miRNA genes, including the loop and other sequences, are represented by the other area, which includes the mature miRNA and miRNA*, that portion is changeable (Bullini & Coluzzi, 1972).

1.5   THE ORIGIN OF PLANT MIRNA GENES

There is currently extremely minimal data concerning the origin of miRNA genes in the genomes of plants. The unique stem-loop structure and how miRNAs couple with their target genes functionally support the theory that miRNA genes were associated with their target genes before they were created (Allen et al., 2004). Despite this, instances of similarity between miRNAs and their target genes are still uncommon and have only been noted in non-conserved miRNA genes, such as the *Arabidopsis* miR161 and miR163 genes. These miRNA genes typically have a single copy, unlike conserved miRNA genes, which frequently contain many copies. Pentatricopeptide repeat proteins and S-adenosylmethionine-dependent methyltransferases are the targets of miR161 and miR163, respectively. The areas outside the mature miRNA and its pairing sequence (miRNA*) of these two genes showed significant homology to their target genes. A potential mechanism for the origin and development of miRNA genes with specific target specificities has been proposed (Allen et al., 2004). The proposed model postulates that tail-to-tail or head-to-head gene configurations resulted from inverted gene duplication events from a single founder gene, either with or without the promoter of the founder gene. At the inverted duplication locus, sequence divergence took place under restrictions to preserve the fold-back structure and DCL1 recognition. Eventually, only the miRNA-complementary or miRNA sequences that matched the founding gene sequence (Allen et al., 2004) were kept in place due to sequence degeneration.

However, no relationship between the target sites and their conserved miRNA genes has been discovered, a model of this sort may only apply to plant non-conserved miRNA genes. Furthermore, since animal precursors are too short to include information on their founder genes, it would be impossible to explain how animal miRNAs are produced. It is believed that animal miRNA regulatory mechanisms are gained by interactions between target genes and their miRNAs or "gain-of-interaction" events. The majority of animal miRNAs control the translation of protein-coding genes by binding to the untranslated regions at the 3'-end, where more mismatches are tolerated during the pairing process (Bartel & Chen, 2004). Contrarily, it has been found that plant miRNAs and their target gene regions are almost similar and that this matching frequently results in the target gene transcript cleavage. The various functional modes in the genomes of both plants and animals imply that there were different origination mechanisms for miRNA genes in the two kingdoms.

1.6   EVOLUTION OF PLANT MIRNAS

The evolutionarily ancient occurrence of gene silencing by complementary RNA molecules seems to have been lost in certain lineages of eukaryotes, including yeast in the process of evolution. For instance, there is debate on the veracity of two miRNAs, miR854 and miR855, which are shared by the kingdom Animalia and Plantae (Arteaga-Vázquez et al., 2006; Cuperus et al., 2011; Pashkovskiy & Ryazansky, 2013). However, caution should be used when drawing evolutionary inferences about conserved and non-conserved miRNA families since, aside from a few model species, miRNAs are not well-defined in many plants, and several of those identified thus far still need more proof to be confirmed. Taylor and his colleagues examined all 6,172 miRNA genes deposited in miRBase version 20 in depth to illustrate this issue and found that almost one-third of these miRNAs lacked sufficient evidence to be labelled as genuine miRNAs. The same study also notes that there are significant evolutionary gaps across species for which data on miRNA content are available. For instance, the moss lineage's only representation in miRBase *is Physcomitrella patens* (Taylor et al., 2014). As the miRNA catalogs of new species are revealed, our knowledge of the evolutionarily non-conserved and conserved miRNA families across evolutionary lineages is susceptible to changes. However, it was shown that eight miRNA families, including miR156, miR159/319, miR160, miR166, miR171, miR408, miR390/391, and miR395, were preserved in the ancestor Embryophyta (Cuperus et al., 2011). Of them, it is determined that miR156 and miR166 are conserved miRNAs that play roles in flower formation throughout the plant kingdom (Luo et al., 2013). Additionally, all vascular plants included members of the miR396 family, and all seed plants contained members of the miR397 and miR398 families. The angiosperm miR403, miR828, and miR2111 families have been identified as being unique to eudicots so far (Cuperus et al., 2011). AGO2 and AGO3 proteins are interesting targets of the miR403 family, suggesting a feedback mechanism in miRNA synthesis (Jagtap & Shivaprasad, 2014). Syntenic connections may confirm that miRNAs conserved across two or more species are, in fact, evidence of a real evolutionary conservation event. Lists of conserved and non-conserved miRNA families are subject to alteration in response to discoveries. For instance, miR1133 and miR167 discovered on the short arm of *Triticum aestivum's* chromosome 5D have homologs on the syntenic *Brachypodium distachyon* chromosome 4, strongly suggesting that these miRNAs are likely conserved across both species (Kurtoglu et al., 2013).

1.7   MIRNA GENE CLUSTERS (THE FORMATION OF MIRNA GENE FAMILIES)

Both plant and animal genomes have been shown to include collections of miRNA genes. Numerous miRNA genes on certain clusters can be transcribed as a single polycistron because they are so closely packed together (Guddeti et al., 2005; Lee et al., 2002). When most miRNA gene families were formed, the development of miRNA gene clusters may have represented broader evolutionary events. For instance, the animal miR-17 gene cluster consists of the following miRNAs: miR-17, miR-18, miR-19a, miR-19b, miR-20, miR-25, miR-92, miR-93, miR-106a, and miR-106b. Despite being linked to evolution, several of these genes are not homologous. The E2F1 gene, which is involved in human cell cycle development, is negatively regulated by miR-17 genes, which impart crucial roles (O'Donnell et al., 2005). According to phylogenetic reconstruction, this cluster's history was shaped by a first phase of local (tandem) duplications, a string of cluster duplications, and a subsequent loss of specific miRNA genes from the resultant paralogous clusters. It seems that the early evolution of the vertebrate lineages (Tanzer & Stadler, 2004; Tanzer & Stadler, 2006; Tanzer et al., 2005) is intimately related to the convoluted history of the miR-17 gene family. The clusters are present in all vertebrates, including teleost fish and humans. The fact that they have been fixed in several contemporary animal genomes suggests that these miRNA gene family arrangements have benefited from natural selection.

In plant genomes, the observation is different. Even though plant miRNA gene families are substantially bigger, only a small portion of their members have been discovered to be grouped in a region of several kilobases (Maher et al., 2006; Guddeti et al., 2005). The fact that miRNA genes from the same family are frequently dispersed across the genome suggests that plant genomes have undergone substantial reorganization since the amplification of these families. The miR395 gene family stands out as one particular miRNA gene family. Several plant genomes include the miR395 gene family in clusters of varying cluster sizes and intergenic spacing. Plant miRNA gene clusters, in contrast to animal miRNA gene clusters, consist of homologous components (Guddeti et al., 2005; Jones-Rhoades & Bartel, 2004; Zhang et al., 2006). No reports of clusters in plant genomes that comprise non-homologous but evolutionary-connected miRNA gene members have been made as of yet. Since many functionally related genes can be concurrently regulated by the co-transcription of non-homologous miRNA genes in animal miRNA gene clusters, a dosage effect would result from the co-transcription of similar or identical miRNA genes in a plant gene cluster.

1.8   FUNCTIONS OF MIRNA

miRNAs have been implicated in numerous biological processes, including development and growth, genome integrity maintenance, transduction of signals, hormone signaling pathways, innate immunity, hormone homeostasis, and response to various environmental biotic and abiotic stresses (Navarro et al., 2006; Sun, 2012; Xie et al., 2014; Zhang & Wang, 2015). RNA polymerase II (pol II) often converts them into pri-miRNAs. After being cut by a group of RNase-III nucleases known as "Dicer-like proteins," these pri-miRNAs join with proteins from the ARGONAUTE (AGO) family to create RNA-induced silencing complexes (RISCs). Then, RISCs participate in the expression and control of target genes (Song et al., 2019). MiRNAs control things like blooming patterns, leaf polarity, and meristem features. MiRNAs are crucial for coordinating plant growth as evidenced by the various impacts that mutations in miRNA transcription or processing complexes often have on plant structure and function. For instance, the miR156/miR172 pathway and the HD-ZIP III-miR165/166 pathways both play crucial roles in the vascular development, meristem development, and leaf polarity as well as time of flowering and its pattern (D'Ario et al., 2017; Ramachandran et al., 2017; Du et al., 2020; Ma et al., 2014; Lian et al., 2021). Endogenous miRNAs regulate gene expression throughout plant development by directing their attention to relevant target genes. Numerous miRNAs interact with hormones to carry out their functions. MiRNAs have several targets that are involved in hormone signaling, and these targets work together with the miRNAs to quickly and efficiently control plant growth, development, and differentiation. To modulate hormone responses or to regulate particular miRNA levels, this signaling uses miRNAs as intermediates (Jodder, 2020; Yu & Wang, 2020). Evidence suggests that miRNAs can act as diffuse inhibitor signals in tissues, where they play a complex role in the differentiation of the tissues (Chen & Rechavi, 2021).

1.8.1 ROLE OF MIRNAS IN SIGNAL TRANSDUCTION

MiRNAs have recently been found to control important hormone signaling pathway constituents, as well as hormone homeostasis and associated developmental processes, according to recent findings from several different laboratories (Sorin et al., 2005; Mallory et al., 2005; Guo et al., 2005). Numerous miRNAs have an impact on hormone signaling pathways, particularly in signal transduction. Plant hormones, notably auxin, have a significant role in controlling the growth and development of the plant, especially cell division, cell elongation, and differentiation (Palme et al., 1991; Woodward & Bartel, 2005). Additionally, the growth hormones of the plant are crucial for organ development and adaptation to varied environmental conditions (Palme et al., 1991; Woodward & Bartel, 2005). Zhang et al. (2005) discovered that numerous miRNAs (miR159, miR160, miR164, and miR167) were derived from tissues stimulated by gibberellic acid (GA), abscisic acid (ABA), salicylic acid (SA), jasmonic acid (JA), and other plant hormones after analyzing a significant number of ESTs. Through the miRNA-guided cleavage of mRNA encoding GAMYB-related proteins, which are transcription factors controlling LEAFY, miR159 modulates GAMYB (Achard et al., 2004). According to Achard et al. (2004), the plant hormone gibberellin positively controlled the miR159 gene (GA). Reduced LEAFY mRNAs, a later blooming period, and altered floral development were caused by miR159 overexpression (Achard et al., 2004).

An essential hormone in plants is called auxin, which is mostly indole-3-acetic acid. By modulating AUXIN RESPONSE FACTORS (ARF), a class of DNA-binding proteins unique to plants impacts a variety of aspects in the growth and development of plants (Hagen & Guilfoyle, 2002; Weijers & Jurgens, 2005). ARFs bind auxin response promoters (AREs), which control the production of auxin-inducible genes such as auxin/indole-3-acetic acid (Aux/IAA) and GH3 (Hagen & Guilfoyle, 2002). Approximately 23 ARFs have so far been found in *Arabidopsis* as a whole among which at least five of those ARFs contain miRNA complementary sites. According to Rhoades et al. (2002) and Mallory et al. (2005), ARF10, ARF16, and ARF17 are prospective targets of miR160, whereas ARF6 and ARF8 are potential targets of miR167. Mallory et al. (2005) provided evidence that miR160 function disruption raised ARF10, ARF16, and ARF17 mRNA levels changed the expression of GH3-like genes and resulted in serious developmental abnormalities. Additionally, it has been noted that ARF plants that are resistant to miR160 exhibit severe developmental defects (Mallory et al., 2005; Sorin et al., 2005).

NAC1 is a transcriptional activator that controls the transport inhibitor response 1 (TIR1) to transduce signals of auxin for lateral root initiation (Xie et al., 2002; Guo et al., 2005). Loss-of-function miR164 mutants were shown by Guo et al. (2005) to disrupt the auxin signaling pathway, increasing NAC1 mRNA levels and lateral root development. On the other hand, miR164 overexpression inhibited lateral root initiation and downregulated the expression of the NAC1 gene (Guo et al., 2005). The expression of miR164 and NAC1 can be controlled by auxin through a signaling pathway (Guo et al., 2005). Auxin signaling was reduced and ended by miRNA-guided NAC1 mRNA cleavage in an autoregulatory loop, as evidenced by the induction of miR164 at least partially by auxin (Guo et al., 2005). During lateral root growth, auxin signals are downregulated by miR164-guided NAC mRNA cleavage.

Other miRNAs, such as miR393, may be implicated in signaling pathways via modulating TIR1 in addition to miR160, miR164, and miR167. In reaction to auxin, an SCF E3 ubiquitin ligase that contains TIR1 breaks down Aux/IAA proteins (Gray et al., 2001; Mallory et al., 2005). MiR393 is currently anticipated to target TIR1 and the three F-box proteins that are most closely linked to it (Sunkar & Zhu, 2004; Jones-Rhoades & Bartel, 2004). This shows that miRNA can modify the function of the E3 enzyme and target proteins of the F-box.

These findings collectively show that miRNAs have a role in signal transduction related to hormone mediation. MiRNAs can target either NAC1 and ARF directly or can get activated by phytohormones.

1.8.2   ROLE OF miRNAs IN GENE REGULATION

MiRNAs control the expression of their target genes either by translational repression or mRNA cleavage. The target's identity influences the post-transcriptional processes that are chosen. Once integrated into a cytoplasmic RISC, the miRNA will either specify cleavage if the mRNA has enough complementarity to the miRNA to be cleaved, or it will suppress translation if the mRNA lacks enough complementarity to be cleaved but has an appropriate cluster of miRNA-binding sites.

1.8.2.1   miRNA-Directed Target mRNA Cleavage

mRNA targets that are fully or nearly complementary are recognized by plant miRNAs, which then trigger endonucleolytic mRNA cleavage (Rhoades et al., 2002; Llave et al., 2002; Tang et al., 2003; Voinnet, 2009). The AGOs catalyze the cleavage, which takes place in between the 10th and 11th position nucleotides, opposite the miRNA strand, and the nonspecific nucleases destroy the resultant 5' and 3' mRNA fragments freshly formed from the 5' and 3' ends (Souret et al., 2004). Cleavage still occurs even if the pairing is not perfect with its 5' terminus of miRNA (Kasschau et al., 2003; Palatnik et al., 2003). The cut site, therefore, seems to be determined concerning miRNA residues rather than miRNA: target base pairs. The miRNA is still there after the mRNA is cut, and it can direct the cutting of new mRNAs (Hutvágner & Zamore, 2002; Tang et al., 2003). These investigations showed that a key mechanism of miRNA-based control in plants is the endonucleolytic cleavage of highly complementary targets.

1.8.2.2   miRNA-Directed Translational Repression of Target Genes

Translation repression in plants has been linked to miR156/miR157, targeting the SBP-box gene SPL3, and miR172, targeting APETALA 2 (Aukerman & Sakai, 2003; Chen, 2004; Gandikota et al., 2007). The production of GFP protein was greater in mutants of *A. thaliana* incapable of silencing a GFP reporter in the binding site of 3'-UTR with a miR171, but the mRNA levels were not recovered. This indicated that the alterations prevented mRNA cleavage but not miRNA-mediated translational repression. However, the findings imply that plant miRNAs may still suppress translation without degradation of the target (Brodersen et al., 2008).

The eDC4, or Enhancer of decapping 4 (sometimes referred to as HeDlS and Ge1), known as vARICOSe in *A. thaliana* and in animals, is required for miRNA-mediated mRNA decay but not for translational repression (Eulalio et al., 2007). However, in *Drosophila melanogaster* cells, eDC4 is known to be a suppressor of miRNA-mediated gene silencing; however, vARICOSe mutants in plants showed increased levels of protein expression but not mRNA levels. This is because eDC4 inhibits the miRNA-mediated degradation of mRNA, suppressed by eDC4, and restores transcript levels (Brodersen et al., 2008).

1.8.2.3 miRNA-Directed DNA Methylation

Additionally, miRNAs alter the pattern of gene expression and are implicated in DNA methylation, a well-known epigenetic control. Variants of miRNAs that participate in RNA-directed DNA methylation are linked to AGO proteins (Law & Jacobsen, 2010). According to research on *A. thaliana*, AGO4, AGO6, and AGO9 bind to 24-nt siRNAs with 5′ adenosine made by DCL3 from dsRNAs produced by the actions of RDR2 and DNA-dependent RNA polymerase IV to drive the methylation of target DNA (Qi et al., 2006; Mi et al., 2008; Havecker et al., 2010; Law & Jacobsen, 2010). Rice DCL3a has been demonstrated to process multiple miRNA foldbacks, resulting in 24-nt siRNAs that bind to rice AGO4a and AGO4b preferentially and function similarly to 24-nt siRNAs in directing the methylation of target genes (Wu et al., 2010). The buildup of miRNA-target RNA duplexes and hypermethylation of the target RNA-encoding genes resulted in gene silence in the moss *Physcomitrella patens* dcl1b mutants (Khraiwesh et al., 2010). Therefore, it may be inferred that miRNA and miRNA-AGO complexes can work at the transcriptional silencing level in addition to their role in mRNA repression.

1.8.3 miRNA Role in miRNA and siRNA Biogenesis and Function

One of the crucial enzymes in the production of mature miRNA is DCL1 (Papp et al., 2003; Bartel, 2004). The miR162 target is DCL1 mRNA (Xie et al., 2003). According to Xie et al. (2003), DCL1 mRNA levels were much greater in hen1 mutant plants than in wild-type *Arabidopsis* plants. Xie and colleagues used transgenic *Arabidopsis* plants expressing P1/HC-Pro protein, a virus-encoded suppressor of RNA silencing that prevents miRNA-guided targeted mRNA degradation (Kasschau et al., 2003), to study the function of miR162. This was done to further test the hypothesis that DCL1 mRNA is negatively regulated by miRNA-guided mRNA cleavage. Their findings demonstrated that the transgenic plants' DCL1 mRNAs were overexpressed. These findings suggest that miR162 inhibits the regulation of DCL1 mRNA.

MiRNA-guided gene regulation necessitates that miRNA form a complex with RISC, regardless of the regulatory method (mRNA cleavage or translational repression). Since RISC is made up of several components, each component's loss of functionality has an impact on both RISC and miRNA function. A crucial part of RISC is made up of ARGONAUTE proteins, sometimes referred to as PPT proteins (Hammond et al., 2001; Caudy et al., 2003; Vaucheret et al., 2004). A single-stranded RNA binding site can be found in ARGONAUTE proteins (Yan et al., 2003). Because of this property, ARGONAUTE proteins can interact directly with miRNAs both before and after RISC detects their mRNA targets (Vaucheret et al., 2004). The *Arabidopsis* ARGONAUTE protein family consists of 10 proteins (Morel et al., 2002). AGO1 (ARGONAUTE1) appears to be necessary for stem cell activity and organ polarity, according to several tests (Kidner & Martienssen, 2005b). Only ARGONAUTE 1 (AGO1) mRNA is the target of miR168, according to computational analysis and 5VRACE tests (Rhoades et al., 2002; Vazquez et al., 2004; Kidner & Martienssen, 2005a). Important for the post-transcriptional gene regulation of the ago1 gene is the miR168-guided cleavage of AGO1 mRNA (Vazquez et al., 2004). Vaucheret et al. (2004) showed that altering AGO1 mRNA's nucleotides to make them less complementary with miR168 caused an increase in the amount of AGO1 mRNA, which in turn led to developmental defects resembling those of the *dcl1* or *hen1* mutants (Xie et al., 2003; Vaucheret et al., 2004; Kidner & Martienssen, 2005a, 2005b). Increased target mRNA suggested that mutant plants developed non-functional RISCs (Vaucheret et al., 2004). By producing a compensatory miRNA that is corresponding to the mutant AGO1 mRNA, these abnormalities were prevented (Vaucheret et al., 2004).

In recent times, it was discovered that five ta-siRNA-producing transcripts were both complementary to miR173 or miR390 and targets of these two miRNAs (Allen et al., 2005; Bartel, 2005). Four ta-siRNA primary transcripts were in phase processed by miR173, according to Allen et al. (2005), but trans-acting siRNA3 (TAS3) ta-siRNA primary transcripts were processed by miR390. Allen and colleagues suggested a hypothesis for miRNA-directed initiation of ta-siRNA synthesis in light of these discoveries. According to their concept, miRNAs helped build a 5V or 3V terminal inside the pre-ta-siRNA transcripts, which was followed by RDR6-dependent dsRNA production and Dicer-like processing to produce phased tasiRNAs that adversely affect other genes (Allen et al., 2005). In addition to directing the expression of the genes responsible for plant growth, miRNA also controls its synthesis and/or function. Currently, it is known that at least five miRNAs (miR162, miR168, miR173, miR390, and miR398) influence the synthesis or function of miRNAs.

1.8.4 MIRNA ROLE IN DEVELOPMENTAL PROCESSES

The growth of plants is a highly regulated process that is managed on many different levels. The identification of miRNAs and their functions in plant growth has opened up new avenues for research into the processes underpinning this control. Major pathways in the development of plants like the development of leaf, polarity and patterning, development of flower and floral identity, time of flowering, transition to developmental phase, development of root and shoot, hormone signaling, and responses to stress all have miRNA target genes that are primarily proteins of F-box and transcription factors (Jones-Rhoades et al., 2006).

Mutants of *A. thaliana* dcl1, AGO1, HYL1, HST, and HEN1, have developmental abnormalities that provide information related to the essential roles in the development of plants by the miRNAs (Lu & Fedoroff, 2000; Chen et al., 2002). Loss-of-function dcl1 gene was associated with reduced levels of miRNA and several developmental defects (Park et al., 2002; Reinhart et al., 2002; Dugas & Bartel, 2004). Reduced levels of HASTY, a protein necessary for miRNA-miRNA* duplex transport from the nucleus to the cytoplasm, led to a variety of pleiotropic traits, including altered flower and leaf morphology, decrease in fertility, rapid shift in phase of vegetation, and altered inflorescence phyllotaxis (Bollman et al., 2003). These results unequivocally show that miRNAs play a role in several plant developmental processes.

1.8.5 ROLE IN STRESS

miRNAs are involved in several biological processes in eukaryotic cells and complex regulatory networks, and they have an impact on how plants grow and respond to biotic and abiotic problems (Pegler et al., 2019; Secic et al., 2021). Numerous research on miRNAs in recent years have shown that plants must have miRNA-directed gene expression control to respond to their environment (Sun et al., 2019). By controlling the expression of specific miRNAs or creating new miRNAs, plants that are under abiotic stress can increase their tolerance to environmental stimuli and thrive while doing so. A range of environmental factors, such as high and low temperatures, drought, saline conditions, heavy metals, and oxidation, can also activate certain miRNAs in plants. The expression of the associated miRNAs can be upregulated or downregulated depending on the kind of stress (Jian et al., 2010). Stress-response-based studies on miRNAs can offer significant insights into the expression of genes and stress resistance breeding in plants, a potent method for unexplored knowledge of plants' adaptive mechanisms.

The miRNA responses to one or more environmental stimuli can vary as their responses to plant development and growth. The miRNAs have been studied for their roles in the control of growth and development. Previous research has demonstrated that miRNAs control the growth of roots, floral organs and flowers, leaves, embryogenesis, fruit ripening, and its development in a plant (Gao et al., 2017; Correa et al., 2018; Plotnikova et al., 2019; Siddiqui et al., 2019).

Recent noteworthy results have shown that miRNAs and their targets may regulate agronomic properties in agricultural plants, suggesting that miRNAs are crucial for the enhancement of agricultural traits that contribute to better quality and yield, such as the height of the plant, branching of panicle, size, and quality of the grain, the yield of the crop, tillering, male sterility, and others (Wang et al., 2012; Zhang et al., 2013; Yang et al., 2019).

1.9   ARTIFICIAL MIRNAS (AMIRNAS)

To create short RNAs (sRNAs) that drive gene silencing in either plants or animals, the artificial microRNA (amiRNA) technique uses endogenous miRNA precursors (Alvarez et al., 2006; Niu et al., 2006; Schwab et al., 2006). The miRNA-miRNA* duplex is produced preferentially by miRNA precursors. MiRNA with the desired sequence will accumulate when both sequences are changed without modifying structural elements like mismatches or bulges. The amiRNAs were created and tested first in human cell lines (Zeng et al., 2002) and then in *Arabidopsis* (Parizotto et al., 2004), where they were successfully employed to obstruct the expression of reporter genes. Later, it was shown that amiRNAs may target endogenous genes as well as reporter genes, and they appear to function with a comparable level of efficacy in other plant species (Alvarez et al., 2006; Schwab et al., 2006).

According to the characteristics of plant miRNAs, amiRNA sequences are created so that the chosen 21 mere precisely silences the target genes. When creating amiRNAs, three main considerations must be made. First, they should begin with a U (found in the majority of plant and animal miRNAs); second, they should show 5′ instability in comparison to their amiRNA* (the nucleotides pairing to the miRNA in the precursor); and third, their tenth nucleotide should either be an A (Reynolds et al., 2004) for synthetic siRNAs or a U (Mallory et al., 2004), which is overrepresented in plant miRNA.

The natural precursor structures of Ath-miR159a, Ath-miR164b, Ath-miR172a, Ath-miR319a, and Osa-miR528 have served as the foundation for the majority of amiRNAs in plants to date. A straightforward amiRNA vector (pAmiR169d) was created by Liu et al. (2010) based on the structure of the *Arabidopsis* miR169d precursor (pre-miR169d). AmiRNAs work well when expressed from either tissue-specific or constitutive promoters. amiRNAs have demonstrated their potential for creating virus-resistant plants. Cotton farming is severely hampered by cotton leaf curl disease, which is brought on by single-stranded DNA viruses of the genus *Begomovirus* (family Geminiviridae). Two amiRNA constructs to change *Nicotiana benthamiana*: (P1C), which had the miR169a sequence except for the replaced 21 nt of the V2 gene sequence of Cotton leaf curl Burewala virus (CLCuBuV), and (P1D), which had the backbone sequence of the miRNA169a that were modified to partially restore the hydrogen bonding of the mature miRNA duplex. Compared to P1D plants, P1C transgenic plants showed better overall resistance to challenge all tested viruses, especially against CLCuBuV (Ali et al., 2013). A potential method for preventing the transmission and proliferation of geminiviruses in host plants is provided by amiRNAs. The overlapping area of the AV1 and AV2 (pre-coat protein) transcripts, as well as the central region of the AV1 (coat protein) transcript, are the targets of amiRNAs (amiR-AV1–1). Transgenic tomato plants expressing amiR-AV1–1 were highly resistant to the Tomato leaf curl New Delhi virus (ToLCNDV), which targets the coat protein of geminivirus, Tomato leaf curl virus (ToLCV) (Vu et al., 2013). The amiRNA strategy may effectively protect plants against viral infection, and this has the potential to offer broad-spectrum defence against a variety of viruses.

1.10   CONCLUSION AND FUTURE PROSPECTS

A family of non-coding, short-sized RNAs known as microRNAs control gene expression and are involved in a variety of biological activities. As the bulk of plant miRNA targets is transcription factors and other regulatory proteins, hence miRNAs are crucial regulators of gene expression. For functional genomics to more fully comprehend the fundamental processes behind the diverse plant developmental events, miRNAs offer a potential technique. The miRNA biogenesis gene mutations have resulted in severe pleiotropic disorders. A variety of strange abnormalities are displayed by miRNA overexpression and miRNA-resistant forms of certain miRNA targets (Jones-Rhoades et al., 2006).

The miRNAs are crucial regulators of developmental processes, and they are also thought to have a role in plants' reactions to biotic and abiotic stress. During biotic stress, various miRNA families are known to perform significant functions, including viral, bacterial, fungal, insect, and nematode assault, as well as under abiotic stress which includes heat, cold, salt, drought, etc. (Sunkar et al., 2012). A better understanding of the responses of plants to stress under harsh environmental circumstances can be aided by knowledge of the stress-responsive genes regulated by miRNAs, which may also be used to fine-tune gene expression to increase stress tolerance in plants. MiRNAs have been a significant tool for crop development throughout time. Successful efforts in this regard have already been made with plants like tomato and *Arabidopsis*. Breeding new crop cultivars with superior agronomic features may assist to minimise food insecurity.

While transgenic plants may experience unwanted phenotypic alterations, miRNAs are still promising tools for agricultural development. To properly implement transgenic techniques and achieve desired phenotypes with a minimum amount of off-target effects, it is imperative to understand the regulation of basic processes mediated by miRNA in plants.

REFERENCES

Achard, P., Herr, A., Baulcombe, D. C., & Harberd, N. P. (2004). Modulation of floral development by a gibberellin-regulated microRNA. *Development*, *131*, 3357–3365. https://doi.org/10.1242/dev.01206

Adai, A., Johnson, C., Mlotshwa, S., Archer-Evans, S., Manocha, V., Vance, V., & Sundaresan, V. (2005). Computational prediction of miRNAs in Arabidopsis thaliana. *Genome Research*, *15*, 78–91. https://doi.org/10.1101/gr.2908205

Agarwal, V., Bell, G. W., Nam, J. W., & Bartel, D. P. (2015). Predicting effective microRNA target sites in mammalian mRNAs. *Elife*, *4*, e05005. https://doi.org/10.7554/eLife.05005

Ali, I., Amin, I., Briddon, R. W., & Mansoor, S. (2013). Artificial microRNA-mediated resistance against the monopartite begomovirus cotton leaf curl Burewala virus. *Virology Journal*, *10*(1), 1–8. https://doi.org/10.1186/1743-422X-10-231

Allen, E., Xie, Z., Gustafson, A. M., & Carrington, J. C. (2005). microRNA-directed phasing during transacting siRNA biogenesis in plants. *Cell*, *121*(2), 207–221. https://doi.org/10.1016/j.cell.2005.04.004

Allen, E., Xie, Z., Gustafson, A. M., Sung, G. H., Spatafora, J. W., & Carrington, J. C. (2004). Evolution of microRNA genes by inverted duplication of target gene sequences in Arabidopsis thaliana. *Nature Genetics*, *36*(12), 1282–1290. https://doi.org/10.1038/ng1478

Alvarez, J. P., Pekker, I., Goldshmidt, A., Blum, E., Amsellem, Z., & Eshed, Y. (2006). Endogenous and synthetic microRNAs stimulate simultaneous, efficient, and localized regulation of multiple targets in diverse species. *Plant Cell*, *18*(5), 1134–1151. https://doi.org/10.1105/tpc.105.040725

Arteaga-Vázquez, M., Caballero-Pérez, J., & Vielle-Calzada, J. P. (2006). A family of microRNAs present in plants and animals. *Plant Cell*, *18*(12), 3355–3369. https://doi.org/10.1105/tpc.106.044420

Aukerman, M. J., & Sakai, H. (2003). Regulation of flowering time and floral organ identity by a microRNA and its APETALA2-like target genes. *Plant Cell*, *15*(11), 2730–2741. https://doi.org/10.1105/tpc.016238

Axtell, M. J., & Bartel, D. P. (2005). Antiquity of microRNAs and their targets in land plants. *The Plant Cell*, *17*(6), 1658–1673. https://doi.org/10.1105/tpc.105.032185

Axtell, M. J., & Meyers, B. C. (2018). Revisiting criteria for plant microRNA annotation in the era of big data. *Plant Cell*, *30*, 272–284. https://doi.org/10.1105/tpc.17.00851

Axtell, M. J., Snyder, J. A., & Bartel, D. P. (2007). Common functions for diverse small RNAs of land plants. *The Plant Cell*, *19*(6), 1750–1769. https://doi.org/10.1105/tpc.107.051706

Bartel, B. (2005). MicroRNAs directing siRNA biogenesis. *Nature Structural & Molecular Biology*, *12*(7), 569–571. https://doi.org/10.1038/nsmb0705-569

Bartel, D. P. (2004). MicroRNAs: Genomics, biogenesis, mechanism, and function. *Cell*, *116*(2), 281–297. https://doi.org/10.1016/S0092-8674(04)00045-5

Bartel, D. P., & Chen, C. Z. (2004). Micromanagers of gene expression: The potentially widespread influence of metazoan microRNAs. *Nature Reviews Genetics*, *5*(5), 396–400. https://doi.org/10.1038/nrg1328

Bollman, K. M., Aukerman, M. J., Park, M. Y., Hunter, C., Berardini, T. Z., & Poethig, R. S. (2003). HASTY, the Arabidopsis ortholog of exportin 5/MSN5, regulates phase change and morphogenesis. *Development*, *130*(8), 1493–1504. https://doi.org/10.1242/dev.00362

Bologna, N. G., Iselin, R., Abriata, L. A., Sarazin, A., Pumplin, N., Jay, F., Grentzinger, T., Dal Peraro, M., & Voinnet, O. (2018). Nucleo-cytosolic shuttling of ARGONAUTE1 prompts a revised model of the plant microRNA Pathway. *Molecular Cell*, *69*, 709–719. https://doi.org/10.1016/j.molcel.2018.01.007

Bologna, N. G., Mateos, J. L., Bresso, E. G., & Palatnik, J. F. (2009). A loop-to-base processing mechanism underlies the biogenesis of plant microRNAs miR319 and miR159. *The EMBO Journal*, *28*, 3646–3656. https://doi.org/10.1038/emboj.2009.292

Bologna, N. G., Schapire, A. L., Zhai, J., Chorostecki, U., Boisbouvier, J., Meyers, B. C., & Palatnik, J. F. (2013). Multiple RNA recognition patterns during microRNA biogenesis in plants. *Genome Research*, *23*(10), 1675–1689. https://doi.org/10.1101/gr.153387.112

Bologna, N. G., & Voinnet, O. (2014). The diversity, biogenesis, and activities of endogenous silencing small RNAs in Arabidopsis. *Annual Review of Plant Biology*, *65*, 473–503. https://doi.org/10.1146/annurev-arplant-050213-035728

Brodersen, P., Sakvarelidze-Achard, L., Bruun-Rasmussen, M., Dunoyer, P., Yamamoto, Y. Y., Sieburth, L., & Voinnet, O. (2008). Widespread translational inhibition by plant miRNAs and siRNAs. *Science*, *320*(5880), 1185–1190. https://doi.org/10.1126/science.1159151

Budak, H., & Akpinar, B. A. (2015). Plant miRNAs: Biogenesis, organization and origins. *Functional & Integrative Genomics*, *15*, 523–531. https://doi.org/10.1007/s10142-015-0451-2

Bullini, L., & Coluzzi, M. (1972). Natural selection and genetic drift in protein polymorphism. *Nature*, *239*(5368), 160–161. https://doi.org/10.1038/239160a0

Caudy, A. A., Ketting, R. F., Hammond, S. M., Denli, A. M., Bathoorn, A. M., Tops, B. B., Silva, J. M., Myers, M. M., Hannon, G. J., & Plasterk, R. H. (2003). A micrococcal nuclease homologue in RNAi effector complexes. *Nature*, *425*(6956), 411–414. https://doi.org/10.1038/nature01956

Chen, P. Y., Manninga, H., Slanchev, K., Chien, M., Russo, J. J., Ju, J., Sheridan, R., John, B., Marks, D. S., Gaidatzis, D., & Sander, C. (2005). The developmental miRNA profiles of zebrafish as determined by small RNA cloning. *Genes & Development*, *19*(11), 1288–1293. https://doi.org/10.1101/gad.1310605

Chen, X. (2004). A microRNA as a translational repressor of APETALA2 in Arabidopsis flower development. *Science*, *303*(5666), 2022–2025. https://doi.org/10.1126/science.1088060

Chen, X., Liu, J., Cheng, Y., & Jia, D. (2002). HEN1 functions pleiotropically in Arabidopsis development and acts in C function in the flower. *Development*, *129*,1085–1094. https://doi.org/10.1242/dev.129.5.1085

Chen, X., & Rechavi, O. (2021). Plant and animal small RNA communications between cells and organisms. *Nature Reviews Molecular Cell Biology*, *23*, 185–203. https://doi.org/10.1038/s41580-021-00425-y

Correa, J. P. O., Silva, E. M., & Nogueira, F. T. S. (2018). Molecular control by non-coding RNAs during fruit development: From gynoecium patterning to fruit ripening. *Frontiers in Plant Science*, *9*, 1760. https://doi.org/10.3389/fpls.2018.01760

Creasey, K. M., Zhai, J., Borges, F., Van Ex, F., Regulski, M., Meyers, B. C., & Martienssen, R. A. (2014). miRNAs trigger widespread epigenetically activated siRNAs from transposons in Arabidopsis. *Nature*, *508*(7496), 411–415. https://doi.org/10.1038/nature13069

Cuperus, J. T., Fahlgren, N., & Carrington, J. C. (2011). Evolution and functional diversification of MIRNA genes. *Plant Cell*, *23*(2), 431–442. https://doi.org/10.1105/tpc.110.082784

D'Ario, M., Griffiths-Jones, S., & Kim, M. (2017). Small RNAs: Big impact on plant development. *Trends in Plant Science*, *22*, 1056–1068. https://doi.org/10.1016/j.tplants.2017.09.009

Deng, P. C., Muhammad, S., Cao, M., & Wu, L. (2018). Biogenesis and regulatory hierarchy of phased small interfering RNAs in plants. *Plant Biotechnology Journal*, *16*, 965–975. https://doi.org/10.1111/pbi.12882

Dolata, J., Bajczyk, M., Bielewicz, D., Niedojadlo, K., Niedojadlo, J., Pietrykowska, H., Walczak, W., Szweykowska-Kulinska, Z., & Jarmolowski, A. (2016). Salt stress reveals a new role for ARGONAUTE1 in miRNA biogenesis at the transcriptional and posttranscriptional levels. *Plant Physiology*, *172*, 297–312. https://doi.org/10.1104/pp.16.00830

Dong, Z., Han, M. H., & Fedoroff, N. (2008). The RNA-binding proteins HYL1 and SE promote accurate in vitro processing of pri-miRNA by DCL1. *Proceedings of the National Academy of Sciences of the United States of America*, *105*, 9970–9975. https://doi.org/10.1073/pnas.0803356105

Du, F., Gong, W., Bosca, S., Tucker, M., & Vaucheret, H. (2020). Dose-dependent AGO1-mediated inhibition of the miRNA165/166 pathway modulates stem cell maintenance in arabidopsis shoot apical meristem. *Plant Communication*, *1*, 100002. https://doi.org/10.1016/j.xplc.2019.100002

Dugas, D. V., & Bartel, B. (2004). MicroRNA regulation of gene expression in plants. *Current Opinion in Plant Biology*, *7*(5), 512–520. https://doi.org/10.1016/j.pbi.2004.07.011

Eulalio, A., Rehwinkel, J., Stricker, M., Huntzinger, E., Yang, S. F., Doerks, T., Dorner, S., Bork, P., Boutros, M., & Izaurralde, E. (2007). Target-specific requirements for enhancers of decapping in miRNA-mediated gene silencing. *Genes & Development*, *21*(20), 2558–2570. https://doi.org/10.1101/gad.44310

Fahlgren, N., Howell, M. D., Kasschau, K. D., Chapman, E. J., Sullivan, C. M., Cumbie, J. S., Givan, S. A., Law, T. F., Grant, S. R., Dangl, J. L., & Carrington, J. C. (2007). High-throughput sequencing of Arabidopsis microRNAs: Evidence for frequent birth and death of MIRNA genes. *PLoS One*, *2*(2), e219. https://doi.org/10.1371/journal.pone.0000219

Fahlgren, N., Jogdeo, S., Kasschau, K. D., Sullivan, C. M., Chapman, E. J., Laubinger, S., Smith, L. M., Dasenko, M., Givan, S. A., Weigel, D., & Carrington, J. C. (2010). MicroRNA gene evolution in Arabidopsis lyrata and Arabidopsis thaliana. *Plant Cell*, *22*(4), 1074–1089. https://doi.org/10.1105/tpc.110.073999

Fang, Y. D., & Spector, D. L. (2007). Identification of nuclear dicing bodies containing proteins for microRNA biogenesis in living Arabidopsis plants. *Current Biology*, *17*, 818–823. https://doi.org/10.1016/j.cub.2007.04.005

Fei, Q. L., Xia, R., & Meyers, B. C. (2013). Phased, secondary, small interfering RNAs in posttranscriptional regulatory networks. *Plant Cell*, *25*, 2400–2415. https://doi.org/10.1105/tpc.113.114652

Floyd, S., & Bowman, J. (2004). Ancient microRNA target sequences in plants. *Nature*, *428*, 485–486. https://doi.org/10.1038/428485a

Gandikota, M., Birkenbihl, R. P., Höhmann, S., Cardon, G. H., Saedler, H., & Huijser, P. (2007). The miRNA156/157 recognition element in the 3′ UTR of the Arabidopsis SBP box gene SPL3 prevents early flowering by translational inhibition in seedlings. *The Plant Journal*, *49*(4), 683–693. https://doi.org/10.1111/j.1365-313X.2006.02983.x

Gao, R., Wang, Y., Gruber, M. Y., & Hannoufa, A. (2017). miR156/SPL10 modulates lateral root development, branching and leaf morphology in Arabidopsis by silencing AGAMOUS-LIKE 79. *Frontiers in Plant Science*, *8*, 2226. https://doi.org/10.3389/fpls.2017.02226

Giraldez, A. J., Cinalli, R. M., Glasner, M. E., Enright, A. J., Thomson, J. M., Baskerville, S., Hammond, S. M., Bartel, D. P., & Schier, A. F. (2005). MicroRNAs regulate brain morphogenesis in zebrafish. *Science*, *308*(5723), 833–838. https://doi.org/10.1126/science.1109020

Gray, W. M., Kepinski, S., Rouse, D., Leyser, O., & Estelle, M. (2001). Auxin regulates SCF (TIR1)-dependent degradation of AUX/IAA proteins. *Nature*, *414*, 271–276. https://doi.org/10.1038/35104500

Griffiths-Jones, S., Grocock, R. J., Van Dongen, S., Bateman, A., & Enright, A. J. (2006). miRBase: microRNA sequences, targets and gene nomenclature. *Nucleic Acids Research*, *34*(suppl_1), D140–D144. https://doi.org/10.1093/nar/gkj112

Guddeti, S., Zhang, D. C., Li, A. L., Leseberg, C. H., Kang, H., Li, X. G., Zhai, W. X., Johns, M. A., & Mao, L. (2005). Molecular evolution of the rice miR395 gene family. *Cell Research*, *15*(8), 631–638. https://doi.org/10.1038/sj.cr.7290333

Guo, H. S., Xie, Q., Fei, J. F., & Chua, N. H. (2005). MicroRNA directs mRNA cleavage of the transcription factor NAC1 to downregulate auxin signals for Arabidopsis lateral root development. *Plant Cell*, *17*, 1376–1386. https://doi.org/10.1105/tpc.105.030841

Hagen, G., & Guilfoyle, T. (2002). Auxin-responsive gene expression: Genes, promoters and regulatory factors. *Plant Molecular Biology*, *49*, 373–385. https://doi.org/10.1023/A:1015207114117

Hammond, S. M., Boettcher, S., Caudy, A. A., Kobayashi, R., & Hannon, G. J. (2001). Argonaute2, a link between genetic and biochemical analyses of RNAi. *Science*, *293*(5532), 1146–1150. https://doi.org/10.1126/science.1064023

Havecker, E. R., Wallbridge, L. M., Hardcastle, T. J., Bush, M. S., Kelly, K. A., Dunn, R. M., Schwach, F., Doonan, J. H., & Baulcombe, D. C. (2010). The Arabidopsis RNA-directed DNA methylation argonautes functionally diverge based on their expression and interaction with target loci. *The Plant Cell*, *22*(2), 321–334. https://doi.org/10.1105/tpc.109.072199

Heisel, S. E., Zhang, Y., Allen, E., Guo, L., Reynolds, T. L., Yang, X., Kovalic, D., & Roberts, J. K. (2008). Characterization of unique small RNA populations from rice grain. *PLoS One*, *3*(8), e2871. https://doi.org/10.1371/journal.pone.0002871

Helwak, A., Kudla, G., Dudnakova, T., & Tollervey, D. (2013). Mapping the human miRNA interactome by CLASH reveals frequent noncanonical binding. *Cell*, *153*, 654–665. https://doi.org/10.1016/j.cell.2013.03.043

Hutvágner, G., & Zamore, P. D. (2002). RNAi: Nature abhors a double-strand. *Current Opinion in Genetics & Development*, *12*(2), 225–232. https://doi.org/10.1016/S0959-437X(02)00290-3

Jagtap, S., & Shivaprasad, P. V. (2014). Diversity, expression and mRNA targeting abilities of Argonaute-targeting miRNAs among selected vascular plants. *BMC Genomics*, *15*, 1049. https://doi.org/10.1186/1471-2164-15-1049

Jian, X., Zhang, L., Li, G., Zhang, L., Wang, X., Cao, X., Fang, X., & Chen, F. (2010). Identification of novel stress-regulated microRNAs from Oryza sativa L. *Genomics*, *95*(1), 47–55. https://doi.org/10.1016/j.ygeno.2009.08.017

Jodder, J. (2020). MiRNA-mediated regulation of auxin signaling pathway during plant development and stress responses. *Journal of Biosciences*, *45*, 91. https://doi.org/10.1007/s12038-020-00062-1

Jones-Rhoades, M. W., & Bartel, D. P. (2004). Computational identification of plant microRNAs and their targets, including a stress-induced miRNA. *Molecular Cell*, *14*, 787–799. https://doi.org/10.1016/j.molcel.2004.05.027

Jones-Rhoades, M. W., Bartel, D. P., & Bartel, D. P. (2006). MicroRNAs and their regulatory roles in plants. *Annual Review of Plant Biology*, *57*, 19–53. https://doi.org/10.1146/annurev.arplant.57.032905.105218

Kasschau, K. D., Xie, Z., Allen, E., Llave, C., Chapman, E. J., Krizan, K. A., & Carrington, J. C. (2003). P1/HC-Pro, a viral suppressor of RNA silencing, interferes with Arabidopsis development and miRNA function. *Developmental Cell*, *4*(2), 205–217. https://doi.org/10.1016/S1534-5807(03)00025-X

Khraiwesh, B., Arif, M. A., Seumel, G. I., Ossowski, S., Weigel, D., Reski, R., & Frank, W. (2010). Transcriptional control of gene expression by microRNAs. *Cell*, *140*(1), 111–122. https://doi.org/10.1016/j.cell.2009.12.023

Kidner, C. A., & Martienssen, R. A. (2005a). The developmental role of microRNA in plants. *Current Opinion in Plant Biology*, *8*(1), 38–44. https://doi.org/10.1016/j.pbi.2004.11.008

Kidner, C. A., & Martienssen, R. A. (2005b). The role of ARGONAUTE1 (AGO1) in meristem formation and identity. *Developmental Biology*, *280*(2), 504–517. https://doi.org/10.1016/j.ydbio.2005.01.031

Kurtoglu, K. Y., Kantar, M., Lucas, S. J., & Budak, H. (2013). Unique and conserved microRNAs in wheat chromosome 5D revealed by nextgeneration sequencing. *PLoS One*, *8*, e69801. https://doi.org/10.1371/journal.pone.0069801

Law, J. A., & Jacobsen, S. E. (2010). Establishing, maintaining and modifying DNA methylation patterns in plants and animals. *Nature Reviews Genetics*, *11*(3), 204–220. https://doi.org/10.1038/nrg2719

Lee, R. C., Feinbaum, R. L., & Ambros, V. (1993). The C. elegans heterochronic gene lin-4 encodes small RNAs with antisense complementarity to lin-14. *Cell*, *75*(5), 843–854. https://doi.org/10.1016/0092-8674(93)90529-Y

Lee, Y., Jeon, K., Lee, J. T., Kim, S., & Kim, V. N. (2002). MicroRNA maturation: Stepwise processing and subcellular localization. *The EMBO Journal*, *21*, 4663–4670. https://doi.org/10.1093/emboj/cdf476

Lelandais-Brière, C., Naya, L., Sallet, E., Calenge, F., Frugier, F., Hartmann, C., Gouzy, J., & Crespi, M. (2009). Genome-wide Medicago truncatula small RNA analysis revealed novel microRNAs and isoforms differentially regulated in roots and nodules. *The Plant Cell*, *21*(9), 2780–2796. https://doi.org/10.1105/tpc.109.068130

Lewis, B. P., Burge, C. B., & Bartel, D. P. (2005). Conserved seed pairing, often flanked by adenosines, indicates that thousands of human genes are microRNA targets. *Cell*, *120*(1), 15–20. https://doi.org/10.1016/j.cell.2004.12.035

Lian, H., Wang, L., Ma, N., Zhou, C. M., & Han, L. (2021). Redundant and specific roles of individual MIR172 genes in plant development. *PLoS Biology*, *19*, e3001044. https://doi.org/10.1371/journal.pbio.3001044

Lim, L. P., Glasner, M. E., Yekta, S., Burge, C. B., & Bartel, D. P. (2003). Vertebrate microRNA genes. *Science*, *299*(5612), 1540–1540. https://doi.org/10.1101/gad.1074403

Liu, C., Xin, Y., Xu, L., Cai, Z., Xue, Y., Liu, Y., Xie, D., Liu, Y., & Qi, Y. (2018). Arabidopsis Argonaute 1 binds chromatin to promote gene transcription in response to hormones and stresses. *Developmental Cell*, *44*, 348–361.e7. https://doi.org/10.1016/j.devcel.2017.12.002

Liu, C., Zhang, L., Sun, J., Wang, M. B., & Liu, Y. (2010). A simple artificial microRNA vector based on ath-miR169d precursor from Arabidopsis. *Molecular Biology Reports*, *37*(2), 903–909. https://doi.org/10.1007/s11033-009-9713-1

Llave, C., Xie, Z., Kasschau, K. D., & Carrington, J. C. (2002). Cleavage of scarecrow-like mRNA targets directed by a class of Arabidopsis miRNA. *Science*, *297*(5589), 2053–2056. https://doi.org/10.1126/science.1076311

Lu, C., & Fedoroff, N. A. (2000). Mutation in the Arabidopsis HYL1 gene encoding a dsRNA binding protein affects responses to abscisic acid, auxin and cytokinin. *Plant Cell*, *12*, 2351–2366. https://doi.org/10.1105/tpc.12.12.2351

Lu, C., Jeong, D. H., Kulkarni, K., Pillay, M., Nobuta, K., German, R., Thatcher, S. R., Maher, C., Zhang, L., Ware, D., & Liu, B. (2008a). Genome-wide analysis for discovery of rice microRNAs reveals natural antisense microRNAs (nat-miRNAs). *Proceedings of the National Academy of Sciences of the United States of America*, *105*(12), 4951–4956. https://doi.org/10.1073/pnas.0708743105

Lu, C., Kulkarni, K., Souret, F. F., MuthuValliappan, R., Tej, S. S., Poethig, R. S., Henderson, I. R., Jacobsen, S. E., Wang, W., Green, P. J., & Meyers, B. C. (2006). MicroRNAs and other small RNAs enriched in the Arabidopsis RNA-dependent RNA polymerase-2 mutant. *Genome Research*, *16*(10), 1276–1288. https://doi.org/10.1101/gr.5530106

Luo, Y., Guo, Z., & Li, L. (2013). Evolutionary conservation of microRNA regulatory programs in plant flower development. *Developmental Biology, 380,* 133–144. https://doi.org/10.1016/j.ydbio.2013.05.009

Ma, Z., Coruh, C., & Axtell, M. J. (2010). Arabidopsis lyrata small RNAs: Transient MIRNA and small interfering RNA loci within the Arabidopsis genus. *The Plant Cell, 22*(4), 1090–1103. https://doi.org/10.1105/tpc.110.073882

Ma, Z., Hu, X., Cai, W., Huang, W., & Zhou, X. (2014). Arabidopsis miR171-targeted scarecrow-like proteins bind to GT cis-elements and mediate gibberellin-regulated chlorophyll biosynthesis under light conditions. *PLoS Genetics, 10,* e1004519. https://doi.org/10.1371/journal.pgen.1004519

Maher, C., Stein, L., & Ware, D. (2006). Evolution of Arabidopsis microRNA families through duplication events. *Genome Research, 16,* 510–519. https://doi.org/10.1101/gr.4680506

Mallory, A. C., Bartel, D. P., & Bartel, B. (2005). MicroRNA-directed regulation of Arabidopsis Auxin Response Factor 17 is essential for proper development and modulates expression of early auxin response genes. *Plant Cell, 17*(5), 1360–1375. https://doi.org/10.1105/tpc.105.031716

Mallory, A. C., Reinhart, B. J., Jones-Rhoades, M. W., Tang, G., Zamore, P. D., Barton, M. K., & Bartel, D. P. (2004). MicroRNA control of Phabulosa in leaf development: Importance of pairing to the microRNA 5' region. *The EMBO Journal, 23*(16), 3356–3364. https://doi.org/10.1038/sj.emboj.7600340

Mallory, A., & Vaucheret, H. (2006). Functions of microRNAs and related small RNAs in plants. *Nature Genetics, 38*(Suppl 6), S31–S36. https://doi.org/10.1038/ng1791

Mateos, J. L., Bologna, N. G., Chorostecki, U., & Palatnik, J. F. (2010). Identification of microRNA processing determinants by random mutagenesis of Arabidopsis MIR172a precursor. *Current Biology, 20,* 49–54. https://doi.org/10.1016/j.cub.2009.10.072

Mette, M. F., van der Winden, J., Matzke, M., & Matzke, A. J. (2002). Short RNAs can identify new candidate transposable element families in Arabidopsis. *Plant Physiology, 130*(1), 6–9. https://doi.org/10.1104/pp.007047

Mi, S., Cai, T., Hu, Y., Chen, Y., Hodges, E., Ni, F., Wu, L., Li, S., Zhou, H., Long, C., & Chen, S. (2008). Sorting of small RNAs into Arabidopsis argonaute complexes is directed by the 5' terminal nucleotide. *Cell, 133*(1), 116–127. https://doi.org/10.1016/j.cell.2008.02.034

Millar, A. A., & Waterhouse, P. M. (2005). Plant and animal microRNAs: Similarities and differences. *Functional & Integrative Genomics, 5,* 129–135. https://doi.org/10.1007/s10142-005-0145-2

Morel, J. B., Godon, C., Mourrain, P., Béclin, C., Boutet, S., Feuerbach, F., Proux, F., & Vaucheret, H. (2002). Fertile hypomorphic ARGONAUTE (ago1) mutants impaired in post-transcriptional gene silencing and virus resistance. *Plant Cell, 14*(3), 629–639. https://doi.org/10.1105/tpc.010358

Navarro, L., Dunoyer, P., Jay, F., Arnold, B., Dharmasiri, N., Estelle, M., Voinnet, O., & Jones, J. D. (2006). A plant miRNA contributes to antibacterial resistance by repressing auxin signaling. *Science, 312,* 436–439. https://doi.org/10.1126/science.1126088

Niu, Q. W., Lin, S. S., Reyes, J. L., Chen, K. C., Wu, H. W., Yeh, S. D., & Chua, N. H. (2006). Expression of artificial microRNAs in transgenic Arabidopsis thaliana confers virus resistance. *Nature Biotechnology, 24*(11), 1420–1428. https://doi.org/10.1038/nbt1255

Nozawa, M., Miura, S., & Nei, M. (2012). Origins and evolution of microRNA genes in plant species. *Genome Biology and Evolution, 4,* 230–239. https://doi.org/10.1093/gbe/evs002

O'Brien, J., Hayder, H., Zayed, Y., & Peng, C. (2018). Overview of microRNA biogenesis, mechanisms of actions, and circulation. *Frontiers in Endocrinology, 9,* 402. https://doi.org/10.3389/fendo.2018.00402

O'Donnell, K. A., Wentzel, E. A., Zeller, K. I., Dang, C. V., & Mendell, J. T. (2005). c-Myc-regulated microRNAs modulate E2F1 expression. *Nature, 435,* 839–843. https://doi.org/10.1038/nature03677

Palatnik, J. F., Allen, E., Wu, X., Schommer, C., Schwab, R., Carrington, J. C., & Weigel, D. (2003). Control of leaf morphogenesis by microRNAs. *Nature, 425*(6955), 257–263. https://doi.org/10.1038/nature01958

Palme, K., Hesse, T., Moore, I., Campos, N., Feldwisch, J., Garbers, C., Hesse, F., & Schell, J. (1991). Hormonal modulation of plant growth: The role of auxin perception. *Mechanisms of Development, 33,* 97–106. https://doi.org/10.1016/0925-4773(91)90076-I

Papp, I., Mette, M. F., Aufsatz, W., Daxinger, L., Schauer, S. E., Ray, A., Van Der Winden, J., Matzke, M., & Matzke, A. J. (2003). Evidence for nuclear processing of plant micro RNA and short interfering RNA precursors. *Plant Physiology, 132*(3), 1382–1390. https://doi.org/10.1104/pp.103.021980

Parizotto, E. A., Dunoyer, P., Rahm, N., Himber, C., & Voinnet, O. (2004). In vivo investigation of the transcription, processing, endonucleolytic activity, and functional relevance of the spatial distribution of a plant miRNA. *Genes & Development, 18*(18), 2237–2242. https://doi.org/10.1101/gad.307804

Park, M. Y., Wu, G., Gonzalez-Sulser, A., Vaucheret, H., & Poethig, R. S. (2005). Nuclear processing and export of microRNAs in Arabidopsis. *Proceedings of the National Academy of Sciences of the USA, 102,* 3691–3696.

Park, W., Li, J., Song, R., Messing, J., & Chen, X. (2002). CARPEL FACTORY, a Dicer homolog, and HEN1, a novel protein, act in microRNA metabolism in Arabidopsis thaliana. *Current Biology*, *12*(17), 1484–1495. https://doi.org/10.1016/S0960-9822(02)01017-5

Pashkovskiy, P. P., & Ryazansky, S. S. (2013). Biogenesis, evolution, and functions of plant microRNAs. *BiochemBiokhimii a*, *78*, 627–637. https://doi.org/10.1134/S0006297913060084

Pegler, J. L., Grof, C. P. L., & Eamens, A. L. (2019). The plant microRNA pathway: The production and action stages. *Methods in Molecular Biology*, *1932*, 15–39. https://doi.org/10.1007/978-1-4939-9042-9_2

Plotnikova, A., Kellner, M. J., Schon, M. A., Mosiolek, M., & Nodine, M. D. (2019). MicroRNA dynamics and functions during Arabidopsis embryogenesis. *Plant Cell*, *31*, 2929–2946. https://doi.org/10.1105/tpc.19.00395

Qi, Y., He, X., Wang, X. J., Kohany, O., Jurka, J., & Hannon, G. J. (2006). Distinct catalytic and non-catalytic roles of ARGONAUTE4 in RNA-directed DNA methylation. *Nature*, *443*(7114), 1008–1012. https://doi.org/10.1038/nature05198

Rajagopalan, R., Vaucheret, H., Trejo, J., & Bartel, D. P. (2006). A diverse and evolutionarily fluid set of microRNAs in Arabidopsis thaliana. *Genes & Development*, *20*, 3407–3425. https://doi.org/10.1101/gad.1476406

Ramachandran, P., Carlsbecker, A., & Etchells, J. P. (2017). Class III HD-ZIPs govern vascular cell fate: An HD view on patterning and differentiation. *Journal of Experimental Botany*, *68*, 55–69. https://doi.org/10.1093/jxb/erw370

Reinhart, B. J., Slack, F. J., Basson, M., Pasquinelli, A. E., Bettinger, J. C., Rougvie, A. E., Horvitz, H. R., & Ruvkun, G. (2000). The 21-nucleotide let-7 RNA regulates developmental timing in Caenorhabditis elegans. *Nature*, *403*(6772), 901–906. https://doi.org/10.1038/35002607

Reinhart, B. J., Weinstein, E. G., Rhoades, M. W., Bartel, B., & Bartel, D. P. (2002). MicroRNAs in plants. *Genes & Development*, *16*(13), 1616–1626. https://doi.org/10.1101/gad.1004402

Reynolds, A., Leake, D., Boese, Q., Scaringe, S., Marshall, W. S., & Khvorova, A. (2004). Rational siRNA design for RNA interference. *Nature Biotechnology*, *22*(3), 326–330. https://doi.org/10.1038/nbt936

Rhoades, M. W., Reinhart, B. J., Lim, L. P., Burge, C. B., Bartel, B., & Bartel, D. P. (2002). Prediction of plant microRNA targets. *Cell*, *110*, 513–520. https://doi.org/10.1016/S0092-8674(02)00863-2

Rogers, K., & Chen, X. M. (2013). Biogenesis, turnover, and mode of action of plant MicroRNAs. *Plant Cell*, *25*, 2383–2399. https://doi.org/10.1105/tpc.113.113159

Schwab, R., Ossowski, S., Riester, M., Warthmann, N., & Weigel, D. (2006). Highly specific gene silencing by artificial microRNAs in Arabidopsis. *The Plant Cell*, *18*(5), 1121–1133. https://doi.org/10.1105/tpc.105.039834

Schwab, R., Palatnik, J. F., Riester, M., Schommer, C., Schmid, M., & Weigel, D. (2005). Specific effects of microRNAs on the plant transcriptome. *Developmental Cell*, *8*, 517–527. https://doi.org/10.1016/j.devcel.2005.01.018

Secic, E., Kogel, K. H., & Ladera-Carmona, M. J. (2021). Biotic stress-associated microRNA families in plants. *Journal of Plant Physiology*, *263*, 153451. https://doi.org/10.1016/j.jplph.2021.153451

Siddiqui, Z. H., Abbas, Z. K., Ansari, M. W., & Khan, M. N. (2019). The role of miRNA in somatic embryogenesis. *Genomics*, *111*, 1026–1033. https://doi.org/10.1016/j.ygeno.2018.11.022

Song, L., Axtell, M. J., & Fedoroff, N. V. (2010). RNA secondary structural determinants of miRNA precursor processing in Arabidopsis. *Current Biology*, *20*, 37–41. https://doi.org/10.1016/j.cub.2009.10.076

Song, X., Li, Y., Cao, X., & Qi, Y. (2019). MicroRNAs and their regulatory roles in plant-environment interactions. *Annual Review of Plant Biology*, *70*, 489–525. https://doi.org/10.1146/annurev-arplant-050718-100334

Sorin, C., Bussell, J. D., Camus, I., Ljung, K., Kowalczyk, M., Geiss, G., McKhann, H., Garcion, C., Vaucheret, H., Sandberg, G., & Bellini, C. (2005). Auxin and light control of adventitious rooting in Arabidopsis require ARGONAUTE1. *Plant Cell*, *17*, 1343–1359. https://doi.org/10.1105/tpc.105.031625

Souret, F. F., Kastenmayer, J. P., & Green, P. J. (2004). AtXRN4 degrades mRNA in Arabidopsis and its substrates include selected miRNA targets. *Molecular Cell*, *15*(2), 173–183. https://doi.org/10.1016/j.molcel.2004.06.006

Subramanian, S., Fu, Y., Sunkar, R., Barbazuk, W. B., Zhu, J. K., & Yu, O. (2008). Novel and nodulation-regulated microRNAs in soybean roots. *BMC Genomics*, *9*(1), 1–14. https://doi.org/10.1186/1471-2164-9-160

Sun, G. (2012). MicroRNAs and their diverse functions in plants. *Plant Molecular Biology*, *80*, 17–36. https://doi.org/10.1007/s11103-011-9817-6

Sun, X., Lin, L., & Sui, N. (2019). Regulation mechanism of microRNA in plant response to abiotic stress and breeding. *Molecular Biology Reports*, *46*, 1447–1457. https://doi.org/10.1007/s11033-018-4511-2

Sunkar, R., Li, Y. F., & Jagadeeswaran, G. (2012). Functions of microRNAs in plant stress responses. *Trends in Plant Science*, *17*(4), 196–203. https://doi.org/10.1016/j.tplants.2012.01.010

Sunkar, R., Zhou, X., Zheng, Y., Zhang, W., & Zhu, J. K. (2008). Identification of novel and candidate miRNAs in rice by high throughput sequencing. *BMC Plant Biology*, *8*(1), 1–17. https://doi.org/10.1186/1471-2229-8-25

Sunkar, R., & Zhu, J. K. (2004). Novel and stress-regulated microRNAs and other small RNAs from Arabidopsis. *Plant Cell*, *16*, 2001–2019. https://doi.org/10.1105/tpc.104.022830

Szittya, G., Moxon, S., Santos, D. M., Jing, R., Fevereiro, M. P., Moulton, V., & Dalmay, T. (2008). High-throughput sequencing of Medicago truncatula short RNAs identifies eight new miRNA families. *BMC Genomics*, *9*(1), 1–9. https://doi.org/10.1186/1471-2164-9-593

Tang, G., Reinhart, B. J., Bartel, D. P., & Zamore, P. D. (2003). A biochemical framework for RNA silencing in plants. *Genes & Development*, *17*(1), 49–63. https://doi.org/10.1101/gad.1048103

Tanzer, A., & Stadler, P. F. (2004). Molecular evolution of a microRNA cluster. *Journal of Molecular Biology*, *339*(2), 327–335. https://doi.org/10.1016/j.jmb.2004.03.065

Tanzer, A., & Stadler, P. F. (2006). Evolution of microRNAs. In S. Y. Yig (Ed.), *microRNA protocols, methods in molecular biology* (pp. 335–350). Humana Press Inc. https://doi.org/10.1385/1-59745-123-1:335

Tanzer, A., Amemiya, C. T., Kim, C. B., & Stadler, P. F. (2005). Evolution of microRNAs located within Hox gene clusters. *The Journal of Experimental Zoology – B: Molecular and Developmental Evolution*, *304B*, 75–85. https://doi.org/10.1002/jez.b.21021

Taylor, R. S., Tarver, J. E., Hiscock, S. J., & Donoghue, P. C. J. (2014). Evolutionary history of plant microRNAs. *Trends in Plant Science*, *19*, 175–182. https://doi.org/10.1016/j.tplants.2013.11.008

Tuskan, G. A., Difazio, S., Jansson, S., Bohlmann, J., Grigoriev, I., Hellsten, U., Putnam, N., Ralph, S., Rombauts, S., Salamov, A., & Schein, J. (2006). The genome of black cottonwood, Populus trichocarpa (Torr & Gray). *Science*, *313*, 1596–1604. https://doi.org/10.1126/science.1128691

Van Vu, T., Choudhury, N. R., & Mukherjee, S. K. (2013). Transgenic tomato plants expressing artificial microRNAs for silencing the pre-coat and coat proteins of a begomovirus, tomato leaf curl New Delhi virus, show tolerance to virus infection. *Virus Research*, *172*(1–2), 35–45. https://doi.org/10.1016/j.virusres.2012.12.008

Vaucheret, H., Vazquez, F., Crété, P., & Bartel, D. P. (2004). The action of ARGONAUTE1 in the miRNA pathway and its regulation by the miRNA pathway are crucial for plant development. *Genes & Development*, *18*(10), 1187–1197. https://doi.org/10.1101/gad.1201404

Voinnet, O. (2009). Origin, biogenesis, and activity of plant microRNAs. *Cell*. *136*(4), 669–687. https://doi.org/10.1016/j.cell.2009.01.046

Vazquez, F., Gasciolli, V., Crété, P., & Vaucheret, H. (2004). The nuclear dsRNA binding protein HYL1 is required for microRNA accumulation and plant development, but not posttranscriptional transgene silencing. *Current Biology*, *14*(4), 346–351. https://doi.org/10.1016/j.cub.2004.01.035

Wang, S., Wu, K., Yuan, Q., Liu, X., Liu, Z., Lin, X., Zeng, R., Zhu, H., Dong, G., Qian, Q., & Zhang, G. (2012). Control of grain size, shape and quality by OsSPL16 in rice. *Nature Genetics*, *44*(8), 950–954. https://doi.org/10.1038/ng.2327

Weijers, D., & Jurgens, G. (2005). Auxin and embryo axis formation: The ends in sight? *Current Opinion in Plant Biology*, *8*, 32–37. https://doi.org/10.1016/j.pbi.2004.11.001

Werner, S., Wollmann, H., Schneeberger, K., & Weigel, D. (2010). Structure determinants for accurate processing of miR172a in Arabidopsis thaliana. *Current Biology*, *20*, 42–48. https://doi.org/10.1016/j.cub.2009.10.073

Woodward, A. W., & Bartel, B. (2005). Auxin: Regulation, action, and interaction. *Annals of Botany*, *95*, 707–735. https://doi.org/10.1093/aob/mci083

Wu, L., Zhou, H., Zhang, Q., Zhang, J., Ni, F., Liu, C., & Qi, Y. (2010). DNA methylation mediated by a microRNA pathway. *Molecular Cell*, *38*(3), 465–475. https://doi.org/10.1016/j.molcel.2010.03.008

Xie, F., Stewart Jr., C. N., Taki, F. A., He, Q., Liu, H., & Zhang, B. (2014). Highthroughput deep sequencing shows that microRNAs play important roles in switch grass responses to drought and salinity stress. *Plant Biotechnology Journal*, *12*, 354–366. https://doi.org/10.1111/pbi.12142

Xie, Q., Guo, H. S., Dallman, G., Fang, S., Weissman, A. M., & Chua, N. H. (2002). SINAT5 promotes ubiquitin-related degradation of NAC1 to attenuate auxin signals. *Nature*, *419*, 167–170. https://doi.org/10.1038/nature00998

Xie, Z., Kasschau, K. D., & Carrington, J. C. (2003). Negative feedback regulation of Dicer-Like1 in Arabidopsis by microRNA-guided mRNA degradation. *Current Biology*, *13*(9), 784–789. https://doi.org/10.1016/S0960-9822(03)00281-1

Xie, Z. X., Allen, E., Fahlgren, N., Calamar, A., Givan, S. A., & Carrington, J. C. (2005). Expression of Arabidopsis MIRNA genes. *Plant Physiology*, *138*, 2145–2154. https://doi.org/10.1104/pp.105.062943

Yan, K. S., Yan, S., Farooq, A., Han, A., Zeng, L., & Zhou, M. M. (2003). Structure and conserved RNA binding of the PAZ domain. *Nature*, *426*(6965), 469–474. https://doi.org/10.1038/nature02129

Yang, G., Li, Y., Wu, B., Zhang, K., Gao, L., & Zheng, C. (2019). MicroRNAs transcriptionally regulate promoter activity in Arabidopsis thaliana. *Journal of Integrative Plant Biology*, *61*(11), 1128–1133. https://doi.org/10.1111/jipb.12775

Yang, L., Liu, Z. Q., Lu, F., Dong, A. W., & Huang, H. (2006). SERRATE is a novel nuclear regulator in primary microRNA processing in Arabidopsis. *The Plant Journal*, *47*, 841–850. https://doi.org/10.1111/j.1365-313X.2006.02835.x

Ying, S. Y., Chang, D. C., & Lin, S. L. (2008). The microRNA (miRNA): Overview of the RNA genes that modulate gene function. *Molecular Biotechnology*, *38*(3), 257–268. https://doi.org/10.1007/s12033-007-9013-8

Yu, B., Yang, Z., Li, J., Minakhina, S., Yang, M., Padgett, R. W., Steward, R., & Chen, X. (2005). Methylation as a crucial step in plant microRNA biogenesis. *Science*, *307*, 932–935. https://doi.org/10.1126/science.1107130

Yu, S., & Wang, J. W. (2020). The crosstalk between microRNAs and gibberellin signaling in plants. *Plant and Cell Physiology*, *61*, 1880–1890. https://doi.org/10.1093/pcp/pcaa079

Zeng, Y., Wagner, E. J., & Cullen, B. R. (2002). Both natural and designed micro RNAs can inhibit the expression of cognate mRNAs when expressed in human cells. *Molecular Cell*, *9*(6), 1327–1333. https://doi.org/10.1016/S1097-2765(02)00541-5

Zhang, B., Pan, X., Cannon, C. H., Cobb, G. P., & Anderson, T. A. (2006). Conservation and divergence of plant microRNA genes. *The Plant Journal*, *46*, 243–259. https://doi.org/10.1111/j.1365-313X.2006.02697.x

Zhang, B., & Wang, Q. (2015). MicroRNA-based biotechnology for plant improvement. *Journal of Cellular Physiology*, *230*, 1–15. https://doi.org/10.1002/jcp.24685

Zhang, B. H., Pan, X. P., Wang, Q. L., Cobb, G. P., & Anderson, T. A. (2005). Identification and characterization of new plant microRNAs using EST analysis. *Cell Research*, *15*, 336–360. https://doi.org/10.1038/sj.cr.7290302

Zhang, H., Xia, R., Meyers, B. C., & Walbot, V. (2015). Evolution, functions, and mysteries of plant ARGONAUTE proteins. *Current Opinion in Plant Biology*, *27*, 84–90. https://doi.org/10.1016/j.pbi.2015.06.011

Zhang, L., Hou, D., Chen, X., Li, D., Zhu, L., Zhang, Y., Li, J., Bian, Z., Liang, X., Cai, X., & Yin, Y. (2012). Exogenous plant MIR168a specifically targets mammalian LDLRAP1: Evidence of cross-kingdom regulation by microRNA. *Cell Research*, *22*(1), 107–126. https://doi.org/10.1038/cr.2011.158

Zhang, L. W., Song, J. B., Shu, X. X., Zhang, Y., & Yang, Z. M. (2013). miR395 is involved in detoxification of cadmium in Brassica napus. *Journal of Hazardous Materials*, *250–251*, 204–211. https://doi.org/10.1016/j.jhazmat.2013.01.053

Zhu, Q. H., Spriggs, A., Matthew, L., Fan, L., Kennedy, G., Gubler, F., & Helliwell, C. (2008). A diverse set of microRNAs and microRNA-like small RNAs in developing rice grains. *Genome Research*, *18*(9), 1456–1465. https://doi.org/10.1101/gr.075572.107

2 Plant miRNAs
Biogenesis, Mode of Action, and Their Role

Bipin Maurya, Lakee Sharma, Nidhi Rai, Vishnu Mishra, Ashish Kumar, and Shashi Pandey Rai

2.1   INTRODUCTION

Small RNAs such as siRNAs, miRNA, snc RNA, and piRNA are usually short nucleotide sequences having a non-coding region. Through post-transcriptional gene silencing (PTGS) and chromatin-dependent (CDGS) gene silencing, these short non-coding RNA have recently emerged as a key regulator of gene expression. Small RNAs may use RNA activation (RNAa) mechanism to increase the expression of the targeted genes (Li et al., 2006; Pushparaj et al., 2008). In terms of origin, structure, and biological roles, small RNAs are classified into three main categories: miRNA (microRNA), siRNA (small interfering RNA), and piRNA (Piwi-interacting RNAs) 21–35 nucleotides, whereas miRNA and siRNA molecules have a length of approximately 21–22 nucleotides (Guleria et al., 2011). The initial process of the origin of miRNAs is transcription completed inside the nucleus, which transcribes a large pri-miRNA. It is further processed to nearly 70 nucleotides pre-miRNAs in the nucleus. These miRNAs are transported to the cytoplasm and behave like mature miRNAs (Jiang et al., 2009).Various pieces of research suggest that small RNAs are a critical component for regulating the developmental and physiological processes of organisms. They also play a role in the regulation of many cellular functions, including cell growth, differentiation, apoptosis, metabolism, migration, and defence-related processes (Stojadinovic et al., 2007). In plants, small RNAs have two major categories, that is, miRNA and siRNA. The miRNA originated from single-stranded RNAs in the nucleus, while siRNA originated from double-stranded RNA (Voinnet, 2009). The miRNA regulates various biological processes of plants which are highly conserved and species-specific.

Victor Ambros and his associates discovered the first short RNA, lin-4, in *Caenorhabditis elegans* in 1993. The developmentally important gene lin-14 protein level is repressed by lin-4. This lin-4 gene does not encode any protein but rather makes a pair of small RNAs like one shorter RNA fragment (21 nt long) and a second longer RNA fragment (60 nt long). The longer fragment of RNA makes the stem-loop structure by folding and is anticipated as an ancestor of 21-mer RNA, and the shorter lin-4 RNA was known as today's miRNA. After seven years of disclosure of lin-4, one other miRNA, let-7, was identified in the *Caenorhabditis elegans* (Reinhart et al., 2000). This miRNA let-7 regulates the expression of the lin-41 gene, which play important role in delayed larval stage into adult cell stage (Nelson & Ambros, 2021). Almost all animal phyla that contain miRNA let-7 have a defining biological significance.

Afterward, several researchers cloned the miRNAs from different organisms. Cloning of small RNAs suggested the pervasive existence of small RNAs in plants has been disclosed by cloning of small RNAs. The gene silencing phenomenon, which causes the expression of chalcone synthase, a pigment-producing gene, in variegated flowers in place of a deep purple flower, was accidentally discovered in the petunia flower. Subsequently, a co-suppression phenomenon was also observed because transgene and homologous endogenous gene was suppressed (Campbell & Choy, 2005). In

DOI: 10.1201/9781003248453-2

the fungus *Neurospora crassa* a similar process has been observed, referred to as quelling, and a similar mechanism has also been seen in animals known as RNA interference (Mourelatos et al., 2002).

The first plant miRNA was discovered in *Arabidopsis thaliana* in year 2002. Lave and his co-workers cloned a set of 21–24 nt length of miRNAs from *Arabidopsis*. Interestingly, in the genome of *Arabidopsis thaliana* a variety of miRNAs were identified that are present in the transposable elements (Oliver et al., 2021). In *Arabidopsis*, a differential expression motif of 16 miRNAs was detected in which 8 were also found from the rice genome showing its conserved nature (Tien et al., 2021). In *Arabidopsis thaliana*, a family of RNase III enzyme Dicer named CARPEL FACTORY was identified, which upon mutation inhibited the accumulation of miRNAs. Several shreds of evidence confirm the mechanism present in the plant's system that regulates this processing of miRNA is similar to animals (Pandita, 2019). There are more than 38,186 miRNA loci predicted by different biological techniques such as genomic screening, cloning of small RNAs, computational approaches, and ESTs. There are several plant miRNAs known which are evolutionarily conserved at the species level from angiosperm to mosses. Functionally small RNAs may act as novel biomarkers in gene regulation processes and act as therapeutic targets for disease identification and management. Recently there have been more studies on small RNAs that suggest the role of miRNAs in various critical cellular mechanisms with novel coordination with other small molecules (B. K. Sun & Tsao, 2008).

A small RNA sequence controls the expression of multiple genes because small RNA binds to its complementary target genes either imperfectly or perfectly (Ha & Kim, 2014). Therefore, small RNAs are equally crucial as transcription factors for gene expression (K. Chen et al., 2008). In the cell more than 30% of the genes may be directly regulated by small RNAs (Saucier et al., 2019). In humans about 850 mature miRNA sequences have been identified, among which several miRNAs are highly conserved in various species. There are various functions of small RNAs at the cellular level involved in cell differentiation, proliferation, migration, death/apoptosis, growth, cell defence, and metabolism. It also participates in animal development, such as early embryonic development, cardiac development, neural development, and germline development. At the time of development, the function of small RNAs occurs in a spatiotemporal, cell-specific, and tissue-specific fashion that shows the important role in cell differentiation. The first evidence about the function of small RNAs in differentiation of embryonic stem cell was to find out the alterations in the expression of miRNAs in embryonic stem cell differentiation, in which various miRNAs were either upregulated or downregulated (Houbaviy et al., 2003). An experiment has been conducted for future verification of the critical role of miRNAs on mice by gene knockout encoding Dicer, which has been a critical enzyme in the biogenesis of small RNAs (Bhaskaran & Mohan, 2014).

2.2   ORIGIN AND EVOLUTION OF PLANT MIRNAS

Research evidence clearly shows the evolution of newly originated miRNA genes originated from inversion of duplication of target genes that reveal the resemblance to those flanking regions where the miRNA complementary site in target genes (Cui et al., 2017). Inverted repeats possibly make hairpin-like structures that produce many small RNAs, that is, miRNAs, siRNAs. At the time of evolution, a few parts of the hairpin structure continue to develop the MIR gene that further encodes the main small RNA. Various examples exhibit the sequence similarity between the target gene and MIR genes promoting inverted gene duplication model to the genesis of miRNAs gene (Axtell et al., 2011; Liang et al., 2014; Liu et al., 2008; H. Zhang et al., 2015). An additional root for the origin of miRNA are MITEs (miniature inverted-repeat transposable elements), which seem like the precursors for miRNA when MITEs RNAs fold into a stem-loop structure. It has been reported that 10 and 38 miRNAs are derived from MITEs in *Arabidopsis* and rice, respectively (Piriyapongsa & Jordan, 2008; Xu et al., 2011). Nevertheless, MITEs can give rise to various small RNAs, that can be well classified as siRNAs (Piriyapongsa & Jordan, 2008). So the hypothesis for the origin of MITEs is still unclear. The example which supports this hypothesis is TamiR1123, which regulates the vernalization process in wheat, and was detected to be originated from the locus of MITE (Jingyi Wang et al., 2021). Nearly half of the total MIR gene families of *Arabidopsis lyrata* that can attach to protein-coding genes yield miRNA that targets the homologous genes (Nozawa et al., 2012). The inverted gene duplication hypothesis cannot describe this statement, and it indicates that there should be another origin for MIR genes in plants. It can be predicted that the source of miRNAs by hairpin regions dispersed throughout the genome can be the precursors of miRNAs (Nozawa et al., 2012). For example, in *Arabidopsis thaliana*, mpss 05 presents a premiRNA as foldback that may align at two domains and possibly have arisen by duplication of a chromosomal segment (Fenselau de Felippes, 2010).

The origin of miRNA probably comes through gene silencing by complementary RNA sequences which has been lost in some eukaryotic members. There are two miRNAs, miR854 and miR855, that are commonly present in both plant and animal kingdoms, but their authenticity is controversial (Budak & Akpinar, 2015). However, phylogenetic studies of miRNA families containing conserved and non-conserved regions should be done with caution. This is because miRNAs have not been extensively characterized in some plants, and additional review of their evolutionary evidence is needed.

To address this question, Taylor and his collaborators probed the 6,172 miRNA genes deposited in miRBase, and approximately one-third of these miRNAs are genuine miRNAs. There is not enough evidence to say there is. Moreover, there are large evolutionary gaps among plant species for which miRNA information is available. *Physcomitrella patens* is therefore only representative of the moss lineage miRBase (Taylor et al., 2014). Nevertheless, his eight miRNA families of miR156, miR159/319, miR160, miR166, miR171, miR408, miR390/391, and miR395 are conserved in embryonic plants (Cuperus et al., 2011). Both miR156 and miR166 are conserved throughout the plant kingdom and play an important role in flowering (Y. Luo et al., 2013). In addition, the miR396 family is found in vascular plants, and miR397 and miR398 are found in all seed plants. Other miRNA families such as miR403, miR828, and miR2111 are specific to eudicots (Cuperus et al., 2011). Interestingly, target proteins such as AGO2 and AGO3 of the miR403 family exhibit

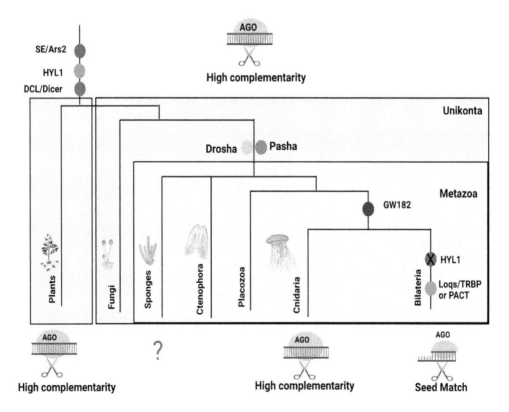

FIGURE 2.1   A tentative flow chart describes the evolution of miRNA in plant system and animal system.

miRNA biogenesis feedback. The dates of conserved and un-conserved miRNA families are subject to change based on new knowledge, syntenic relationships. In summary, evolutionary patterns of miRNA conservation can be misleading unless additional evidence supports relationships between miRNAs themselves and species.

2.3   COMPARISON BETWEEN ANIMAL AND PLANT MIRNAS

As we have discussed previously, miRNA belongs to a different class of short endogenous RNAs that came into the limelight over a few years as it plays a crucial role in gene regulation at the level of post-transcriptional and functional roles in developmental and physiological processes in several eukaryotic lineages of plant system and animal system. There is an assumption that miRNAs have developed alone in both lineages; nevertheless the likelihood from current evidence from plants, non-bilaterian metazoans, and most of the algae shows that already the last ancestor of these families might possess miRNA pathways of post-transcriptional gene regulation. In plants and animals, miRNA play a significant role in a biological process that is crucial for normal development (Bartel, 2004; X. Chen et al., 2009; Giraldez et al., 2005; Voinnet, 2009), that has been clear after the discovery of the function of miRNA in *Caenorhabditis elegans* two decades ago (Lee et al., 1993; Wightman et al., 1993).

There is a possibility that miRNAs must have role in animal complexity and also involved in evolution of animal complexity and developmental process (Grimson et al., 2008; Peterson et al., 2009). A missing homology of sequence between animals and plants miRNA families in miRNA mode of action and biogenesis indicated that miRNAs had arisen alone in both lineages from a siRNA mechanism present earlier from a descendent of common ancestors of all eukaryotes. Still, questions arise related to plant and animal miRNAs. There are numerous primary similarities among plant and animal miRNAs because, in each lineage, the miRNAs play a critical function in the development, disease, ageing, and regulation of transcription ranges of the diverse molecular process.

Both miRNAs of plants and animals appear to mainly exert their genetic and transcriptomic influences through the gene expression at the level of mRNA stability and translation repression. There are some miRNA species that have been reported that may be active in both plants and animals (miRNA- 155 and miRNA-168), which are the members of the family miRNA-854, and this indicates a common origin and function on the basis of the course of evolution, for example, *Homo sapiens* and *Arabidopsis thaliana* diverged 1.5 billion years ago (Zhao et al., 2018). Apart from the similarities, more attention is paid to the variations between the animal and plant miRNAs. It counts the gene structure, size of mature miRNAs, 5' nucleotide sequences, and most importantly the missing of any genetic conservation between two groups. It has been proven by mechanistic studies that both groups significantly differ in their functions and mode of biogenesis. The conclusion is that both groups have originated from dual origins in their evolutionary lineages. Moreover, miRNAs play fundamental roles in both systems at the level of development, and they predominantly control the expression of the regulatory genes. The differences are also counted in both plant and animal miRNAs; in the primary step of biogenesis in animals, there is the involvement of Drosha, whereas, in plants, this step is taken by DCL1 (DICER). Most plant miRNAs originated from a single primary transcript whose loci are present in intergenic regions; however, the polycistronic transcript was derived from intergenic areas of the chromosome. Plant miRNAs are predominantly controlled by their targets by cleavage in RNA coding regions. On the other hand in animals miRNAs are primarily regulated by translational inhibition using targets at 3'-UTR. So the differences in plant and animal miRNA systems mentioned here present a clear-cut picture. The reason for the differences between the miRNAs' system in both plants and animals is that they evolved individually, but it may probably be from the same type of ancestor after the divergence of animals and plants, which exemplifies convergent evolution.

2.4   MIRNA GENE TRANSCRIPTION

Transcriptional process is the potential checkpoint in expression of miRNA. Micro RNAs belong to a large family of small non-coding RNA, and their primary function is to act as co-factors in diverse gene silencing pathways. By the transcription process, enzyme RNA polymerase II (pol II) produces pre-miRNA. The unique character of pri-miRNAs (miRNAs transcript) is having cap structure and a poly(A) tails belonging to the class II gene transcript. Many primary transcripts have been analyzed in different organisms, including plant and animals. The full length of the primary transcript is generally longer than 1kb and has an N^7-methyl guanosine cap. The promoters that manage the miRNA transcription also contain the site of pol II promoters. The discovery of RNA pol II plays a key role in miRNA transcription, effectively answering the question of why miRNA is different in their expression level, because RNA polymerase II promoters are great in strength and highly regulated.

Like transcription, other processes in coding genes like 5' capping, polyadenylation, and splicing indicate that approximately the known regulatory mechanisms for miRNA transcription probably apply to gene transcriptions encoding miRNA. During pre-miRNA transcription, many transcription factors are involved, like MED 20A (Mediater20A), MED17 (Mediator17), and MED18 (Mediator 18). They affect the transcription by interacting with RNA pol II (S. Zhang et al., 2015). MED20A, MED17, and MED18, which specify the attachment of RNA polymerase II at the promoter of the gene, formed a transcription complex before the start of transcription. In the *Arabidopsis* plant, NOT 2_3_5 domain protein, that is, NOT2, binds with polymerase II and accelerates the transcription of miRNA (Bevilacqua et al., 2016). A protein CDC5 contains the SANT domain of MYB protein, which identifies and attaches to the promoter and affects transcriptional regulation. The transcription elongation process is critical, and the elongation factor plays a crucial role in the transcription. ELP2 and ELP5 are the conserved transcription factors involved in the plant miRNA transcription (S. Zhang et al., 2015). The disruption of the transcript elongation decreases RNA polymerase II occupancy of miRNA loci and downregulates the pre-miRNA transcription process. The miRNA genes form pri-miRNA with the help of pol II, cycling DOF transcription factor (CDF2), and other proteins, which upregulates the miRNA transcription. miRNA transcription plays a crucial role in the eventual synthesis of miRNA and processing and other modifications after transcription is completed.

2.5   POST-TRANSCRIPTIONAL CONTROL OF MIRNA BIOGENESIS

In plants, biogenesis of initiates by the transcription of gene encoding miRNA by the activity of pol II–like protein results in the synthesis of 1kb pri-miRNA. Further processing of generated pri-miRNA is performed mainly by DCL1 (Dicer-like1) protein to produce precursor pri-miRNA. Mature miRNA has been developed from pri-miRNA by several complex alterations (Figure 2.2). The length of the stem-loop (pri-miRNA) structure varies, and it is responsible for different cleavage forms. The direction of pre-miRNA from base to loop affects the bidirectional processing and direct involvement of DCL1 in the fining of pri-miRNA. It is a homolog to the Drosophila Dicer and resembles another Dicer gene of humans. It has different conserved domains like N-terminal helicase, PAZ, RNA ribonuclease, and 1–2 dsRNA binding domains at the carboxyl terminus. PAZ domain is bound with dsRNA, which has two–nucleotide protuberance at the 3' end. The cleaving activity of the DCL1 protein, necessary for the processing and modifications of a specific type of miRNAs, is provided by the ATPase activity of the helicase domain. DDL protein binds to the RNA and performs miRNA processing. The conclusion is that the DCL1 plays a defining role in the whole process of biogenesis and processing.

A complex processing unit containing DCL1 and other proteins cut the pri-miRNA transcript and formed mature and processed miRNA. There are plenty of other such protein residues in the processing complex that binds with the DCL1 protein and control its capacity to perform precise processing. A protein factor HYL1 (HYPONASTIC LEAVES1) binds with DCL1 to increase the accuracy of cleavage of pri-miRNA (Achkar et al., 2018). The N-terminal of HYL1 protein has two domains that bind to dsRNA and a protein interaction binding domain present at C-terminus.

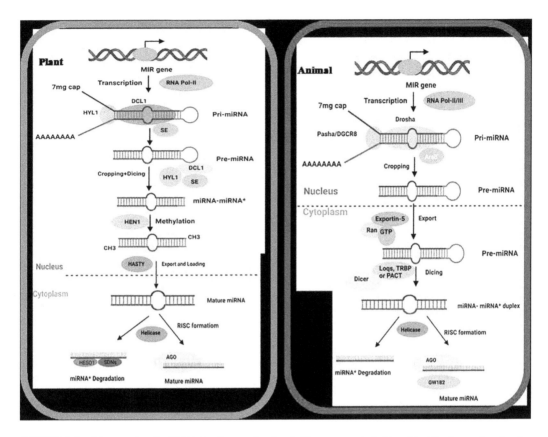

FIGURE 2.2 Biogenesis of miRNA in plant and animal. The process is initiated inside the nucleus by the conversion of pri-miRNA into pre-miRNA (70 bp long), after the involvement of the DROSHA/DGCR8 or DCL1 complex. Through the nuclear pore complex, synthesized pri-miRNA is transported in the cytoplasm and further processed by DICER protein by forming small duplex fragments; duplex miRNA binds with AGO protein and forms a RISC complex that mediates recognition for mRNA to be targeted.

A domain DsRBD2 is present in HYL1, which was previously believed to affect cleavage site in pri-miRNA during processing, while interference in this homodimer does not affect proteins like DCL1 and SE (SERRATE). SE binds with DCL1, and together, these proteins perform recognition and cleavage of pri-miRNA. SE localizes in the nucleus, and it coordinates with HYL1, and their core region may give a base so that HYL1 and DCL1 can interact with each other. The interacting site of HE is the middle of the N-terminal domain rather than the zinc finger region. Mainly, it provides a platform for the binding of more than one protein (nature of scaffold protein) and relays the signals from the cell wall to the nucleus, thus affecting the specificity of miRNA processing by interacting with DCL1 and HYL1 with pri-miRNA. At post-transcriptional level of RNA processing, NOT2 protein also performs scaffold activity and enhances the formation of miRNA during transcription and the processing of miRNA.

All pol II transcripts are tagged with the cap of CBC (cap binding protein complex). CBP80 and CBP 20 play a key role in miRNA processing. If CBP80 and CBP 20 got mutated, it leads to a higher level of pri-miRNA and less miRNA (Kim et al., 2008). Both CBP 80/20 act as SE, to combinedly bind RNA with DCL1 protein and upregulate the processing of miRNA. TGH is a type of RNA binding protein with domains like G-patch and SWAP domain. This protein has occurred in the DCL1-HYL1-SE processing complex, and it controls the level of miRNAs by regulating the efficacy of the division of pri-miRNA (L. Song et al., 2010; Yang et al., 2014). TGH governs the amount of miRNA by promoting DCL1 cleavage efficiency without affecting processing accuracy.

Recent studies exhibit that the MAC complex contains CDC5 and PRL1 protein factors involved in miRNA processing. CDC5 affects the miRNA gene during transcription along with impacting processing steps. An evolutionary conserved WD-40 protein is PRL1, which interacts with CDC DCL1 and SE to enhance the miRNA processing and strength of pri-miRNA (O'Brien et al., 2018). It assists the functionality of CDC5 protein factor in the MAC complex and acts as part of the DCL1 protein complex to increase pri-miRNA processing. Another protein CDF2 affects the miRNA at the post-transcriptional level by combining with DCL1 (Gao et al., 2021). Different types of proteins are present which affect the functionality of the DCL1 protein complex (processing complex). Phosphorylation has an impotent role in the processing and stability of HYL1 that affects the miRNA processing at the C-terminal domain only when phosphatase-like 1 protein (CPL1/FRY2) binds with DCL1 processing complex at HYL1 and SE protein site (Gao et al., 2021). CPL1 continues the phosphorylation of HYL1 to confirm the fidelity of pre-miRNA cleavage, and SE protein function as a scaffold to provide a platform for the attachment of HYL1 and CPL1 together. Mitogen-activated protein (MAPK3) increases the capacity of HYL1 to get phosphorylated and ultimately decreases its activity. *Arabidopsis thaliana*, K homology (KH) domain proteins cofactor for phosphatase CPL1 and CPL2. HYL1 dephosphorylation is promoted by RCF3 through interacting with CPL1 and CPL2 (Jodder, 2021). The protein phosphatase PP4 (Protein Phosphatase 4) and a protein complex, that is, SMEK1 (Suppressor of MEK1) start the dephosphorylation of HYL1 and thus enhances the processing miRNA by MAPK signaling cascade and also mediates stability of HYL (Z. Sun et al., 2018). SMEK1 can decrease phosphorylation by suppressing the activity of MAPK and connect the PP4 protein to the dephosphorylation of HYL1 to ensure miRNA biogenesis. Regulatory subunit 3 (PP4R3A) present in the PP4 complex having phosphatase activity binds with two of the reluctant genes harbouring two catalytic subunits, PPX1 and PPX2, promoting the formation of miRNA (Chendrimada et al., 2005). Several results attest to this through the mutation analysis. If PP4R3A is mutated, it decreases the synthesis of miRNA, pri-miRNA processing, and activity of the HYL1 protein factor.

The matured pri-miRNA generally has a cap-like structure at 5' end, a poly-A tail at 3' end, and a loop-like steady neck structure. The stability of pri-miRNA is governed by CDKF-1 (cyclin-dependent protein kinase) (O'Brien et al., 2018). In the *Arabidopsis* plant, STABILIZD1 (STA1) is a pre-miRNA that is a homolog of processing factor-6 and is known to have involvement in the processing of miRNA (Ben Chaabane et al., 2013). Like other miRNA processing mutants, mutated STA1 decreases the level of mature miRNAs. STA1 involves miRNA processing by governing pri-miRNA cleavage and enhancing the activity of DCL1. Nuclear localization protein coded by CMA33, that is, XAP5 CIRCADIAN TIMEKEEPER (XCT), acts as TF to control the miRNA biogenesis after initiating transcription of DCL1 (Yan et al., 2017).

The activity of the DCL1 processing complex is influenced by SIC (proline-rich protein) by interacting with HYL1. If loss of function mutation occurs in the SIC protein, the level of miRNA effectively decreases. In recent years it is discovered that importin-β-protein (KETCH1) simplifies the processing and splicing of pri-miRNA by importing HYL1 inside nucleus and helps in the formation of the DCL1 processing complex. MOS2 protein function is the joining of the G patch with the KOW domains during binding with RNA. Mutated MOS2 have a lower level of mature miRNA. SMaLL1 involves miRNA synthesis by altering the rate of intron excision present in the Mir gene (X. Wu et al., 2013). It is a DEAD-box pre-mRNA splicing factor Prp28 homolog. EPL2 and EPL5 are the transcriptional elongation protein that influences miRNA biogenesis during transcription. It involves miRNA processing by binding with DCL1 to direct DCL1 toward the dicer-like complex. A RING-finger E3 ligase, that is, constitutive photomorphogenic1 (COP1), plays a crucial role in photomorphogenesis (Cho et al., 2014; Stepien et al., 2017; Junli Wang et al., 2019). Many light-governed photoreceptors and transcription factors are regulated and destabilized by these RING-finger ligases. With the help of protease, under the influence of high light intensity, it can initiate the degradation of HYL1 in the cytosol and thus regulate miRNA processing. A protein kinase named SnRK2 acts as abscisic acid (ABA) and osmotic stress response factor (Kobayashi et al., 2005). It lowers miRNA storage and phosphorylates HYL1 and SE in the DCL1 processing complex. FCA binds with RNA to enhance the level of some miRNAs by affecting miRNA processing with the help of temperature variability. In the chromosome remodeling complex (SWI/SNF complex), CHR2/BRM is an ATPase chr2 mutant with multiple phenotypes, including alteration in miRNA expression (Z. T. Song et al., 2021). It is found that CHR2 is involved in the miRNA synthesis at the transcriptional level and decreases the accumulation of post-transcriptional miRNA by interacting with the SE protein factor. A mutation changes the secondary conformation of pri-miRNA in chr2. CHR2 influences miRNA processing by remodeling their secondary structure. It is responsible for the controlled functioning of miRNA in plants to balance the whole related plant systems (Gao et al., 2021).

2.6   FORMATION AND FUNCTION OF MIRNA EFFECTOR COMPLEX

Emergence of RISC complex (RNA-induced silencing complex) and miRNA charging, miRNA regulates the expression of the desired target gene by repression in the rate of translation and post-translational cleavage. RISC complex was first discovered in humans, and miRNA is loaded on this complex to combine translational reparation and post-transcriptional silencing (Filipowicz et al., 2005). Like animals, miRNA and AGO proteins interact in plants to make the RISC complex (Filipowicz et al., 2005). First, pre-miRNA interacts with the AGO protein to form a pre-RISC complex, which further develops into a mature RISC protein complex. The most important factor of RISC complex is the AGO protein, which can regulate the silencing activity of the protein complex. There are ten different types of AGO protein in *Arabidopsis*; among them AGO1, 2, 4, 7, and 10 contain cleavage activity that can directly affect the movement of small RNAs (Bologna & Voinnet, 2014; Ossowski et al., 2008). AGO protein joins with the PAZ domain (jointed with miRNA), PIWI domain (a cap-binding-like domain; MC domain), and a Mid domain. PIWI domain has endonuclease activity while a divalent cation helps the Mid domain to bind with the 5' phosphate of ssRNA. AGO1 protein is involved in the cleavage of miRNAs and transcriptional suppression. These functions of the AGO1 protein are possible when miRNA and polyribosomes are present.

The RNA-binding protein complex DRB1/HYL1 binds with the AGO1 protein and plays a role in miRNA synthesis (Yang et al., 2021). This is also responsible for strand selection of miRNA duplex and its directional loading in the RISC complex for passenger strand degradation. Molecular chaperone Hsp90 interacts with AGO1, affects the RISC function, and alters topology. Hsp90 binds with AGO1 and directs double-stranded RNA degradation by producing energy. SQN is an orthologue of immunophilin and cyclophilin 40 (Cyp40) directly involved in miRNA silencing in the *Arabidopsis* (Smith et al., 2009; G. Wu, 2013). SQN works with AGO1 and interacts with Hsp90, HYL1, DCL1, and HEN1 to govern the level of miRNAs. MiRNA Active 1 (EMA1) and Transporter (TRN1) code two different importin-β family proteins. RISC loading process is regulated by EMA1 and TRN1, negatively and positively, respectively. EMA1 inhibits the miRNA loading on the AGO1 complex. Researchers proved this statement as they purified most miRNA from the effector complex in ema1 mutants. The activity of miRNA decreases by the ema1 and trn1 mutants without affecting the level of miRNA, and the trn1 mutant never interferes with miRNA and AGO1 but intercepts the interaction between miRNA and AGO1.

2.7   TRANSLATIONAL REPRESSION

Protein translational repression initially works through ALTERED MERISTEM PROGRAM1 (AMP1), inhibiting the appropriate target mRNA passage into polyribosome, and AMP1 restricted on endoplasmic reticulum (Iwakawa & Tomari, 2015). The miRNA translational reparation process occurs on the endoplasmic reticulum, denoted by AGO1 and miRNAs, where the ER surface serves as the protein translation site (Gao et al., 2021). Additionally, other protein factors are available to control the miRNA mediated translational repression. This process is regulated by the microtubule-severing enzyme KATANIN1 (KTN1) and SUO domain-containing GW protein that promotes translational repression (Won, 2013; Yu et al., 2020). If a mutation occurs in this gene, a higher level of the target protein is acuminated without affecting the mRNA level. The decapping complex is made up of DSP1, DSP2, and VCS by co-localization in the P-bodies, and this process occurs in the cytoplasm.

2.8 ROLE OF MIRNAS IN PLANT DEVELOPMENT

Researchers have performed mutations in specific genes associated in the formation of miRNA-related proteins to understand the function of these regulatory factors in plant organogenesis. The main functions of miRNAs in several plant development processes studied as vasculature, meristem, root, vegetative growth, floral patterning, and organ polarity by mediating the response of auxin and other plant hormones (Table 2.1) (Gautam et al., 2017; Singh et al., 2018). In this context,

TABLE 2.1
miRNAs' role in plant developmental processes.

Plant part	miRNA type	Mechanism of action	References
SAM (shoot apical meristem)	miR165/166	Together with AGO proteins, miR165/166 regulates HD-ZIP III transcripts. AGO targets the abaxial surface of leaf primordia and SAM, which is the accumulation area of miR165/166.	(Zhang et al., 2012)
	miR394	miR394 downregulates the F-box protein LCR (LEAF CURLING RESPONSIVENESS) an inhibitor for stem development.	(Knauer e t al., 2013)
	miR156	miR156 represses *SPL3/4/5 (SQUAMOSA PROMOTER BINDING PROTEIN-LIKE* family members) responsible for the transition of the vegetative shoot in the flowering shoot.	(G. Wu & Poethig, 2006)
	miR171	miR171 causes the termination of the premature vegetative shoot and participates in shoot development.	(Fouracre & Poethig, 2016)
	miR166	miR166 regulates HD-ZIPIII, which in turn regulates tasiR-ARF, which finally regulates ETT/ARF.	(Nagasaki et al., 2007)
	miR159	miR159 family redundantly controls the regulation of MYB33 and MYB65 and causes defects in pleiotropic development, which also includes a reduction in the apical dominance.	(Palatnik et al., 2007)
	miR169	miR169 acts as a nuclear factor Y (NF-Y) and actively participates in the development of shoot apical meristem.	(Du et al., 2017)
Leaf	miR319	miR319 regulates LANCEOLATE and thus changes morphology of tomato leaves.	(Ori et al., 2007).
	miR319	miR319 regulates the BRANCHED/*CYCLOIDEA/ PCF (TCP)* genes and is responsible for leaf development.	(Warthmann et al., 2008)
	miR396	miR396 of *Medicago truncatula* downregulates the expression of *MtGRF* genes and bHLH79 genes.	(Bazin et al., 2013)
	miR159	miR159 family play role in regulation of genes like MYB33 and MYB65 and exerts pleiotropic developmental defects that include the genesis of curled leaves.	(Bologna et al., 2009)

(Continued)

TABLE 2.1 (Continued)
miRNAs role in plant developmental processes

Plant part	miRNA type	Mechanism of action	References
Root	miR393	miR393 negatively regulates the functioning of F-box auxin receptors TIR1, AFB2, and AFB3, role in root development.	(M. Luo et al., 2014)
	miR393/AFB3 complex	miR393/*AFB3* complex acts as nitrogen responsive element and also acts in auxin synthesis and root development.	(Vidal et al., 2010)
	OsmiR393	OsmiR393 of *Oryza sativa* targets *TIR1*, *OsTIR1*, and *OsAFB2* and enhances root emergence.	(Bian et al., 2012)
	miR847	miR847 cleaves the transcriptional repressor of the IAA/ARF complex by binding through complementary sequences and thus promotes root development.	(J.-J. Wang & Guo, 2015)
	miR160	It targets the ARF10, ARF17, and ARF16 and thus suppresses the development of primary roots while promoting lateral roots.	(J.-W. Wang et al., 2005)
	miR167	miR167 targets ARF6 and ARF8 family genes and is responsible for root development.	(Gutierrez et al., 2009).
	miR173	miR173 cleaves the transcript from TAS1 and TAS2 and induces adventitious root development.	(Axtell et al., 2006)
	miR390	miR390 targets the transcripts from *TAS3* and regulates root development.	(Howell et al., 2007)
	miR828	MiR828 targets the transcripts from TAS4 and regulates root development.	(Montgomery et al., 2008)
Flowering	miR159	miR159 in *Arabidopsis* regulates photoperiod, flowering time, and development of anther in the SD Plants (short-day) via altering the expression of several MYB transcription factors.	(Achard et al., 2004)
	miR172	It is targeted by miR156 which regulates the *SQUAMOSA PROMOTER BINDING PROTEIN-LIKE*9 (*SPL9*) that promotes miR172. miR172 promotes flowering in plants, and its mutation causes vegetative shoot development.	(Zhu & Helliwell, 2011)
	miR166/165	Double mutant plants showed that this pair of miRNA mediates a signaling pathway parallel to the WUS-CLV pathway for the development of flowering.	(Jung & Park, 2007)

specific genes are pleiotropic for miRNA development. Kidner and Martienssen (2004) observed that miRNAs play an important role in leaf polarity with the help of the AGRONAUTE 1 protein. Production of radial organs, seedlings with loose meristematic zones, and even sterile plants were obtained after mutation in the AGO1 gene. Bohmert et al., (1998) suggested that AGO1 may be the critical locus for leaf development in *Arabidopsis*, and its mutation may cause the loss of function in PIWI and PAZ proteins (Bohmert et al., 1998). Another pair of genes, that is, PNH/ZLL closely resembles the AGO1 locus, play a key role in embryonic development in SAM (shoot apical meristem); however, PNH/ZLL mutant (png/zll) plants did not cause any defect in the polarity of plants as AGO1 plants showed (Kidner & Martienssen, 2004, 2005). These findings suggested that PNH/ZLL genes might be involved in vascular development.

Like AGO1 and DCL1 genes, mutations in HEN1 and HYL1 also result in some loss of functions but not as severe as AGO1, and DCL1 plants have (Vaucheret et al., 2004). This indicates that HEN1 and HYL1 genes play an important role in miRNAs' regulatory processes and biogenesis. Several reports indicate the indispensable role of many miRNAs like miR165/miR166 in plant growth and development. CUC1 and CUC2 (CUP-SHAPED COTYLEDONS) genes of NAC protein are known for floral organ separation by acting as a suppressor of cell proliferation and specifying boundaries of organ primordia. Mallory et al. suggested that miR165/miR166 regulates the functioning of NAC domains by targeting the CUC1 and CUC2 genes and thus ultimately causing alterations in the patterning and development of embryonic, floral, and vegetative organs (Mallory et al., 2004). Interestingly other miRNAs like miR319 are responsible for the regulation of leaf development. Besides these, miR156 and miR172 are involved in regulating SQUAMOSA PROMOTER BINDING LIKE (SPL family) and genes encoding AP2 proteins (floral organ identity) respectively. Despite this, several miRNA families are associated in the regulation of endosperm development (Djami-Tchatchou et al., 2017). To attest to the significant role of miRNAs in hormone signaling, many workers have performed experiments in this direction, some of which are discussed here. Bonnet et al. (2004) have identified several conserved miRNAs in *Oryza sativa* and *Arabidopsis* and also their target genes. Their computational approach suggests that miR160/miR167 targets the auxin response factor (ARF), while TRANSPORT INHIBITOR RESPONSE1 (TIR1) is the locus for the binding of miR393 (Couzigou & Combier, 2016). The effect of hormones on the functioning of miRNAs comes into light after the above outcomes. Achard and his co-workers have attested that the seed germinating hormone, that is, gibberellic acid (GA3) positively regulates (enhances the Function) miR159. This microRNA is well-known for regulating the flowering period and male organ development by mRNA cleavage.

2.9   CONCLUDING REMARKS

We conclude that miRNA plays a major role in regulation of gene expression. Functions as the fundamental regulators of cell type differentiation, proliferation, and survival, miRNAs have "taken over" from proteins. Changes in miRNA expression are connected to the progression of various developmental processes in plants and several human diseases, including cancer. However, the molecular intricacies of their expression regulation, biogenesis, and transcriptional regulation, as well as the signaling cues that modulate their expression, are still being researched. These mechanistic features must be investigated to gain a better understanding of their involvement in cell physiology and illness. It has been found that the machineries responsible for transcription, splicing, and pri-miRNA processing interact, demonstrating the relationship between DCL1-mediated processing and other aspects of pri-miRNA maturation. Understanding the functional significance of these relationships is the challenging part. All of these mechanisms have the ability to integrate miRNA synthesis into cellular responses to stress and development, including individual MIR locus transcriptional regulation, phospho-protein binding within the DCL1 complex, and processing connections with transcription and splicing elements. The universe of short RNAs has long been divided into plant life and animal life, as well as between miRNA and siRNA. One of the main causes of the separation between plants and animals was the difference in repression mechanisms. It is increasingly clear that plants regularly decrease the expression of miRNA target genes through translational inhibition. Early studies often overlooked protein levels, possibly due to the need for epitope-tagged transgenes, and the full degree of plant miRNA-mediated translational suppression is yet unknown.

2.10   FUTURE PROSPECTS

Several players in miRNA transcription and processing have been identified, which has provided insight into pri-miRNA co-transcriptional splicing, modification, and processing. However, there are still significant hurdles in determining the linkages between these actors and their precise biochemical contributions. At the sub-cellular level, the development, make-up, and role of the dicing body in co-transcriptional pri-miRNA processing are all of great interest. Innovative methods like single-cell biology and in vitro restoration of the dicing apparatus would be needed to address these problems. Despite the fact that 30 modifications, targets, and AGOs all have an impact on miRNA stability, little is known about the underlying mechanisms. Future studies should identify and define other enzymes and modulators involved in these processes. Understanding the biological effects of global or sequence-specific miRNA degradation throughout developmental transitions and in response to environmental cues will also be essential.

2.10.1 Acknowledgement

Authors are thankful to the DST-PURSE, DST-FIST, and CAS (Botany) for providing funds to the Department of Botany, Banaras Hindu University, Varanasi, India. VM is thankful to the Department of Biotechnology (DBT), India (DBT/JRF/15/AL/223) and NIPGR for fellowship.

2.10.1.1 Conflict of Interest
There is no conflict of interest between authors.

2.10.1.2 Author Contributions
BM and LS conceptualized the manuscript. NR, VM, and AK helped in writing, and SP gave the critical suggestions and scientific feedback.

REFERENCES

Achard, P., Herr, A., Baulcombe, D. C., & Harberd, N. P. (2004). Modulation of floral development by a gibberellin-regulated microRNA. *Development (Cambridge, England), 131*(14), 3357–3365.

Achkar, N. P., Cho, S. K., Poulsen, C., Arce, A. L., Re, D. A., Giudicatti, A. J., Karayekov, E., Ryu, M. Y., Choi, S. W., Harholt, J., Casal, J. J., Yang, S. W., & Manavella, P. A. (2018). A quick HYL1-dependent reactivation of microRNA production is required for a proper developmental response after extended periods of light deprivation. *Developmental Cell, 46*(2), 236–247.e236.

Axtell, M. J., Jan, C., Rajagopalan, R., & Bartel, D. P. (2006). A two-hit trigger for siRNA biogenesis in plants. *Cell, 127*(3), 565–577.

Axtell, M. J., Westholm, J. O., & Lai, E. C. (2011). Vive la différence: Biogenesis and evolution of microRNAs in plants and animals. *Genome Biology, 12*(4), 1–13.

Bartel, D. P. (2004). MicroRNAs: Genomics, biogenesis, mechanism, and function. *Cell, 116*(2), 281–297.

Bazin, J., Khan, G. A., Combier, J. P., Bustos-Sanmamed, P., Debernardi, J. M., Rodriguez, R., Sorin, C., Palatnik, J., Hartmann, C., Crespi, M., & Lelandais-Brière, C. (2013). miR396 affects mycorrhization and root meristem activity in the legume Medicago truncatula. *The Plant Journal, 74*(6), 920–934.

Ben Chaabane, S., Liu, R., Chinnusamy, V., Kwon, Y., Park, J. H., Kim, S. Y., Zhu, J. K., Yang, S. W., & Lee, B. H. (2013).STA1, an Arabidopsis pre-mRNA processing factor 6 homolog, is a new player involved in miRNA biogenesis. *Nucleic Acids Research, 41*(3), 1984–1997.

Bevilacqua, P. C., Ritchey, L. E., Su, Z., & Assmann, S. M. (2016). Genome-wide analysis of RNA secondary structure. *Annual Review of Genetics, 50*, 235–266.

Bhaskaran, M., & Mohan, M. (2014). MicroRNAs: History, biogenesis, and their evolving role in animal development and disease. *Veterinary Pathology, 51*(4), 759–774.

Bian, H., Xie, Y., Guo, F., Han, N., Ma, S., Zeng, Z., Wang, J., Yang, Y., & Zhu, M. (2012).Distinctive expression patterns and roles of the miRNA393/TIR1 homolog module in regulating flag leaf inclination and primary and crown root growth in rice (Oryza sativa). *New Phytologist, 196*(1), 149–161.

Bohmert, K., Camus, I., Bellini, C., Bouchez, D., Caboche, M., & Benning, C. (1998). AGO1 defines a novel locus of Arabidopsis controlling leaf development. *The EMBO Journal, 17*(1), 170–180.

Bologna, N. G., Mateos, J. L., Bresso, E. G., & Palatnik, J. F. (2009). A loop-to-base processing mechanism underlies the biogenesis of plant microRNAs miR319 and miR159. *The EMBO Journal, 28*(23), 3646–3656.

Bologna, N. G., & Voinnet, O. (2014). The diversity, biogenesis, and activities of endogenous silencing small RNAs in Arabidopsis. *Annual Review of Plant Biology, 65*, 473–503.

Bonnet, E., Wuyts, J., Rouzé, P., & Van de Peer, Y. (2004). Detection of 91 potential conserved plant microRNAs in Arabidopsis thaliana and Oryza sativa identifies important target genes. *Proceedings of the National Academy of Sciences of the United States of America, 101*(31), 11511–11516.

Budak, H., & Akpinar, B. A. (2015). Plant miRNAs: Biogenesis, organization and origins. *Functional & Integrative Genomics, 15*(5), 523–531.

Campbell, T. N., & Choy, F. Y. (2005). RNA interference: Past, present and future. *Current Issues in Molecular Biology, 7*(1), 1–6.

Chen, K., Song, F., Calin, G. A., Wei, Q., Hao, X., & Zhang, W. (2008). Polymorphisms in microRNA targets: A gold mine for molecular epidemiology. *Carcinogenesis, 29*(7), 1306–1311.

Chen, X., Guo, X., Zhang, H., Xiang, Y., Chen, J., Yin, Y., Cai, X., Wang, K., Wang, G., Ba, Y., Zhu, L., Wang, J., Yang, R., Zhang, Y., Ren, Z., Zen, K., Zhang, J., & Zhang, C. Y. (2009). Role of miR-143 targeting KRAS in colorectal tumorigenesis. *Oncogene, 28*(10), 1385–1392.

Chendrimada, T. P., Gregory, R. I., Kumaraswamy, E., Norman, J., Cooch, N., Nishikura, K., & Shiekhattar, R. (2005). TRBP recruits the Dicer complex to Ago2 for microRNA processing and gene silencing. *Nature, 436*(7051), 740–744.

Cho, S. K., Chaabane, S. B., Shah, P., Poulsen, C. P., & Yang, S. W. (2014). COP1 E3 ligase protects HYL1 to retain microRNA biogenesis. *Nature Communications, 5*(1), 1–10.

Couzigou, J. M., & Combier, J. P. (2016). Plant micro RNA s: Key regulators of root architecture and biotic interactions. *New Phytologist, 212*(1), 22–35.

Cui, J., You, C., & Chen, X. (2017). The evolution of microRNAs in plants. *Current Opinion in Plant Biology, 35*, 61–67.

Cuperus, J. T., Fahlgren, N., & Carrington, J. C. (2011). Evolution and functional diversification of MIRNA genes. *The Plant Cell, 23*(2), 431–442.

Djami-Tchatchou, A. T., Sanan-Mishra, N., Ntushelo, K., & Dubery, I. A. (2017). Functional roles of microRNAs in agronomically important plants—potential as targets for crop improvement and protection. *Frontiers in Plant Science, 8*, 378.

Du, Q., Zhao, M., Gao, W., Sun, S., & Li, W. X. (2017). micro RNA/micro RNA* complementarity is important for the regulation pattern of NFYA 5 by miR169 under dehydration shock in Arabidopsis. *The Plant Journal, 91*(1), 22–33.

Fenselau de Felippes, F. (2010). *Origin, biogenesis and non-cell autonomous effect of small RNAs in Arabidopsis thaliana.* Universität Tübingen.

Filipowicz, W., Jaskiewicz, L., Kolb, F. A., & Pillai, R. S. (2005). Post-transcriptional gene silencing by siRNAs and miRNAs. *Current Opinion in Structural Biology, 15*(3), 331–341.

Fouracre, J. P., & Poethig, R. S. (2016). The role of small RNAs in vegetative shoot development. *Current Opinion in Plant Biology, 29*, 64–72.

Gao, Z., Nie, J., & Wang, H. (2021). MicroRNA biogenesis in plant. *Plant Growth Regulation, 93*(1), 1–12.

Gautam, V., Singh, A., Verma, S., Kumar, A., Kumar, P., Singh, S., Mishra, V., & Sarkar, A. K. (2017). Role of miRNAs in root development of model plant Arabidopsis thaliana. *Indian Journal of Plant Physiology, 22*(4), 382–392.

Giraldez, A. J., Cinalli, R. M., Glasner, M. E., Enright, A. J., Thomson, J. M., Baskerville, S., Hammond, S. M., Bartel, D. P., & Schier, A. F. (2005). MicroRNAs regulate brain morphogenesis in Zebrafish. *Science, 308*(5723), 833–838.

Grimson, A., Srivastava, M., Fahey, B., Woodcroft, B. J., Chiang, H. R., King, N., Degnan, B. M., Rokhsar, D. S., & Bartel, D. P. (2008). Early origins and evolution of microRNAs and Piwi-interacting RNAs in animals. Nature, 455(7217), 1193–1197.

Guleria, P., Mahajan, M., Bhardwaj, J., & Yadav, S. K. (2011). Plant small RNAs: Biogenesis, mode of action and their roles in abiotic stresses. *Genomics, Proteomics & Bioinformatics, 9*(6), 183–199.

Gutierrez, L., Bussell, J. D., Pacurar, D. I., Schwambach, J., Pacurar, M., & Bellini, C. (2009). Phenotypic plasticity of adventitious rooting in Arabidopsis is controlled by complex regulation of AUXIN RESPONSE FACTOR transcripts and microRNA abundance. *The Plant Cell, 21*(10), 3119–3132.

Ha, M., & Kim, V. N. (2014). Regulation of microRNA biogenesis. *Nature Reviews Molecular Cell Biology, 15*(8), 509–524.

Houbaviy, H. B., Murray, M. F., & Sharp, P. A. (2003). Embryonic stem cell-specific microRNAs. *Developmental Cell, 5*(2), 351–358.

Howell, M. D., Fahlgren, N., Chapman, E. J., Cumbie, J. S., Sullivan, C. M., Givan, S. A., Kasschau, K. D., & Carrington, J. C. (2007). Genome-wide analysis of the RNA-DEPENDENT RNA POLYMERASE6/DICER-LIKE4 pathway in Arabidopsis reveals dependency on miRNA-and tasiRNA-directed targeting. *The Plant Cell, 19*(3), 926–942.

Iwakawa, H.-o., & Tomari, Y. (2015). The functions of microRNAs: mRNA decay and translational repression. *Trends in Cell Biology*, *25*(11), 651–665.

Jiang, Q., Wang, Y., Hao, Y., Juan, L., Teng, M., Zhang, X., Li, M., Wang, G., & Liu, Y. (2009). miR2Disease: A manually curated database for microRNA deregulation in human disease. *Nucleic Acids Research*, *37*(suppl_1), D98–D104.

Jodder, J. (2021). Regulation of pri-MIRNA processing: Mechanistic insights into the miRNA homeostasis in plant. *Plant Cell Reports*, 1–16.

Jung, J.-H., & Park, C.-M. (2007). MIR166/165 genes exhibit dynamic expression patterns in regulating shoot apical meristem and floral development in Arabidopsis. *Planta*, *225*(6), 1327–1338.

Kidner, C. A., & Martienssen, R. A. (2004). Spatially restricted microRNA directs leaf polarity through ARGONAUTE1. *Nature*, *428*(6978), 81–84.

Kidner, C. A., & Martienssen, R. A. (2005). The developmental role of microRNA in plants. *Current Opinion in Plant Biology*, *8*(1), 38–44.

Kim, S., Yang, J.-Y., Xu, J., Jang, I.-C., Prigge, M. J., & Chua, N.-H. (2008). Two cap-binding proteins CBP20 and CBP80 are involved in processing primary microRNAs. *Plant and Cell Physiology*, *49*(11), 1634–1644.

Knauer, S., Holt, A. L., Rubio-Somoza, I., Tucker, E. J., Hinze, A., Pisch, M., Javelle, M., Timmermans, M. C., Tucker, M. R., & Laux, T. (2013). A protodermal miR394 signal defines a region of stem cell competence in the Arabidopsis shoot meristem. *Developmental Cell*, *24*(2), 125–132.

Kobayashi, Y., Murata, M., Minami, H., Yamamoto, S., Kagaya, Y., Hobo, T., Yamamoto, A., & Hattori, T. (2005). Abscisic acid-activated SNRK2 protein kinases function in the gene-regulation pathway of ABA signal transduction by phosphorylating ABA response element-binding factors. *The Plant Journal*, *44*(6), 939–949.

Lee, R. C., Feinbaum, R. L., & Ambros, V. (1993). The C. elegans heterochronic gene lin-4 encodes small RNAs with antisense complementarity to lin-14. *Cell*, *75*(5), 843–854.

Li, L. C., Okino, S. T., Zhao, H., Pookot, D., Place, R. F., Urakami, S., Enokida, H., & Dahiya, R. (2006). Small dsRNAs induce transcriptional activation in human cells. *Proceedings of the National Academy of Sciences*, *103*(46), 17337–17342.

Liang, G., He, H., Li, Y., Wang, F., & Yu, D. (2014). Molecular mechanism of microRNA396 mediating pistil development in Arabidopsis. *Plant Physiology*, *164*(1), 249–258.

Liu, N., Okamura, K., Tyler, D. M., Phillips, M. D., Chung, W.-J., & Lai, E. C. (2008). The evolution and functional diversification of animal microRNA genes. *Cell Research*, *18*(10), 985–996.

Luo, M., Gao, J., Peng, H., Pan, G., & Zhang, Z. (2014). MiR393-targeted TIR1-like (F-box) gene in response to inoculation to R. Solani in Zea mays. *Acta Physiologiae Plantarum*, *36*(5), 1283–1291.

Luo, Y., Guo, Z., & Li, L. (2013). Evolutionary conservation of microRNA regulatory programs in plant flower development. *Developmental Biology*, *380*(2), 133–144.

Mallory, A. C., Reinhart, B. J., Jones-Rhoades, M. W., Tang, G., Zamore, P. D., Barton, M. K., & Bartel, D. P. (2004). MicroRNA control of PHABULOSA in leaf development: Importance of pairing to the microRNA 5′ region. *The EMBO Journal*, *23*(16), 3356–3364.

Montgomery, T. A., Howell, M. D., Cuperus, J. T., Li, D., Hansen, J. E., Alexander, A. L., Chapman, E. J., Fahlgren, N., Allen, E., & Carrington, J. C. (2008). Specificity of ARGONAUTE7-miR390 interaction and dual functionality in TAS3 trans-acting siRNA formation. *Cell*, *133*(1), 128–141.

Mourelatos, Z., Dostie, J., Paushkin, S., Sharma, A., Charroux, B., Abel, L., Rappsilber, J., Mann, M., & Dreyfuss, G. (2002). miRNPs: A novel class of ribonucleoproteins containing numerous microRNAs. *Genes & Development*, *16*(6), 720–728.

Nagasaki, H., Itoh, J., Hayashi, K., Hibara, K., Satoh-Nagasawa, N., Nosaka, M., Mukouhata, M., Ashikari, M., Kitano, H., Matsuoka, M., Nagato, Y., & Sato, Y. (2007). The small interfering RNA production pathway is required for shoot meristem initiation in rice. *Proceedings of the National Academy of Sciences*, *104*(37), 14867–14871.

Nelson, C., & Ambros, V. (2021). A cohort of Caenorhabditis species lacking the highly conserved let-7 microRNA. *G3*, *11*(3), jkab022.

Nozawa, M., Miura, S., & Nei, M. (2012). Origins and evolution of microRNA genes in plant species. *Genome Biology and Evolution*, *4*(3), 230–239.

O'Brien, J., Hayder, H., Zayed, Y., & Peng, C. (2018). Overview of microRNA biogenesis, mechanisms of actions, and circulation. *Frontiers in Endocrinology*, *9*, 402.

Oliver, C., Annacondia, M. L., Wang, Z., Jullien, P. E., Slotkin, R. K., Köhler, C., & Martinez, G. (2021). The miRNome function transitions from regulating developmental genes to transposable elements during pollen maturation. *The Plant Cell, 34(2)*, 784–801.

Ori, N., Cohen, A. R., Etzioni, A., Brand, A., Yanai, O., Shleizer, S., Menda, N., Amsellem, Z., Efroni, I., Pekker, I., Alvarez, J. P., Blum, E., Zamir, D., & Eshed, Y. (2007). Regulation of LANCEOLATE by miR319 is required for compound-leaf development in tomato. *Nature Genetics*, *39*(6), 787–791.

Ossowski, S., Schwab, R., & Weigel, D. (2008). Gene silencing in plants using artificial microRNAs and other small RNAs. *The Plant Journal*, *53*(4), 674–690.

Palatnik, J. F., Wollmann, H., Schommer, C., Schwab, R., Boisbouvier, J., Rodriguez, R., Warthmann, N., Allen, E., Dezulian, T., Huson, D., Carrington, J. C., & Weigel, D. (2007). Sequence and expression differences underlie functional specialization of Arabidopsis microRNAs miR159 and miR319. *Developmental Cell*, *13*(1), 115–125.

Pandita, D. (2019). Plant MIRnome: miRNA biogenesis and abiotic stress response. In *Plant abiotic stress response* (pp. 449–474). Springer.

Peterson, K. J., Dietrich, M. R., & McPeek, M. A. (2009). MicroRNAs and metazoan macroevolution: Insights into canalization, complexity, and the Cambrian explosion. *Bioessays*, *31*(7), 736–747.

Piriyapongsa, J., & Jordan, I. K. (2008). Dual coding of siRNAs and miRNAs by plant transposable elements. *RNA*, *14*(5), 814–821.

Pushparaj, P., Aarthi, J., Manikandan, J., & Kumar, S. (2008). siRNA, miRNA, and shRNA: In vivo applications. *Journal of Dental Research*, *87*(11), 992–1003.

Reinhart, B. J., Slack, F. J., Basson, M., Pasquinelli, A. E., Bettinger, J. C., Rougvie, A. E., Horvitz, H. R., & Ruvkun, G. (2000). The 21-nucleotide let-7 RNA regulates developmental timing in Caenorhabditis elegans. *Nature*, *403*(6772), 901–906.

Saucier, D., Wajnberg, G., Roy, J., Beauregard, A. P., Chacko, S., Crapoulet, N., Fournier, S., Ghosh, A., Lewis, S. M., Marrero, A., O'Connell, C., Ouellette, R. J., & Morin, P. J. (2019). Identification of a circulating miRNA signature in extracellular vesicles collected from amyotrophic lateral sclerosis patients. *Brain Research*, *1708*, 100–108.

Singh, A., Gautam, V., Singh, S., Sarkar Das, S., Verma, S., Mishra, V., Mukherjee, S., & Sarkar, A. K. (2018). Plant small RNAs: Advancement in the understanding of biogenesis and role in plant development. *Planta*, *248*(3), 545–558.

Smith, M. R., Willmann, M. R., Wu, G., Berardini, T. Z., Möller, B., Weijers, D., & Poethig, R. S. (2009). Cyclophilin 40 is required for microRNA activity in Arabidopsis. *Proceedings of the National Academy of Sciences*, *106*(13), 5424–5429.

Song, L., Axtell, M. J., & Fedoroff, N. V. (2010). RNA secondary structural determinants of miRNA precursor processing in Arabidopsis. *Current Biology*, *20*(1), 37–41.

Song, Z. T., Liu, J. X., & Han, J. J. (2021). Chromatin remodeling factors regulate environmental stress responses in plants. *Journal of Integrative Plant Biology*, *63*(3), 438–450.

Stepien, A., Knop, K., Dolata, J., Taube, M., Bajczyk, M., Barciszewska-Pacak, M., Pacak, A., Jarmolowski, A., & Szweykowska-Kulinska, Z. (2017). Posttranscriptional coordination of splicing and miRNA biogenesis in plants. *Wiley Interdisciplinary Reviews: RNA*, *8*(3), e1403.

Stojadinovic, O., Lee, B., Vouthounis, C., Vukelic, S., Pastar, I., Blumenberg, M., Brem, H., & Tomic-Canic, M. (2007). Novel genomic effects of glucocorticoids in epidermal keratinocytes: Inhibition of apoptosis, interferon-γ pathway, and wound healing along with promotion of terminal differentiation. *Journal of Biological Chemistry*, *282*(6), 4021–4034.

Sun, B. K., & Tsao, H. (2008). Small RNAs in development and disease. *Journal of the American Academy of Dermatology*, *59*(5), 725–737.

Sun, Z., Li, M., Zhou, Y., Guo, T., Liu, Y., Zhang, H., & Fang, Y. (2018). Coordinated regulation of Arabidopsis microRNA biogenesis and red light signaling through Dicer-like 1 and phytochrome-interacting factor 4. *PLoS Genetics*, *14*(3), e1007247.

Taylor, R. S., Tarver, J. E., Hiscock, S. J., & Donoghue, P. C. (2014). Evolutionary history of plant microRNAs. *Trends in Plant Science*, *19*(3), 175–182.

Tien, V. Q., Duong, N. H., Nhan, D. T., & Vu, P. M. (2021). In silico analysis of Osa-miR164 gene family in rice (Oryza Sativa). *VNU Journal of Science: Natural Sciences and Technology*, *37*(3).

Vaucheret, H., Vazquez, F., Crété, P., & Bartel, D. P. (2004). The action of ARGONAUTE1 in the miRNA pathway and its regulation by the miRNA pathway are crucial for plant development. *Genes & Development*, *18*(10), 1187–1197.

Vidal, E. A., Araus, V., Lu, C., Parry, G., Green, P. J., Coruzzi, G. M., & Gutiérrez, R. A. (2010). Nitrate-responsive miR393/AFB3 regulatory module controls root system architecture in Arabidopsis thaliana. *Proceedings of the National Academy of Sciences*, *107*(9), 4477–4482.

Voinnet, O. (2009). Origin, biogenesis, and activity of plant microRNAs. *Cell*, *136*(4), 669–687.

Wang, J., Li, L., Li, C., Yang, X., Xue, Y., Zhu, Z., Mao, X., & Jing, R. (2021). A transposon in the vacuolar sorting receptor gene TaVSR1-B promoter region is associated with wheat root depth at booting stage. *Plant Biotechnology Journal, 19*(7), 1456–1467.

Wang, J., Mei, J., & Ren, G. (2019). Plant microRNAs: Biogenesis, homeostasis, and degradation. *Frontiers in Plant Science, 10*, 360.

Wang, J.-J., & Guo, H.-S. (2015). Cleavage of INDOLE-3-ACETIC ACID INDUCIBLE28 mRNA by microRNA847 upregulates auxin signaling to modulate cell proliferation and lateral organ growth in Arabidopsis. *The Plant Cell, 27*(3), 574–590.

Wang, J.-W., Wang, L.-J., Mao, Y.-B., Cai, W.-J., Xue, H.-W., & Chen, X.-Y. (2005). Control of root cap formation by microRNA-targeted auxin response factors in Arabidopsis. *The Plant Cell, 17*(8), 2204–2216.

Warthmann, N., Das, S., Lanz, C., & Weigel, D. (2008). Comparative analysis of the MIR319a microRNA locus in Arabidopsis and related Brassicaceae. *Molecular Biology and Evolution, 25*(5), 892–902.

Wightman, B., Ha, I., & Ruvkun, G. (1993). Posttranscriptional regulation of the heterochronic gene lin-14 by lin-4 mediates temporal pattern formation in C. elegans. *Cell, 75*(5), 855–862.

Won, S. Y. (2013). *A study of genes in DNA methylation and transcriptional gene silencing in Arabidopsis.* University of California, Riverside.

Wu, G. (2013). Plant microRNAs and development. *Journal of Genetics and Genomics, 40*(5), 217–230.

Wu, G., & Poethig, R. S. (2006). Temporal regulation of shoot development in Arabidopsis thaliana by miR156 and its target SPL3. *Development (Cambridge, England), 133*(18), 3539–3547.

Wu, X., Shi, Y., Li, J., Xu, L., Fang, Y., Li, X., & Qi, Y. (2013). A role for the RNA-binding protein MOS2 in microRNA maturation in Arabidopsis. *Cell Research, 23*(5), 645–657.

Xu, J., Li, C. X., Li, Y. S., Lv, J. Y., Ma, Y., Shao, T. T., Xu, L. D., Wang, Y. Y., Du, L., Zhang, Y. P., Jiang, W., Li, C. Q., Xiao, Y., & Li, X. (2011). MiRNA–miRNA synergistic network: Construction via co-regulating functional modules and disease miRNA topological features. *Nucleic Acids Research, 39*(3), 825–836.

Yan, J., Wang, P., Wang, B., Hsu, C. C., Tang, K., Zhang, H., Hou, Y. J., Zhao, Y., Wang, Q., Zhao, C., Zhu, X., Tao, W. A., Li, J., & Zhu, J. K. (2017). The SnRK2 kinases modulate miRNA accumulation in Arabidopsis. *PLoS Genetics, 13*(4), e1006753.

Yang, X., Dong, W., Ren, W., Zhao, Q., Wu, F., & He, Y. (2021). Cytoplasmic HYL1 modulates miRNA-mediated translational repression. *The Plant Cell, 33*(6), 1980–1996.

Yang, X., Ren, W., Zhao, Q., Zhang, P., Wu, F., & He, Y. (2014). Homodimerization of HYL1 ensures the correct selection of cleavage sites in primary miRNA. *Nucleic Acids Research, 42*(19), 12224–12236.

Yu, Y., Mo, X., & Mo, B. (2020). Introduction to plant small RNAs. In *Plant small RNA* (pp. 3–35). Elsevier.

Zhang, H., Xia, R., Meyers, B. C., & Walbot, V. (2015). Evolution, functions, and mysteries of plant ARGONAUTE proteins. *Current Opinion in Plant Biology, 27*, 84–90.

Zhang, J., Zhang, S., Han, S., Wu, T., Li, X., Li, W., & Qi, L. (2012). Genome-wide identification of microRNAs in larch and stage-specific modulation of 11 conserved microRNAs and their targets during somatic embryogenesis. *Planta, 236*(2), 647–657.

Zhang, S., Liu, Y., & Yu, B. (2015). New insights into pri-miRNA processing and accumulation in plants. *Wiley Interdisciplinary Reviews: RNA, 6*(5), 533–545.

Zhao, Y., Cong, L., & Lukiw, W. J. (2018). Plant and animal microRNAs (miRNAs) and their potential for inter-kingdom communication. *Cellular and Molecular Neurobiology, 38*(1), 133–140.

Zhu, Q.-H., & Helliwell, C. A. (2011). Regulation of flowering time and floral patterning by miR172. *Journal of Experimental Botany, 62*(2), 487–495.

3 miRNAs and Plant Development

Shah Rafiq, Nasir Aziz Wagay, Abdul Hadi,
Aabida Ishrath, and Zahoor Ahmad Kaloo

3.1 INTRODUCTION

With the discovery of short RNAs, a new class of regulatory genes that have greatly increased our understanding of gene regulation began to proliferate at the beginning of the twenty-first century. Through the RNA interference pathway, they have significant effects on the post-transcriptional regulation of protein-coding genes (Bartel & Bartel, 2003). microRNAs, (miRNAs) which are mostly 20–22 nucleotide (nt), are now becoming a prominent class of post-transcriptional endogenous gene regulators (Voinnet, 2009). Pre-miRNAs, which are stem-loop-structured intermediates, are processed into mature miRNAs from much longer original transcripts called pri-miRNAs. RNA-induced silencing complex is typically where mature miRNA is inserted and binds with target transcripts (Axtell et al., 2011). It is generally known that miRNAs are extensively present in both plants and animals and are essential for many basic biological functions (Voinnet, 2009). Plant responses to environmental stresses and growth are significantly regulated by miRNAs (Samad et al., 2017; Sunkar et al., 2012).

During their existence, plants went through several developmental stages, each of which is marked by the expression of unique physical features or the growth of new organs (Huijser & Schmid, 2011; Jin et al., 2013). In plants, miRNAs have been found to regulate many developmental and physiological processes (Sunkar et al., 2012; Johnson et al., 2017; Zhang et al., 2006). Most of the genes that miRNAs target encode transcription factors, so miRNAs play a critical role in the control of protein DNA and protein-protein interactions during plant development. These signaling pathways also control other essential biological pathways including phase transition and nutrient balance, as well as cell proliferation, growth, and specialization. They, therefore, serve as miRNA effectors in the plant's life cycle (Jones-Rhoades et al., 2006; Bartel & Bartel, 2003). It is challenging to determine miRNAs' exact functions in plants because, despite having a diverse variety of impacts upon several facets of plant growth, their regulatory actions may vary at various stages of growth. Most of the miRNAs recognized to serve a function in plant development are well conserved throughout the plant kingdom, while others are specialized to perform a particular function (Djami-Tchatchou et al., 2017; Cuperus et al., 2011). An miRNA's level of conservation reveals its "age" in the perspective of evolutionary history and is directly correlated with the way it expresses and functions (Cuperus et al., 2011).

3.2 ROLE OF MIRNAS IN PLANT GROWTH AND DEVELOPMENT

Plant growth is a well-organized process that is overseen at many different stages. Research into the mechanisms behind this regulation has found new directions with the discovery of miRNAs and potential roles in plant development. Numerous plant crucial physiological processes such as developmental phase transition, floral identity, shoot and root development, leaf development, hormone

DOI: 10.1201/9781003248453-3

signaling, patterning and polarity, stress response, flower development, and flowering time involve miRNA target genes that are predominantly transcription factors and F-box proteins (Jones-Rhoades et al., 2006). For plants to thrive and grow, miRNA transcription must be in a healthy state.

3.2.1 Role in Vegetative Development

The above and subterranean (shoot and root) tissues make up a plant's vegetative organs. For this the organs must initially develop from meristem cells, then must proliferate throughout time while being closely regulated by processes of cell division, expansion, and differentiation. The generation of organ polarity and the establishment of organ borders are two additional patterning processes necessary for the normal formation of plants. The miRNA-related pathways are engaged in nearly all the processes, making them some of the most significant endogenous signals regulating plant vegetative growth.

3.2.2 Role of miRNA in Shoot Meristem

Throughout their entire life cycle, plants may continually develop new organs, unlike mammals. Their embryonic apical meristem matures and houses a group of stem cells with the ability to differentiate in several directions and replicate themselves. The shoot apical meristem (SAM) controls the formation and growth of the plant's above-ground organs to a large extent. The STM (shoot meristemless)-WUS (Wuschel)-CLV (Clavata) pathway is essential for maintaining meristem activity (Somssich et al., 2016; Gaillochet & Lohmann, 2015). Flower meristems also exhibit some of the same processes.

miRNA targets and regulates many genes in the STM-WUS-CLV signaling pathway, directing the development and maintenance of the SAM. A significant factor in the control of gene regulation networks is miRNA (**Figure 3.1**). The SAM's surface L1 layer generates miR394, which diffuses downstream to the Organising Center (OC; **Figure 3.1**). When Leaf Curling Responsiveness (LCR) transcription in the OC is suppressed, WUS, a gene unique to SAM, is downregulated (Song et al., 2012; Knauer et al., 2013). The miR394 is present in both the L1 and OC layers, but its detrimental impact on LCR only manifests in the OC layer, demonstrating how miR394's functioning in *A. thaliana* depends on the precise level of the gene (Knauer et al., 2013). While this is going on, miR394-LCR plays several different functions in stem cell regulation (Kumar et al., 2019). AGO10 can selectively attach to miR165/166 and hence encourage HD-ZIP III expression. The miR165/166 targets the significant HD-ZIP III transcription factor family, which controls SAM in *A. thaliana*. Plant meristematic tissue is obliterated when miR166/165 is unable to bind to AGO10 or when the AGO10 gene is silenced. AGO10 and AGO1 compete for miR166/165 binding affinity. Plants will reduce the transcription of the HD-ZIP III genes and stop SAM production when miR166/165 binds to AGO1. Recent findings indicate that the connection between AGO10 and miR165/166 is exclusively linked to the structure of those molecules and that AGO10's catalytic activity is not involved (Zhu et al., 2011).

3.2.3 Role of miRNAs in Leaf Development

The formation of leaves is defined as the enlargement of the leaf blades following the separation of the leaf precursor cells (primordium) from SAM. Numerous regulatory components are included in these processes. The SAM's polar transport and distribution of auxin control organogenesis (Veit, 2009). The auxin response factors (ARF) gene family includes the miR160 target genes ARF10, ARF16, and ARF17, which regulate the auxin response. Arf10 and Arf17, two miR160-resistant mutants of the *A. thaliana* plant, exhibit an unequal number of cotyledons and a serrated, upward-curling leaf

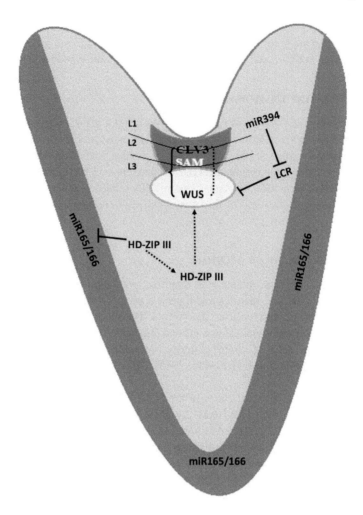

FIGURE 3.1   miRNA functions in embryo.

edge (Liu et al., 2007). The transcription of the MYB transcription factor in leaf primordium is one of many transcription factors that simultaneously control how leaves develop. By blocking the meristematic-specific gene KNOX1, these gene families PHANTASTICA/ASYMMETRIC LEAVES1/ ROUGH SHEATH2 may repress transcription to encourage development and differentiation (Piazza et al., 2005; Hay et al., 2004). In the development of dorsal-ventral polarization in plant leaves, transcription of MYB protein ASYMMETERIC LEAVES1 and the HD-ZIP III dictates the fate of the ventral axis, whilst expression of ARF3, ARF4, and KANADI (KAN) defines the destiny of the dorsal axis. In *A. thaliana*, the YABBY gene functions downstream of the KAN gene and is essential for the dorsal development of the leaf. The role of HD-ZIP III in leaf polarization is widely acknowledged (**Figure 3.2**). Due to miR165/166's inhibition of the HD-ZIP III family on the abaxial side, HD-ZIP III transcription was only kept on the adaxial side (Zhong & Ye, 2004). AGO1 is required for miR165/166 to control and confine PHBOLUSA (PHB) to the adaxial side and to target itself to HDZIP III transcripts in leaves (Kidner & Martienssen, 2004). Inhibiting acellular autonomous miR165/166 function and preventing HD-ZIP III mRNA accumulation require AGO10 on the adaxial leaf surface, much like AGO1 does (Liu et al., 2009).

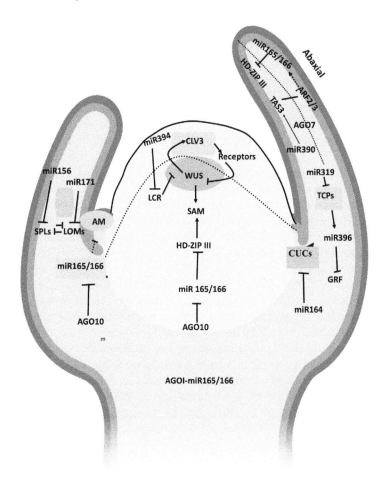

FIGURE 3.2 miRNA role in the shoot apex.

3.2.3.1 Establishment of Adaxial-Abaxial Polarity

Leaf polarization is dependent on miR390 and its activator AGO7 (**Figure 3.2**). TAS3 ta-siRNA which prevents ARF3 and ARF4 expression on the abaxial side of leaves establishes the adaxial side of leaves (Chitwood et al., 2009). In *A. thaliana* and *Z. maize*, the ventral ta-siARF pathway interacts sporadically with the dorsal regulatory components. Moreover, the ta-siARF pathway inhibits miR165/miR166 production and supports HD-ZIP III growth. There is no doubt that maize leaf polarity will change if the ta-siARF pathway is inhibited. According to Wang et al. (2011), miR396 controlled the proliferation of leaf cells by concentrating on growth regulating factors (GRFs), resulting in an impact on how the dorsal-ventral axis polarity developed in leaves.

3.2.3.2 Controlling the Shape and Size of Leaves

Organ primordia expand and divide in several waves in the final stages of shoot development before differentiating into distinct forms. miRNAs control the shape and size of leaves. The development of the leaf lamina and the definition of the leaf edges serve as the model systems for leaf morphogenesis. These patterning processes are regulated by miR319 and the genes in the CYCLOIDEA, PCF (TCP) FAMILYANDTEOSINTE BRANCHED1 that it targets. In contrast to the overexpression of miR319, which produced bigger leaves with curled borders and multiple TCP genes losing function, upregulation of miR319-resistant TCP4 (mTCP4) generates smaller leaves in *Arabidopsis* (Koyama et al., 2010; Efroni et al., 2008). The activation of KIP RELATED PROTEIN1/CYCLIN-DEPENDENT

KINASE INHIBITOR1 (ICK1), a cyclin-dependent kinase inhibitor of the cell cycle that decreases cell proliferation, was revealed to be one of the contributing reasons in the downregulation of TCP4 on leaf development (Wang et al., 2000; Schommer et al., 2014). In addition to these redundant functions in leaf morphogenesis, other miR319-targeted TCP genes also function primarily via controlling the miR164-CUC pathway (Koyama et al., 2010).

The interplay of the proteins CUC2, TCP4, and the miR156 TARGET SQUAMOSAL PROMOTER BINDING PROTEIN like (SPL9) in *Arabidopsis* and the distantly related *Cardamine hirsuta* govern the age-related development of leaf complexity. TCP4 interacts with CUC2 in the early vegetative phases and suppresses dimerization with the homolog CUC3, which suppresses the production of leaf serrations and leads to the generation of simple leaves. As a plant ages, SPL9 protein levels increase, stopping TCP4-CUC connections and allowing CUC2 and CUC3 to dimerize. This action acts as a cue to encourage the formation of compound leaves later on (Rubio-Somoza et al., 2014).

Along with miR164, miR396 also functions downstream of TCP genes (Schommer et al., 2014). Seven GRF family members required for cell division and abaxial-adaxial polarity are the targets of miR396 in *Arabidopsis* (Das Gupta & Nath, 2015; Rodriguez et al., 2010). Because miR396 inhibits plant cell growth, transgenic lines overexpressing it and GRF loss-of-function mutants both produce smaller rosettes (Rodriguez et al., 2010; Horiguchi et al., 2005). This is due to *Arabidopsis*'s slower pace of development from the leaf's proximal to distal ends (Das Gupta & Nath, 2015). Cao et al. (2016) suggest that miR396 regulates GRFs to prevent organ formation in tomato plants. In rice, miR396-targeted OsGRF4 and OsGRF6 are also efficient and beneficial controllers of leaf size and yield attributes (Gao et al., 2015). In numerous species, mioR396 also controls non-GRF targets. By regulating its basic Helix-Loop-Helix (bHLH) transcription factor miR396, bHLH74 aids in the patterning of the leaf edge and veins in the Brassicaceae and Cleomaceae families (Debernardi et al., 2012). In contrast, miR396 controls the downstream gene HaWRKY6 in the sunflower (*Helianthus annuus*), which helps the plant respond to heat stress by promoting growth (Giacomelli et al., 2012). These findings suggest that miR396 can detect new gene transcription as it develops.

Additionally, miR396's activity may be impacted by environmental cues. For instance, UV-B light can activate it, resulting in plants growing leaves with a smaller margin (Casadevall et al., 2013). Similar to miR826, which controls leaf growth in response to nitrogen and phosphate deficiency, NLA is the target of miR827 (Lin et al., 2013; Hewezi et al., 2016). These miRNAs could serve as internal signaling systems for the environmental signals that influence the development of plant organs.

3.2.3.3 Leaf Senescence

Senescence is the last stage of the leaf's development. Several miRNAs have been shown to have essential roles as regulators of this process and are responsible for controlling how phytohormones work. For example, WRKY53 a gene that positively controls several senescence-associated genes (SAG), is directly regulated by miR396-targeted HD-ZIP III genes, which in turn affect leaf senescence (Miao & Zentgraf, 2007; Miao et al., 2004). The miR319-TCP4 module directly binds to LIPOXYGENASE2 (LOX2), a crucial enzyme in the production of Jasmonic acid (JA), to regulate JA-dependent leaf senescence. WRKY53 is also increased in the miR319-resistant rTCP4 transgenic plant (Schommer et al., 2008). Plant hormone ethylene affects the miR164-ORE1 pathway because ETHYLENE-INSENSITIVE3 (EIN3), a crucial gene in ethylene signaling, directly suppresses miR164 and encourages the transcription of ORE1 (Li et al., 2013b; Kim et al., 2014). The intricate interaction between miR164 and the ethylene regulators is demonstrated by the fact that ORE1 stimulates AMINOCYCLOPROPANE-1-CARBOXYLIC ACID (ACC) synthase ACS2, a vital enzyme in the production of ethylene (Qiu et al., 2015). Another important component in the senescence of leaves is auxin (Osborne, 1959). ARF2, which miR390 particularly targets, speeds up leaf withering in *Arabidopsis* by effectively regulating auxin transmission (Lim et al., 2010).

The Agamous Like 16 (AGL16) gene that miR824 targets controls the development of stomata. Overexpression of miR824 reduces stomatal density like that of AGL16 mutant plants. Conversely, stomatal density will rise as miR824's influence in AGL16 is weakened (Kutter et al., 2007). An increase in GL15 (Glossyl5) activity in maize may cause the growth of the reproductive organs to be delayed and more immature leaves to appear. Additionally, by negatively regulating the GL15 mRNAs, miRl72 can speed up the development of maize's immature leaves into mature leaves (Lauter et al., 2005). An increase in GL15 (Glossyl5) activity in maize may cause the growth of the reproductive organs to be delayed and more immature leaves to appear. Additionally, by negatively regulating the GL15 mRNAs, miRl72 can speed up the development of maize's immature leaves into mature leaves (Kutter et al., 2007). A TCP transcription factor is encoded by the Lanceolate gene of the tomato (*Solanum lycopersicum*). Single leaves may emerge in plants with compound leaf structures as a result of mutation or downregulation. Several leaflets may merge into a single leaf as a result of miR319's impact on the LANCEOLATE (LA) gene (Ori et al., 2007).

3.2.4   ROLE OF MIRNAS IN VASCULAR DEVELOPMENT

Vascular plants employ xylem to move water and nutrients absorbed by roots upwards, whereas a phloem is used to move carbohydrates absorbed by leaves downwards. Xylem, cambium/procambium, and phloem are three nicely ordered tissues that make up the vascular bundle (**Figure 3.3**). The HD-ZIP III gene family is highly expressed in the vascular tissue of *A. thaliana*. Overexpressing miR165 in *A. thaliana* can influence plant morphogenesis by regulating the polar differentiation of vascular tissue cells and reducing the expression of all HD-ZIP III family members (Jia et al., 2015; Du & Wang, 2015). The miR166 has been shown to alter the Homeobox 15 protein (ATHB15), thus altering the growth of vascular tissue cells in *A. thaliana* (Kim et al., 2005). Since the miR165/166 target region in the HD-ZIP III gene class is shared in almost all plants, this module is essential for the growth and development of plants (Floyd & Bowman, 2004).

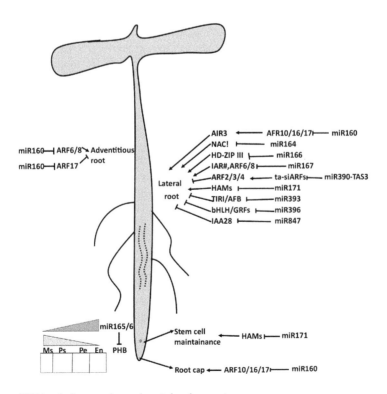

FIGURE 3.3   miRNA role in vascular and root development.

Some miRNAs in plants are also associated with the development of cell walls and fibres (Kim, 2005). The formation of vascular secondary walls is reportedly influenced by a new miRNA (miR857) in a manner that is copper ion dependent. The miR857 regulates the expression of the putative laccase LACCASE7, a gene belonging to the laccase family that has an impact on the amount of lignin (Zhao et al., 2015). Scientific investigations have shown that several elements, including ARF, miR390, and TAS3, that are connected to vascular morphogenesis and leaf orientation are all preserved in land plants. TAS3 ta-siRNA for instance activates ARF in liverworts similarly to angiosperms. The semi-dominant phv (phavoluta) mutant of *Nicotiana tabacum*, which lacks the regulation of miR165, exhibits aberrant radial development of the stem and leaf vascular systems as well as discontinuous vascular tissue at the stem nodes. The discovery showing miRl65 regulates the formation of the vascular cambium shows why plants still retain miR165/166's role in vascular.

3.2.5 MiRNAs and Reproductive Development

The change of vegetative meristems into floral meristems, where flowers develop, grow, and generate seeds to complete the plant's lifetime, occurs when the plant reaches a certain phase or under specific circumstances. To reproduce and transmit their genetic makeup to their progeny, plants must go through this reproductive development process. The miRNAs are important regulators at many stages of plant reproduction.

3.2.5.1 Role of miRNA in Flower Development

3.2.5.1.1 The Transition From the Vegetative Phase to the Reproductive Phase

Numerous studies have discovered the role of miRNA in blossoming. A class of transcriptional regulators (SBP/SPL proteins) with a focus on plants assist *A. thaliana* in entering the vegetative phase. The expression of miR156 and miR157 is suppressed during the developmental stage of the juvenile. The number of SBP/SPL proteins rise, miR156/157 levels fall, and the plant transitions from the vegetative to the reproductive phases (Fouracre et al., 2021). The primary regulator of plant growth cycle transformation, miR156, influences plant phase transition by concentrating on transcriptional regulators that bind to SPL (squamosa promoter binding protein-like) sequences (**Figure 3.4**; He et al., 2018). *A. thaliana*'s delayed blooming phase was brought on by miR156 overexpression, which also resulted in the downregulation of SPL3/5, SPL9, and SPL15. As a result, *A. thaliana* grew more slowly and produced a lot more leaves (Zhang et al., 2019; Xu et al., 2016b). Additionally, it has been discovered that rice miR156 and SPLs have a role in floral development (Xie et al., 2006). According to the latest studies, the strong downward repression of miR156 on SPL3 affects the transcription of the FT gene in *A. thaliana* leaves, which eventually has an impact on how the plant transitions into its flowering phase. This postpones blossoming (Kim et al., 2012). High amounts of miR156 keep plants in their initial phases of growth in a manner that is similar to how insect juvenile hormones work. The progressive decrease in miR156 levels during development facilitates the change from a young to a mature state. Further investigation revealed that the drop in miR156 content was associated with the physiological age of plants instead of with the actual age (i.e., time) of plants (Cheng et al., 2021).

Similar to miR156, miR172 suppresses and degrades its target mRNA to control when plants blossom and how their floral organs grow (Jung et al., 2007). By controlling AP2-like genes including SCHNARCHZAPFEN, TARGET OF EARLY ACTIVATION TAGGED 1/2/3, AND SM-LIKE 2, miR172 controls the transition of plants grown from the embryonic to the blooming stage. Through altering AP2 transcription factors miR172 controls the time of flowering, floral morphology, and plant development (Aukerman & Sakai, 2003). Protein expression of the miR172 gene promotes early blooming in *A. thaliana*, but overexpression of the AP2 gene delays flowering.

Additionally, miR172 and miR156 collaborate in some miRNA-controlled phases of the plant development cycle. While certain SPLs increased the expression of miR172, miR156 suppressed the SPL family's expression. As per earlier studies, the young-to-mature change in *A. thaliana* is

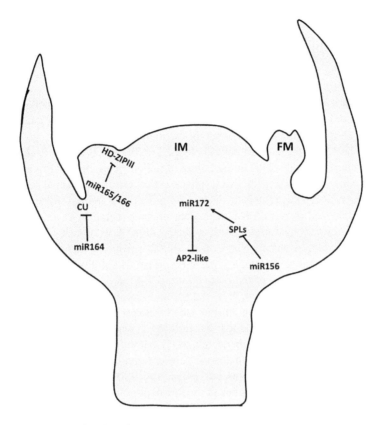

FIGURE 3.4 miRNA role in floral meristem.

primarily regulated by the mir156-SPL-miR172 pathway. The leaf modules and apical meristem modules are subclasses of the miR156-SPL-miR172 pathway, and they both contain different ratios of miR172 and SPL coding gene units. While the SPL15-miR172d unit enhances blooming by boosting transcription of the box genes in the apical meristem, the SPL9-miR172b/c units in leaves decrease the inflorescence time by transcriptional activation of the FT gene. The environment's temperature and photoperiod may also influence the miR172 gene's transcription, and various miR172 genes have various response routes (Lian et al., 2021).

The miRNAs miR159 and miR319 are two additional miRNAs that support floral development. TCP (TCP FAMILY TRANSCRIPTION FACTOR) and MYB transcription factors are their respective target genes. The miR139 and miR159 enhanced production may result in abnormal floral development, particularly delayed flowering (Palatnik et al., 2007). Significant pleiotropic malformations, retarded development, curled leaves, and malformed sepals and anthers are displayed by *A. thaliana's* loss-of-function miR159. MYB33 and MYB65 production is controlled by miR159 (Yu et al., 2012; Tsuji et al., 2006). The miR159 regulates *A. thaliana's* phase transition because it can prevent miR156 from being hyperactive during the vegetative transition stage (Guo et al., 2017). By altering the miR156-SPL pathway's effect on gene expression, the change from young to mature status in *Arabidopsis* is regulated by AB15. The miR159's intended target, MYB33, activates the promoter of ABA INSENSITIVE 5 (ABI5) to boost the transcription of ABI5 (Guo et al., 2021).

3.2.5.1.2 miRNAs: Role in Controlling Floral Patterning

The floral meristem separates into distinct floral organs as it moves from vegetative to reproductive development. The calyx, corolla, androecium, and gynoecium are produced by the four concentric whorls of the growing flower in *Arabidopsis*, and their identities are determined by three groups of floral homeotic transcription factors known as A, B, and C function genes (Bowman et al., 1991). The miR172 controls the post-transcriptional and translational regulation of the crucial class Bgene AP2 (Mlotshwa et al., 2006). The miR172 upregulation causes a significant floral homeotic transition where the reproductive organs replace the perianth organs. The ap2 mutant exhibits the same phenotype because the C-function gene AGAMOUS (AG) is transcribed ectopically in the two outer whorls. AP2 and AG compete for the definition of the second and third whorl boundaries (Wollmann et al., 2010). The miR172 targets a variety of transcription factors from the AP2 family and plays an essential role in both floral imitation and organ differentiation.

Vital miRNA MiR159 interacts with genes belonging to the MYB family to regulate the formation of male reproductive organs. Due to the failure of pollen ejection, MiR159 is overexpressed in the anthers of *Arabidopsis* and rice, and this results in male sterility (Tsuji et al., 2006). It is exciting to learn that tomato miR159 targets a gene distinct from the MYB family, suggesting that miR159's activity may be controlled in an MYB-independent manner (Buxdorf et al., 2010). The miR164 largely helps in the formation of organ borders during floral development, much like its function during the vegetative stage. Constative expression of the miRNA or the loss of function CUC1/2 results in merged sepals, merged stamens, and missing petals (Mallory et al., 2004; Laufs et al., 2004). On the other hand, the loss of the other two miR164 family members significantly worsens the miR164c phenotype, which is brought by the mutation in the primary miR164 gene, mIR164c, and results in reduced organogenesis in all four whorls (Baker et al., 2005). Due to these morphological imperfections, it may be concluded that miR164 genes play dual roles in the growth and extension of the floral organ border (Sieber et al., 2007). Although upregulation of both target genes SINAM2 and GOBLET (GOB) results in aberrant flower boundaries and growth, that is, the crucial role of miR164 is likewise retained in tomatoes (Blein et al., 2008; Berger et al., 2009). The miR319 loss causes a decline in the number of petals in the flower, as opposed to miR164 loss, which is predominantly brought about by the relaxation of TCP4 repression (Nag et al., 2009). TCP4 may have a parallel role in the control of cell proliferation during both the vegetative and reproductive phases because floral-specific upregulation of TCP4 inhibits the growth of petals and stamens (Efroni et al., 2008; Nag et al., 2009).

ARF genes that are controlled by miR160 and miR167 also significantly influence floral development. *Arabidopsis* carpels (foc) with mutant floral organs generate deformed flowers with fragile sepals and petals and poor reproduction rates (Liu et al., 2010). Additionally, the formation of reduced floral organs is caused by mir167 upregulation in *Arabidopsis*, tomato, and rice, demonstrating that miRNA-controlled regulation of auxin signaling acts as a preserved process to regulate the development of floral organs (Wu et al., 2006; Ru et al., 2006). In contrast to some of these evolutionary conservative miRNAs, several many others affect floral development in a family-specific way. Both Antirrhinum BLIND (BL) and Petunia FISTULATA (FIS) genes include miR169, which regulates the location-based transcription of type C genes and activates NF-YA genes (Cartolano et al., 2007). *Arabidopsis* also has the miR169-NFYA pathway, although this pathway does not seem to be engaged in floral organ identification, suggesting that *Antirrhinum* and *Petunia* may well have developed specific uses for this system.

3.2.6 Organ Development

The shoot apical meristem and root apical meristem respectively are the developmental forerunners of plant shoots and roots. The NAC family of transcription factors, which are known to be crucial for cotyledon dissociation and shoot apical meristem development during *Arabidopsis* embryogenesis are encoded by CUC genes. CUC1, CUC2, and NAC1 are among the members of the NAC family that the miR164 family of molecules targets (Schwab et al., 2005; Guo et al., 2005). Similar to NAC gene mutations plants upregulating miR164 exhibit a variety of developmental problems. The lateral root development, however, is improved in loss-of-function miR164 mutants, which also store more

NAC mRNA (Guo et al., 2005). As a result, miR164 destroys the CUC1 and CUC2 mRNAs in the *Arabidopsis* meristem during organ initiation, restricting border expansion.

3.2.7 Role in Other Processes

The control of other developmental stages also depends on miRNA. The auxin response factors ARF16, ARF10, and ARF17 are miR160's targets in *A. thaliana*. Findings show that miR160 is crucial for suppressing ARF10 to encourage seed germination (Liu et al., 2007). According to Llave et al. (2002), the formation of root cap cells occurs when *Arabidopsis* roots expand. By controlling the transcription of RF10, miR160 governs stem cell proliferation at the end of the terminal of the root meristematic area and affects the orientation of root growth (**Figure 3.3**).

The formation of plant root organs, including the creation of root caps, lateral roots, and adventitious roots, is also significantly influenced by miR164 and miR390 (Yoon et al., 2010). The miR164 regulates the development of lateral roots in *A. thaliana*. The miR164 may affect NAC1 expression after auxin induction, altering auxin transport and regulating peripheral root development (**Figure 3.3**; Guo et al., 2005). The miR165/166 in plants is linked to xylem development and cell organization. The miRNA controls developing plant tissue via a complex molecular mechanism (**Figure 3.3**; Carlsbecker et al., 2010). Also, the same miRNA could take different roles in diverse tissues. Like miR165/miR166 is connected to xylem and cell layout as well as leaf polarization (Manuela & Xu, 2020; Tatematsu et al., 2015).

The modulation of morphological structure and agricultural plant production are two other important functions of miRNA. For the regulation of soybean production and morphology, the miR156-SPL gene module is crucial. Axillary bud growth and branching in transgenic soybean overexpressing miR156b have been controlled by downregulating SPL9d transcription. By accelerating grain filling, miR1432 inhibition or overexpression of OsACOT (Acetyl-CoA thioesterase) in rice can significantly enhance yield (Zhao et al., 2019). Genomic investigations suggested miR156f's target gene is OsSPL7, which controls the morphological characteristics of plants including tillering and height of rice (Dai et al., 2018). The miR156f-OsSPL7-OsGH3.8 route is the complete regulatory system for these features in rice, as shown by the finding that OsSPL7 directly attaches to the OsGH3.8 regulators to regulate their translation.

Plant diseases and responses to environmental stress are typically associated with miRNA. Viral infections can have a substantial influence on plant morphology and output. As to increasing evidence, miRNA has a connection to viral illnesses and virus-induced gene silencing (Chapman et al., 2004). Plant viruses have been revealed to contain over 30 RNA silencing silencers also known as pathogenic factors such as p^{19}, p^{21}, p^{25}, and p^{69}. Pathogenic factors can cause the development of numerous plant diseases and deformities throughout development in addition to obstructing the production of siRNA, changing its stability, or interfering with the interaction of siRNA with RISC complexes. High levels of HC-Pro protease (helper-component proteinase) in plants cause the amount of miR171 to decrease, resulting in developmental deletion plants with miR171-related problems including improper branching and an increase in the quantity of short vegetative phytomers and late blooming. The majority of miR171 target mRNAs are increased when the HC-Pro gene is overexpressed in *A. thaliana* which causes virus-mediated illness in plants (Kasschau et al., 2003).

Plants may directly synthesize certain miRNAs in reaction to abiotic stress, while also inducing either reduced or excessive transcription of other miRNAs. These miRNAs have an impact on transcription factors linked to stress tolerance, stress tolerance response protein genes, Plant Growth Regulator 9 response protein genes, and other target genes, allowing plants to react swiftly to environmental changes. The first stress-reported response miRNA in plants was reported in *A. thaliana* (Jones-Rhoades & Bartel, 2004). When *A. thaliana* was exposed to dehydration, low temperatures, salt, or hormones (ABA) treatment, it increased expression of miR393. The miR310 and miR319 transcription levels increased following low-temperature stress, and no responses to drought or NaCl was shown, demonstrating that these two miRNAs exclusively had a role in low-temperature stress, resulting in the downregulation of miR389a (Sunkar & Zhu, 2004).

Low temperatures have been shown to promote miR393, miR397, miR402, miR165/miR166, miR169, and miR172 to improve plant tolerance (Sunkar & Zhu, 2004). While miR399 transcription rose and its target gene PHO2/UBC24 (PHOSPHATE2)'s mRNA level decreased, miR395 transcription level increased in *A. thaliana* due to lack of sulphate (Chiou et al., 2006). The downregulation of miR169 is brought on by drought. Plants having nuclear transcription factor Y subunit A-5 loss of the miR169a seem to be more susceptible to transpiration and are more vulnerable to drought than wild-type plants (Li et al., 2008). By specifically modulating the activity of two superoxide dismutase, COPPER/ZINC SUPEROXIDE DISMUTASE ½, miR398 in grapevine enhances plant biotic stress, leading to excessive salt, heavy metal, UV radiation, drought, etc. (Leng et al., 2017).

Studies on *A. thaliana* and other plants have shown that miRNAs are important in a broad range of biological activities. miRNAs are crucial for the specific control of gene transcription in a tissue-specific manner as opposed to hormones, which have impacts on the whole plant. Additional research is needed to understand how plants control the hormone signaling pathway and influence tissue differentiation by precisely regulating miRNAs. Understanding the role and process of miRNA diffusion between cells and tissues requires further research.

3.3   INTERACTION BETWEEN PLANT HORMONES AND MIRNAS

Plant hormones are crucial regulating substances that are produced by plants. They affect the growth, differentiation, and development of plants on an individual or group level. Auxin (AUX), abscisic acid (ABA), cytokinin (CK), gibberellic acid (GA), brassinosteroid (BR), ethylene (ET), and jasmonic acid (JA) are the main phytohormones. Phytohormones are essential for managing the timing of development, metabolic activity, and stress response throughout the whole plant life cycle because they act as signaling molecules that regulate plant growth and development. Multiple hormones are frequently present throughout various growth stages, allowing plant cells to respond to developmental cues and variations in their internal and external environments in a dynamic fashion (Li et al., 2020). By influencing target genes in hormonal pathways adversely, miRNAs work in conjunction with hormones. The overall quantity of miRNA accumulation was shown to have reduced in the seedlings after the HYL1 mutation, which caused a variety of developmental abnormalities and incorrect sensitivity to ABA, AUX, and CK. This demonstrates that miRNA and these phytohormones' signaling responses are related (Han et al., 2004). The fact that many cis-elements in the miRNA gene promoters respond to stress and phytohormones suggests that they may be used to influence miRNA gene transcription (Ding et al., 2013).

In plants, miRNAs control a series of transcription factors, including auxin receptors. ARF16 and ARF17 gene expression levels increased in *Arabidopsis* when miR160 was suppressed, resulting in aberrant germ formation, an irregular cotyledon shape, sluggish inflorescence growth, root shortening, reduced stamens, root shortening, and other undesirable developmental symptoms. On the other hand, in *Arabidopsis*, overexpressing miR160 decreased the number of lateral roots and prevented the development of the root cap (**Figure 3.3**; Wang et al., 2005; Mallory et al., 2005). These findings suggest that precise miR160 production is required for auxin-related plant growth. Co-regulation of adventitious root development by miR167 and ARf6/8 (Gutierrez et al., 2009). To activate the auxin signaling pathway, miR847 targets and silences IAA28, the AUX/IAA inhibitory protein. The auxin signaling pathway was quickly disinhibited using the miR847/IAA28 mRNA regulation modules and ubiquitination-mediated breakdown of the IAA28 protein (Wang & Guo, 2015). Likewise, miR165/166 targets PHB, an ARF5 activator, directly, and this results in the expression of miR390. As a result, Ta-siRNAs (TasiR-ARF3/4) build up (Dastidar et al., 2019; Muller et al., 2016). The blade's paraxial/distal polarity is likewise determined by the miR165/166-tasiRARFs module.

By downregulating the transcription of SHOOTMERISTEMLESS and BREVIPEDICELLUS, miR159 and miR319 upregulates the activity of IPT (ISOPENTENYL TRANSFERASE) and encourage the synthesis of CK in SAM (Scofield et al., 2014; Rubio-Somoza & Weigel, 2013). By suppressing the B-type ARR genes9 [type B *Arabidopsis* Response Regulators (ARRs)], which are transcription factors that function as positive regulators in the two-component cytokinin signaling pathway, the miR156-SPL9 combination changes cytokinin-related plant regeneration at the level of cytokinin signal transmission (Zhang et al., 2015). Additionally, miRNAs like miR160 and miR165/6 influence the production of cytokinin and the way it is signaled by auxin, which assists in keeping a dynamic balance between auxin and cytokinin (Liu et al., 2016; Dello et al., 2012). The DELLA (aspartic acid-glutamic acid-leucine-leucine-alanine) protein and its partner proteins, such as IDD2 (indeterminate (ID)-domain 2), PHYTOCHROMEINTERACTING FACTOR 4, or SCARECROW-LIKE, can alter the concentrations of some miRNAs in response to another signaling molecule, gibberellin (SCL; Fan et al., 2018; Han et al., 2014). However, a variety of complexes, including miR156-SPL, miR159-GAMYB(L)s, and miR171-SCL modules, can allow miRNAs to directly influence the synthesis of GA and signal transduction (Millar et al., 2019; Sun et al., 2018). Brassinosteroids inhibit the translation of target genes that miRNA has inhibited by interacting with AGO1, a miRNA effector protein, in the endoplasmic reticulum (Wang et al., 2021d).

miRNAs can regulate leaf senescence and seed germination by altering abscisic acid and ethylene levels. miRNA production can be regulated by the signaling hormone abscisic acid and crucial enzyme of the osmotic stress response pathway SnRK2 (SNF1-related protein kinase 2) (Yan et al., 2017). The ABA and ethylene signaling pathways may have an impact on CBP20, a critical protein in the sRNA synthesis (CAP-BINDING PROTEIN 20; Zhang et al., 2016; Li et al., 2013a). Because of this, miRNAs play several roles in the interactions between hormone responses and plant growth.

3.4   CONCLUSION AND FUTURE PERSPECTIVES

Several miRNAs have been found in plants by genetic assessment, high throughput sequencing, and bioinformatics analysis. Determining the biological functions of vital miRNAs in many plant species also made progress. Even though the functions of many miRNAs are still unknown, it is known that these miRNAs collaborate to regulate organ development. As a result, there is still much to be learned about the signaling pathways of miRNAs in plant development. Although species-specific miRNAs typically appear to be more "inert" than conserved miRNAs due to their reduced expression and absence of target genes, the functioning of a miRNA is correlated with its evolutionary conservation status (Cuperus et al., 2011). This emphasizes the need to research this subgroup for novel growth regulators, and it may be crucial for future research to concentrate more on the finer modifications that these miRNAs regulate. Additionally, since distinct conserved miRNAs from vast gene families sometimes perform repetitive tasks, some of them may play roles that are exclusive to certain tissues or organs. It is critical to continue investigating how the multiple members of these conserved miRNA families exhibit and operate in different ways depending on the circumstance and time.

In addition to revealing the signaling pathways of individual miRNAs, an important unanswered subject in research is how several miRNAs related pathways interact. The following questions need to be answered: can a miRNA directly control the transcription of miRNA to initiate a signaling cascade? Can a miRNA affect histone modifications, which frequently affect the expressions of downstream miRNA genes? Can the target gene products of different miRNAs directly interact and perform together? By understanding how these interactions interact in a network to govern a particular developmental process and then how these particular processes link to build the overall pattern of the plant, the findings of these studies may lead us to an even more difficult goal. Therefore, it is highly recommended that miRNA sequencing technology and study methodologies be improved. Several single-cell omics and nanopore sequencing techniques will be utilized on model plants to find more miRNAs, their regulatory pathways, and their mode of action. This will provide a fundamental theoretical foundation for understanding how miRNA regulates plant growth and development, which can then be applied to plants of agricultural importance.

REFERENCES

Aukerman, M. J., & Sakai, H. (2003). Regulation of flowering time and floral organ identity by a microRNA and its APETALA2-like target genes. *Plant Cell, 15*, 2730–2741. https://doi.org/10.1105/tpc.016238

Axtell, M. J., Westholm, J. O., & Lai, E. J. (2011). Vive la difference: Biogenesis and evolution of microRNAs in plants and animals. *Genome Biology, 12*, 221. https://doi.org/10.1186/gb-2011-12-4-221

Baker, C. C., Sieber, P., Wellmer, F., & Meyerowitz, E. M. (2005). The early extra petals1 mutant uncovers a role for microRNA miR164c in regulating petal number in Arabidopsis. *Current Biology, 15*, 303–315. https://doi.org/10.1016/j.cub.2005.02.017

Bartel, B., & Bartel, D. P. (2003). MicroRNAs: At the root of' plant development? *Plant Physiology, 132*(2), 709–717. https://doi.org/10.1104/pp.103.023630

Berger, Y., Harpaz-Saad, S., Brand, A., Melnik, H., Sirding, N., Alvarez, J. P., Zinder, M., Samach, A., Eshed, Y., & Ori, N. (2009). The NAC-domain transcription factor GOBLET specifies leaflet boundaries in compound tomato leaves. *Development, 136*, 823–832. https://doi.org/10.1242/dev.031625

Blein, T., Pulido, A., Vialette-Guiraud, A., Nikovics, K., & Morin, H. (2008). A conserved molecular framework for compound leaf development. *Science, 322*, 1835–1839. https://doi.org/10.1126/science.1166168

Bowman, J. L., Smyth, D. R., & Meyerowitz, E. M. (1991). Genetic interactions among floral homeotic genes of Arabidopsis. *Development, 112*, 1–20. https://doi.org/10.1242/dev.112.1.1

Buxdorf, K., Hendelman, A., Stav, R., Lapidot, M., Ori, N., & Arazi, T. (2010). Identification and characterization of a novel miR159 target not related to MYB in tomato. *Planta, 232*, 1009–1022. https://doi.org/10.1007/s00425-010-1231-9

Cao, D. Y., Wang, J., Ju, Z., Liu, Q. Q., Li, S., Tian, H. Q., Fu, D. Q., Zhu, H. L., Luo, Y. B., & Zhu, B. Z. (2016). Regulations on growth and development in tomato cotyledon, flower and fruit via destruction of miR396 with short tandem target mimic. *Plant Science, 247*, 1–12. https://doi.org/10.1016/j.plantsci.2016.02.012

Carlsbecker, A., Lee, J. Y., Roberts, C. J., Dettmer, J., & Lehesranta, S. (2010). Cell signalling by microRNA165/6 directs gene dose-dependent root cell fate. *Nature, 465*, 316–321. https://doi.org/10.1038/nature08977

Cartolano, M., Castillo, R., Efremova, N., Kuckenberg, M., Zethof, J., Gerats, T., Schwarz-Sommer, Z., & Vandenbussche, M. (2007). A conserved microRNA module exerts homeotic control over Petunia hybrida and Antirrhinum majus floral organ identity. *Nature Genetics, 39*, 901–905. https://doi.org/10.1038/ng2056

Casadevall, R., Rodriguez, R. E., Debernardi, J. M., Palatnik, J. F., & Casati, P. (2013). Repression of growth regulating factors by the micro-RNA396 inhibits cell proliferation by UV-B radiation in Arabidopsis leaves. *Plant Cell, 25*, 3570–3583. https://doi.org/10.1105/tpc.113.117473

Chapman, E. J., Prokhnevsky, A. I., Gopinath, K., Dolja, V. V., & Carrington, J. C. (2004). Viral RNA silencing suppressors inhibit the microRNA pathway at an intermediate step. *Genes & Development, 18*, 1179–1186. https://doi.org/10.1101/gad.1201204

Cheng, Y. J., Shang, G. D., Xu, Z. G., Yu, S., & Wu, L. Y. (2021). Cell division in the shoot apical meristem is a trigger for miR156 decline and vegetative phase transition in Arabidopsis. *Proceedings of the National Academy of Sciences of the United States of America, 118*, e2115667118. https://doi.org/10.1073/pnas.2115667118

Chiou, T. J., Aung, K., Lin, S. I., Wu, C. C., & Chiang, S. F. (2006). Regulation of phosphate homeostasis by microRNA in Arabidopsis. *Plant Cell, 18*, 412–421. https://doi.org/10.1105/tpc.105.038943

Chitwood, D. H., Nogueira, F. T., Howell, M. D., & Montgomery, T. A. (2009). Pattern formation via small RNA mobility. *Genes & Development, 23*, 549–554. https://doi.org/10.1101/gad.1770009

Coen, E. S., & Meyerowitz, E. M. (1991). The war of the whorls: Genetic interactions controlling flower development. *Nature, 353*, 31–37. https://doi.org/10.1038/353031a0

Cuperus, J. T., Fahlgren, N., & Carrington, J. C. (2011). Evolution and functional diversification of MIRNA genes. *Plant Cell, 23*(2), 431–442. https://doi.org/10.1105/tpc.110.082784

Dai, Z., Wang, J., Yang, X., Lu, H., & Miao, X. (2018). Modulation of plant architecture by the miR156f-OsSPL7-OsGH3.8 pathway in rice. *Journal of Experimental Botany, 69*, 5117–5130. https://doi.org/10.1093/jxb/ery273

Das Gupta, M., & Nath, U. (2015). Divergence in patterns of leaf growth polarity is associated with the expression divergence of miR396. *Plant Cell, 27*, 2785–2799. https://doi.org/10.1105/tpc.15.00196

Dastidar, M. G., Scarpa, A., Magele, I., Ruiz-Duarte, P., & von-Born, P. (2019). ARF5/MONOPTEROS directly regulates miR390 expression in the Arabidopsis thaliana primary root meristem. *Plant Direct, 3*, e116. https://doi.org/10.1002/pld3.116

Debernardi, J. M., Rodriguez, R. E., Mecchia, M. A., & Palatnik, J. F. (2012). Functional specialization of the plant miR396 regulatory network through distinct microRNA-target interactions. *PLoS Genetics*, *8*, e1002419. https://doi.org/10.1371/journal.pgen.1002419

Dello, I. R., Galinha, C., Fletcher, A. G., Grigg, S. P., & Molnar, A. (2012). A PHABULOSA/cytokinin feedback loop controls root growth in Arabidopsis. *Current Biology*, *22*, 1699–1704. https://doi.org/10.1016/j.cub.2012.07.005

Ding, Y., Tao, Y., & Zhu, C. (2013). Emerging roles of microRNAs in the mediation of drought stress response in plants. *Journal of Experimental Botany*, *64*, 3077–3086. https://doi.org/10.1093/jxb/ert164

Djami-Tchatchou, A. T., Sanan-Mishra, N., Ntushelo, K., & Dubery, I. A. (2017). Functional roles of microRNAs in agronomically important plants-potential as targets for crop improvement and protection. *Frontiers in Plant Science*, *8*, 378. https://doi.org/10.3389/fpls.2017.00378

Du, Q., & Wang, H. (2015). The role of HD-ZIP III transcription factors and miR165/166 in vascular development and secondary cell wall formation. *Plant Signaling & Behavior*, *10*, e1078955. https://doi.org/10.1080/15592324.2015.1078955

Efroni, I., Blum, E., Goldshmidt, A., & Eshed, Y. (2008). A protracted and dynamic maturation schedule underlies Arabidopsis leaf development. *Plant Cell*, *20*, 2293–2306. https://doi.org/10.1105/tpc.107.057521

Fan, S., Zhang, D., Gao, C., Wan, S., & Lei, C. (2018). Mediation of flower induction by gibberellin and its inhibitor paclobutrazol: MRNA and miRNA integration comprises complex regulatory cross-talk in apple. *Plant and Cell Physiology*, *59*, 2288–2307. https://doi.org/10.1093/pcp/pcy154

Floyd, S. K., Bowman, J. L. (2004). Gene regulation: Ancient microRNA target sequences in plants. *Nature*, *428*, 485–486. https://doi.org/10.1038/428485a

Fouracre, J. P., He, J., Chen, V. J., Sidoli, S., Poethig, R. S. (2021). VAL genes regulate vegetative phase change via miR156-dependent and independent mechanisms. *PLoS Genetics*, *17*, e1009626. https://doi.org/10.1371/journal.pgen.1009626

Gaillochet, C., & Lohmann, J. U. (2015). The never-ending story: From pluripotency to plant developmental plasticity. *Development*, *142*, 2237–2249. https://doi.org/10.1242/dev.117614

Gao, F., Wang, K., Liu, Y., Chen, Y., Chen, P., Shi, Z., Luo, J., Jiang, D., Fan, F., Zhu, Y., & Li, S. (2015). Blocking miR396 increases rice yield by shaping inflorescence architecture. *Nature Plants*, *2*, 15196. https://doi.org/10.1038/nplants.2015.196

Giacomelli, J. I., Weigel, D., Chan, R. L., & Manavella, P. A. (2012). Role of recently evolved miRNA regulation of sunflower HaWRKY6 in response to temperature damage. *New Phytologist*, *195*, 766–773. https://doi.org/10.1111/j.1469-8137.2012.04259.x

Guo, C., Jiang, Y., Shi, M., Wu, X., & Wu, G. (2021). ABI5 acts downstream of miR159 to delay vegetative phase change in Arabidopsis. *New Phytologist*, *231*, 339–350. https://doi.org/10.1111/nph.17371

Guo, C., Xu, Y., Shi, M., Lai, Y., & Wu, X. (2017). Repression of miR156 by miR159 regulates the timing of the juvenile-to-adult transition in arabidopsis. *Plant Cell*, *29*, 1293–1304. https://doi.org/10.1105/tpc.16.00975

Guo, H. S., Xie, Q., Fei, J. F., & Chua, N. H. (2005). MicroRNA directs mRNA cleavage of the transcription factor NAC1 to downregulate auxin signals for Arabidopsis lateral root development. *Plant Cell*, *17*, 1376–1386. https://doi.org/10.1105/tpc.105.030841

Gutierrez, L., Bussell, J. D., Pacurar, D. I., Schwambach, J., & Pacurar, M. (2009). Phenotypic plasticity of adventitious rooting in Arabidopsis is controlled by complex regulation of AUXIN RESPONSE FACTOR transcripts and microRNA abundance. *Plant Cell*, *21*, 3119–3132. https://doi.org/10.1105/tpc.108.064758

Han, J., Fang, J., Wang, C., Yin, Y., & Sun, X. (2014). Grapevine microRNAs responsive to exogenous gibberellin. *BMC Genomics*, *15*, 111. https://doi.org/10.1186/1471-2164-15-111

Han, M. H., Goud, S., Song, L., & Fedoroff, N. (2004). The Arabidopsis double-stranded RNA-binding protein HYL1 plays a role in microRNAmediated gene regulation. *Proceedings of the National Academy of Sciences of the United States of America*, *101*, 1093–1098. https://doi.org/10.1073/pnas.0307969100

Hay, A., Craft, J., & Tsiantis, M. (2004). Plant hormones and homeoboxes: Bridging the gap? *Bioessays*, *26*, 395–404. https://doi.org/10.1002/bies.20016

He, J., Xu, M., Willmann, M. R., McCormick, K., & Hu, T. (2018). Thresholddependent repression of SPL gene expression by miR156/miR157 controls vegetative phase change in Arabidopsis thaliana. *PLoS Genetics*, *14*, e1007337. https://doi.org/10.1371/journal.pgen.1007337

Hewezi, T., Piya, S., Qi, M., Balasubramaniam, M., Rice, J. H., & Baum, T. J. (2016). Arabidopsis miR827 mediates post-transcriptional gene silencing of its ubiquitin E3 ligase target gene in the syncytium of the cyst nematode Heterodera schachtii to enhance susceptibility. *The Plant Journal*, *88*, 179–192. https://doi.org/10.1111/tpj.13238

Horiguchi, G., Kim, G. T., & Tsukaya, H. (2005). The transcription factor AtGRF5 and the transcription coactivator AN3 regulate cell proliferation in leaf primordia of Arabidopsis thaliana. *The Plant Journal*, *43*, 68–78. https://doi.org/10.1111/j.1365-313X.2005.02429.x

Huijser, P., & Schmid, M. (2011). The control of developmental phase transitions in plants. *Development*, *138*, 4117–4129. https://doi.org/10.1242/dev.063511

Jia, X., Ding, N., Fan, W., Yan, J., & Gu, Y. (2015). Functional plasticity of miR165/166 in plant development revealed by small tandem target mimic. *Plant Science*, *233*, 11–21. https://doi.org/10.1016/j.plantsci.2014.12.020

Jin, D., Wang, Y., Zhao, Y., & Chen, M. (2013). MicroRNAs and their crosstalks in plant development. *Journal of Genetics and Genomics*, *40*, 161–170. https://doi.org/10.1016/j.jgg.2013.02.003

Johnson, C. R., Millwood, R. J., Tang, Y., Gou, J., Sykes, R. W., Turner, G. B., Davis, M. F., Sang, Y., Wang, Z.-Y., & Stewart, C. N. (2017). Field-grown miR156 transgenic switchgrass reproduction, yield, global gene expression analysis, and bioconfinement. *Biotechnology for Biofuels and Bioproducts*, *10*, 255. https://doi.org/10.1186/s13068-017-0939-1

Jones-Rhoades, M. W., & Bartel, D. P. (2004). Computational identification of plant microRNAs and their targets, including a stress-induced miRNA. *Molecular Cell*, *14*, 787–799. https://doi.org/10.1016/j.molcel.2004.05.027

Jones-Rhoades, M. W., Bartel, D. P., & Bartel, D. P. (2006). MicroRNAs and their regulatory roles in plants. *Annual Review of Plant Biology*, *57*, 19–53. https://doi.org/10.1146/annurev.arplant.57.032905.105218

Jung, J. H., Seo, Y. H., Seo, P. J., Reyes, J. L., & Yun, J. (2007). The GIGANTEA regulated microRNA172 mediates photoperiodic flowering independent of CONSTANS in Arabidopsis. *Plant Cell*, *19*, 2736–2748. https://doi.org/10.1105/tpc.107.054528

Kasschau, K. D., Xie, Z., Allen, E., Llave, C., & Chapman, E. J. (2003). P1/HC-pro, a viral suppressor of RNA silencing, interferes with Arabidopsis development and miRNA unction. *Developmental Cell*, *4*, 205–217. https://doi.org/10.1016/S1534-5807(03)00025-X

Kidner, C. A., & Martienssen, R. A. (2004). Spatially restricted microRNA directs leaf polarity through ARGONAUTE1. *Nature*, *428*, 81–84. https://doi.org/10.1038/nature02366

Kim, H. J., Hong, S. H., Kim, Y. W., Lee, I. H., Jun, J. H., Phee, B. K., Rupak, T., Jeong, H., Lee, Y., Hong, B. S., Nam, H. G., Woo, H. R., & Lim, P. O. (2014). Gene regulatory cascade of senescence-associated NAC transcription factors activated by ETHYLENE-INSENSITIVE2-mediated leaf senescence signalling in Arabidopsis. *Journal of Experimental Botany*, *65*, 4023–4036. https://doi.org/10.1093/jxb/eru112

Kim, J., Jung, J. H., Reyes, J. L., Kim, Y. S., & Kim, S. Y. (2005). MicroRNAdirected cleavage of ATHB15 mRNA regulates vascular development in Arabidopsis inflorescence stems. *The Plant Journal*, *42*, 84–94. https://doi.org/10.1111/j.1365-313X.2005.02354.x

Kim, J. J., Lee, J. H., Kim, W., Jung, H. S., & Huijser, P. (2012). The microRNA156-SQUAMOSA PROMOTER BINDING PROTEIN-LIKE3 module regulates ambient temperature-responsive flowering via FLOWERING LOCUS T in Arabidopsis. *Plant Physiology*, *159*, 461–478. https://doi.org/10.1104/pp.111.192369

Kim, V. N. (2005). Small RNAs: Classification, biogenesis, and function. *Molecular Cells*, *19*, 1–15.

Knauer, S., Holt, A. L., Rubio-Somoza, I., Tucker, E. J., & Hinze, A. (2013). A protodermal miR394 signal defines a region of stem cell competence in the Arabidopsis shoot meristem. *Developmental Cell*, *24*, 125–132. https://doi.org/10.1016/j.devcel.2012.12.009

Koyama, T., Mitsuda, N., Seki, M., Shinozaki, K., & Ohme-Takagi, M. (2010). TCP transcription factors regulate the activities of ASYMMETRIC LEAVES1 and miR164, as well as the auxin response, during differentiation of LEAVES in Arabidopsis. *Plant Cell*, *22*, 3574–3588. https://doi.org/10.1105/tpc.110.075598

Kumar, A., Gautam, V., Kumar, P., Mukherjee, S., & Verma, S. (2019). Identification and co-evolution pattern of stem cell regulator miR394s and their targets among diverse plant species. *BMC Ecology and Evolution*, *19*, 55. https://doi.org/10.1186/s12862-019-1382-7

Kutter, C., Schob, H., Stadler, M., Meins, F. J., & Si-Ammour, A. (2007). MicroRNA-mediated regulation of stomatal development in Arabidopsis. *Plant Cell*, *19*, 2417–2429. https://doi.org/10.1105/tpc.107.050377

Laufs, P., Peaucelle, A., Morin, H., & Traas, J. (2004). MicroRNA regulation of the CUC genes is required for boundary size control in Arabidopsis meristems. *Development*, *131*, 4311–4322. https://doi.org/10.1242/dev.01320

Lauter, N., Kampani, A., Carlson, S., Goebel, M., & Moose, S. P. (2005). micro-RNA172 down-regulates glossy15 to promote vegetative phase change in maize. *Proceedings of the National Academy of Sciences of the United States of America*, *102*, 9412–9417. https://doi.org/10.1073/pnas.0503927102

Leng, X., Wang, P., Zhu, X., Li, X., & Zheng, T. (2017). Ectopic expression of CSD1 and CSD2 targeting genes of miR398 in grapevine is associated with oxidative stress tolerance. *Functional & Integrative Genomics*, *17*, 697–710. https://doi.org/10.1007/s10142-017-0565-9

Li, T., Gonzalez, N., Inze, D., & Dubois, M. (2020). Emerging connections between small RNAs and phytohormones. *Trends in Plant Science*, *25*, 912–929. https://doi.org/10.1016/j.tplants.2020.04.004

Li, W. X., Oono, Y., Zhu, J., He, X. J., & Wu, J. M. (2008). The Arabidopsis NFYA5 transcription factor is regulated transcriptionally and posttranscriptionally to promote drought resistance. *Plant Cell*, *20*, 2238–2251. https://doi.org/10.1105/tpc.108.059444

Li, Z., Peng, J., Wen, X., & Guo, H. (2013a). Ethylene-insensitive3 is a senescence-associated gene that accelerates age-dependent leaf senescence by directly repressing miR164 transcription in Arabidopsis. *Plant Cell*, *25*, 3311–3328. https://doi.org/10.1105/tpc.113.113340

Li, Z., Peng, J., Wen, X., & Guo, H. (2013b). Ethylene-insensitive3 is a senescence-associated gene that accelerates age-dependent leaf senescence by directly repressing miR164 transcription in Arabidopsis. *Plant Cell*, *25*, 3311–3328. https://doi.org/10.1105/tpc.113.113340

Lian, H., Wang, L., Ma, N., Zhou, C. M., & Han, L. (2021). Redundant and specific roles of individual MIR172 genes in plant development. *PLoS Biology*, *19*, e3001044. https://doi.org/10.1371/journal.pbio.3001044

Lim, P. O., Lee, I. C., Kim, J., Kim, H. J., Ryu, J. S., Woo, H. R., & Nam, H. G. (2010). Auxin response factor 2 (ARF2) plays a major role in regulating auxin-mediated leaf longevity. *Journal of Experimental Botany*, *61*, 1419–1430. https://doi.org/10.1093/jxb/erq010

Lin, W. Y., Huang, T. K., & Chiou, T. J. (2013). Nitrogen limitation adaptation, a target of microRNA827, mediates degradation of plasma membrane-localized phosphate transporters to maintain phosphate homeostasis in Arabidopsis. *Plant Cell*, *25*, 4061–4074. https://doi.org/10.1105/tpc.113.116012

Liu, B., Li, J., Tsykin, A., Liu, L., & Gaur, A. B. (2009). Exploring complex miRNA-mRNA interactions with Bayesian networks by splitting-averaging strategy. *BMC Bioinformatics*, *10*, 408. https://doi.org/10.1186/1471-2105-10-408

Liu, C., Zhang, L., Sun, J., Wang, M. B., & Liu, Y. (2010). A simple artificial microRNA vector based on ath-miR169d precursor from Arabidopsis. *Molecular Biology Reports*, *37*(2), 903–909. https://doi.org/10.1007/s11033-009-9713-1

Liu, P. P., Montgomery, T. A., Fahlgren, N., Kasschau, K. D., & Nonogaki, H. (2007). Repression of AUXIN RESPONSE FACTOR10 by microRNA160 is critical for seed germination and post-germination stages. *The Plant Journal*, *52*, 133–146. https://doi.org/10.1111/j.1365-313X.2007.03218.x

Liu, Z., Li, J., Wang, L., Li, Q., & Lu, Q. (2016). Repression of callus initiation by the miRNA-directed interaction of auxin-cytokinin in Arabidopsis thaliana. *The Plant Journal*, *87*, 391–402. https://doi.org/10.1111/tpj.13211

Llave, C., Xie, Z., Kasschau, K. D., & Carrington, J. C. (2002). Cleavage of Scarecrow-like mRNA targets directed by a class of Arabidopsis miRNA. *Science*, *297*, 2053–2056. https://doi.org/10.1126/science.1076311

Mallory, A. C., Barte, D. P, & Bartel, B. (2005). MicroRNA-directed regulation of Arabidopsis AUXIN RESPONSE FACTOR17 is essential for proper development and modulates expression of early auxin response genes. *Plant Cell*, *17*, 1360–1375. https://doi.org/10.1105/tpc.105.031716

Mallory, A. C., Reinhart, B. J., Jones-Rhoades, M. W., Tang, G., Zamore, P. D., Barton, M. K., & Bartel, D. P. (2004). MicroRNA control of PHABULOSA in leaf development: Importance of pairing to the microRNA 5′ region. *The EMBO Journal*, *23*, 3356–3364 https://doi.org/10.1038/sj.emboj.7600340

Manuela, D., & Xu, M. (2020). Patterning a leaf by establishing polarities. *Frontiers in Plant Science*, *11*, 568730. https://doi.org/10.3389/fpls.2020.568730

Miao, Y., Laun, T., Zimmermann, P., & Zentgraf, U. (2004). Targets of the WRKY53 transcription factor and its role during leaf senescence in Arabidopsis. *Plant Molecular Biology*, *55*, 853–867. https://doi.org/10.1007/s11103-004-2142-6

Miao, Y., & Zentgraf, U. (2007). The antagonist function of Arabidopsis WRKY53 and ESR/ESP in leaf senescence is modulated by the jasmonic and salicylic acid equilibrium. *Plant Cell*, *19*, 819–830. https://doi.org/10.1105/tpc.106.042705

Millar, A. A., Lohe, A., & Wong, G. (2019). Biology and function of miR159 in plants. *Plan. Theory*, *8*, 255. https://doi.org/10.3390/plants8080255

Mlotshwa, S., Yang, Z., Kim, Y., & Chen, X. (2006). Floral patterning defects induced by Arabidopsis APETALA2 and microRNA172 expression in Nicotiana benthamiana. *Plant Molecular Biology*, *61*, 781–793. https://doi.org/10.1007/s11103-006-0049-0

Muller, C. J., Valdes, A. E., Wang, G., Ramachandran, P., & Beste, L. (2016). PHABULOSA mediates an auxin signaling loop to regulate vascular patterning in arabidopsis. *Plant Physiology*, *170*, 956–970. https://doi.org/10.1104/pp.15.01204

Nag, A., King, S., & Jack, T. (2009). miR319a targeting of TCP4 is critical for petal growth and development in Arabidopsis. *Proceedings of the National Academy of Sciences of the United States of America*, *106*, 22534–22539. https://doi.org/10.1073/pnas.0908718106

Ori, N., Cohen, A. R., Etzioni, A., Brand, A., & Yanai, O. (2007). Regulation of LANCEOLATE by miR319 is required for compound-leaf development in tomato. *Nature Genetics*, *39*, 787–791. https://doi.org/10.1038/ng2036

Osborne, D. J. (1959). Control of leaf senescence by auxins. *Nature*, *183*, 1459–1460. https://doi.org/10.1038/1831459a0

Palatnik, J. F., Wollmann, H., Schommer, C., Schwab, R., & Boisbouvier, J. (2007). Sequence and expression differences underlie functional specialization of Arabidopsis microRNAs miR159 and miR319. *Developmental Cell*, *13*, 115–125. https://doi.org/10.1016/j.devcel.2007.04.012

Piazza, P., Jasinski, S., & Tsiantis, M. (2005). Evolution of leaf developmental mechanisms. *New Phytologist*, *167*, 693–710. https://doi.org/10.1111/j.1469-8137.2005.01466.x

Qiu, K., Li, Z., Yang, Z., Chen, J., Wu, S., Zhu, X., Gao, S., Gao, J., Ren, G., Kuai, B., & Zhou, X. (2015). EIN3 and ORE1 accelerate degreening during ethylene-mediated leaf senescence by directly activating chlorophyll catabolic genes in Arabidopsis. *PLoS Genetics*, *11*, e1005399. https://doi.org/10.1371/journal.pgen.1005399

Rodriguez, R. E., Mecchia, M. A., Debernardi, J. M., Schommer, C., & Weigel, D. (2010). Control of cell proliferation in Arabidopsis thaliana by microRNA miR396. *Development*, *137*, 103–112. https://doi.org/10.1242/dev.043067

Ru, P., Xu, L., Ma, H., & Huang, H. (2006). Plant fertility defects induced by the enhanced expression of microRNA167. *Cell Research*, *16*, 457–465. https://doi.org/10.1038/sj.cr.7310057

Rubio-Somoza, I., & Weigel, D. (2013). Coordination of flower maturation by a regulatory circuit of three microRNAs. *PLoS Genetics*, *9*, e1003374. https://doi.org/10.1371/journal.pgen.1003374

Rubio-Somoza, I., Zhou, C. M., Confraria, A., Martinho, C., von Born, P., Baena-Gonzalez, E., Wang, J. W., & Weigel, D. (2014). Temporal control of leaf complexity by miRNA-regulated licensing of protein complexes. *Current Biology*, *24*, 2714–2719. https://doi.org/10.1016/j.cub.2014.09.058

Samad, A. F. A., Sajad, M., Nazaruddin, N., Fauzi, L. A., Murad, A. M. A., Zainal, Z., & Ismail, I. (2017). MicroRNA and transcription factor: Key players in plant regulatory network. *Frontiers in Plant Science*, *8*, 565. https://doi.org/10.3389/fpls.2017.00565

Schommer, C., Debernardi, J. M., Bresso, E. G., Rodriguez, R. E., & Palatnik, J. F. (2014). Repression of cell proliferation by miR319-regulated TCP4. *Molecular Plant*, *7*, 1533–1544. https://doi.org/10.1093/mp/ssu084

Schommer, C., Palatnik, J. F., Aggarwal, P., Chetelat, A., Cubas, P., Farmer, E. E., Nath, U., & Weigel, D. (2008). Control of jasmonate biosynthesis and senescence by miR319 targets. *PLoS Biology*, *6*, e230. https://doi.org/10.1371/journal.pbio.0060230

Schwab, R., Palatnik, J. F., Riester, M., Schommer, C., & Schmid, M. (2005). Specific effects of microRNAs on the plant transcriptome. *Developmental Cell*, *8*, 517–527. https://doi.org/10.1016/j.devcel.2005.01.018

Scofield, S., Dewitte, W., & Murray, J. A. (2014). STM sustains stem cell function in the Arabidopsis shoot apical meristem and controls KNOX gene expression independently of the transcriptional repressor AS1. *Plant Signaling & Behavior*, *9*, e28934. https://doi.org/10.4161/psb.28934

Sieber, P., Wellmer, F., Gheyselinck, J., Riechmann, J. L., & Meyerowitz, E. M. (2007). Redundancy and specialization among plant microRNAs: Role of the MIR164 family in developmental robustness. *Development*, *134*, 1051–1060. https://doi.org/10.1242/dev.02817

Somssich, M., Je, B. I., Simon, R., & Jackson, D. (2016). CLAVATA-WUSCHEL signaling in the shoot meristem. *Development*, *143*, 3238–3248. https://doi.org/10.1242/dev.133645

Song, J. B., Huang, S. Q., Dalmay, T., & Yang, Z. M. (2012). Regulation of LEAF morphology by microRNA394 and its target LEAF CURLING RESPONSIVENESS. *Plant and Cell Physiology*, *53*, 1283–1294. https://doi.org/10.1093/pcp/pcs080

Sun, Z., Li, M., Zhou, Y., Guo, T., & Liu, Y. (2018). Coordinated regulation of Arabidopsis microRNA biogenesis and red light signaling through Dicerlike 1 and phytochrome-interacting factor 4. *PLoS Genetics*, *14*, e1007247. https://doi.org/10.1371/journal.pgen.1007247

Sunkar, R., Li, Y. F., & Jagadeeswaran, G. (2012). Functions of microRNAs in plant stress responses. *Trends in Plant Science*, *17*, 196–203. https://doi.org/10.1016/j.tplants.2012.01.010

Sunkar, R., & Zhu, J. K. (2004). Novel and stress-regulated microRNAs and other small RNAs from Arabidopsis. *Plant Cell, 16*, 2001–2019. https://doi.org/10.1105/tpc.104.022830

Tatematsu, K., Toyokura, K., & Okada, K. (2015). Requirement of MIR165A primary transcript sequence for its activity pattern in Arabidopsis leaf primordia. *Plant Signaling & Behavior, 10*, e1055432. https://doi.org/10.1080/15592324.2015.1055432

Tsuji, H., Aya, K., Ueguchi-Tanaka, M., Shimada, Y., & Nakazono, M. (2006). GAMYB controls different sets of genes and is differentially regulated by microRNA in aleurone cells and anthers. *The Plant Journal, 47*, 427–444. https://doi.org/10.1111/j.1365-313X.2006.02795.x

Veit, B. (2009). Hormone mediated regulation of the shoot apical meristem. *Plant Molecular Biology, 69*, 397–408. https://doi.org/10.1007/s11103-008-9396-3

Voinnet, O. (2009). Origin, biogenesis, and activity of plant microRNAs. *Cell, 136*, 669–687. https://doi.org/10.1016/j.cell.2009.01.046

Wang, H., Zhou, Y., Gilmer, S., Whitwill, S., & Fowke, L. C. (2000). Expression of the plant cyclin-dependent kinase inhibitor ICK1 affects cell division, plant growth and morphology. *The Plant Journal, 24*, 613–623. https://doi.org/10.1046/j.1365-313x.2000.00899.x

Wang, J. J., & Guo, H. S. (2015). Cleavage of INDOLE-3-ACETIC ACID INDUCIBLE28 mRNA by microRNA847 upregulates auxin signaling to modulate cell proliferation and lateral organ growth in Arabidopsis. *Plant Cell, 27*, 574–590. https://doi.org/10.1105/tpc.15.00101

Wang, J. W., Wang, L. J., Mao, Y. B., Cai, W. J., & Xue, H. W. (2005). Control of root cap formation by MicroRNA-targeted auxin response factors in Arabidopsis. *Plant Cell, 17*, 2204–2216. https://doi.org/10.1105/tpc.105.033076

Wang, T., Zheng, Y., Tang, Q., Zhong, S., & Su, W. (2021d). Brassinosteroids inhibit miRNA-mediated translational repression by decreasing AGO1 on the endoplasmic reticulum. *Journal of Integrative Plant Biology, 63*, 1475–1490. https://doi.org/10.1111/jipb.13139

Wang, Y., Itaya, A., Zhong, X., Wu, Y., Zhang, J., van der Knaap, E., Olmstead, R., Qi, Y., & Ding, B. (2011). Function and evolution of a micro-RNA that regulates a Ca2+- ATPase and triggers the formation of phased small interfering RNAs in tomato reproductive growth. *Plant Cell, 23*, 3185–3203. https://doi.org/10.1105/tpc.111.088013

Wollmann, H., Mica, E., Todesco, M., Long, J. A., & Weigel, D. (2010). On reconciling the interactions between APETALA2, miR172 and AGAMOUS with the ABC model of flower development. *Development, 137*, 3633–3642. https://doi.org/10.1242/dev.036673

Wu, M. F., Tian, Q., & Reed, J. W. (2006). Arabidopsis microRNA167 controls patterns of ARF6 and ARF8 expression, and regulates both female and male reproduction. *Development, 133*, 4211–4218. https://doi.org/10.1242/dev.02602

Xie, K., Wu, C., & Xiong, L. (2006). Genomic organization, differential expression, and interaction of SQUAMOSA promoter-binding-like transcription factors and microRNA156 in rice. *Plant Physiology, 142*, 280–293. https://doi.org/10.1104/pp.106.084475

Xu, M., Hu, T., Zhao, J., Park, M. Y., & Earley, K. W. (2016b). Developmental functions of miR156-regulated SQUAMOSA PROMOTER BINDING PROTEIN-LIKE (SPL) genes in Arabidopsis thaliana. *PLoS Genetics, 12*, e1006263. https://doi.org/10.1371/journal.pgen.1006263

Yan, J., Wang, P., Wang, B., Hsu, C. C., & Tang, K. (2017). The SnRK2 kinases modulate miRNA accumulation in Arabidopsis. *PLoS Genetics*, 13, e1006753. https://doi.org/10.1371/journal.pgen.1006753

Yoon, E. K., Yang, J. H., Lim, J., Kim, S. H., & Kim, S. K. (2010). Auxin regulation of the microRNA390-dependent transacting small interfering RNA pathway in Arabidopsis lateral root development. *Nucleic Acids Research, 38*, 1382–1391. https://doi.org/10.1093/nar/gkp1128

Yu, S., Galvao, V. C., Zhang, Y. C., Horrer, D., & Zhang, T. Q. (2012). Gibberellin regulates the Arabidopsis floral transition through miR156-targeted QUAMOSA promoter binding-like transcription factors. *Plant Cell, 24*, 3320–3332. https://doi.org/10.1105/tpc.112.101014

Zhang, B., Pan, X., Cobb, G. P., & Anderson, T. A. (2006). Plant microRNA: A small regulatory molecule with big impact. *Developmental Biology, 289*, 3–16. https://doi.org/10.1016/j.ydbio.2005.10.036

Zhang, F., Wang, L., Lim, J. Y., Kim, T., & Pyo, Y. (2016). Phosphorylation of CBP20 links microRNA to root growth in the ethylene response. *PLoS Genetics, 12*, e1006437. https://doi.org/10.1371/journal.pgen.1006437

Zhang, H., Zhang, L., Han, J., Qian, Z., & Zhou, B. (2019). The nuclear localization signal is required for the function of squamosa promoter binding protein-like gene 9 to promote vegetative phase change in Arabidopsis. *Plant Molecular Biology, 100*, 571–578. https://doi.org/10.1007/s11103-019-00863-5

Zhang, T. Q., Lian, H., Tang, H., Dolezal, K., & Zhou, C. M. (2015). An intrinsic microRNA timer regulates progressive decline in shoot regenerative capacity in plants. *Plant Cell*, *27*, 349–360. https://doi.org/10.1105/tpc.114.135186

Zhao, Y., Lin, S., Qiu, Z., Cao, D., & Wen, J. (2015). MicroRNA857 is involved in the regulation of secondary growth of vascular tissues in arabidopsis. *Plant Physiology*, *169*, 2539–2552. https://doi.org/10.1104/pp.15.01011

Zhao, Y. F., Peng, T., Sun, Z., Teotia, S., & Wen, H. L. (2019). MiR1432-OsACOT (acyl-CoA thioesterase) module determines grain yield via enhancing grain filling rate in rice. *Plant Biotechnology Journal*, *17*, 712–723. https://doi.org/10.1111/pbi.13009

Zhongc, R., & Ye, Z. H. (2004). Amphivasal vascular bundle 1, a gain-offunction mutation of the IFL1/REV gene, is associated with alterations in the polarity of leaves, stems and carpels. *Plant and Cell Physiology*, *45*, 369–385. https://doi.org/10.1093/pcp/pch051

Zhu, H., Hu, F., Wang, R., Zhou, X., & Sze, S. H. (2011). Arabidopsis Argonaute10 specifically sequesters miR166/165 to regulate shoot apical meristem development. *Cell*, *145*, 242–256. https://doi.org/10.1016/j.cell.2011.03.024

4 Dynamic Function of miRNAs in Sensing and Signaling Nutrient Stress in Plants

Samina Mazahar, Yasheshwar, and Shahid Umar

4.1 INTRODUCTION

The growth, development, and yield of the plant depend on a continuous supply of nutrients present in the soil sphere around its root spread. Nutrient availability at the sub-optimum level affects multiple vital functions in plants and hence reduces crop yield significantly. Plants have adapted to the low availability of nutrients and thereby sending a signal to facilitate ion transport via interaction between signaling molecules and transporter genes. The mechanism of uptake and transport of mineral ions is positively correlated with transporter genes involved and adequate quantum of nutrients. This maintains nutrient stability by the engaging system to express transporter genes and altering morphological and anatomical contours of roots thereby resulting in ease of nutrient uptake. MicroRNA (miRNA) is an important and indispensable small RNA signaling molecule that controls post-transcriptional gene expression (Bartel, 2018).

In plants, miRNAs are the master controllers of several biochemical and physiological attributes like growth, nutrition, flowering, and also responses towards environmental fluctuation including biotic and abiotic (Shriram et al., 2016; Li & Zhang, 2016; Brant & Budak, 2018). For the regulation of target genes, the spatiotemporal expression of miRNAs comes under multilevel control (Wang et al., 2019). The origin of miRNA is through inverted duplications and random sequences in the genome (Voinnet, 2009). The processing of miRNA is done by mRNA Dicer-like protein. It is affined into RISC (RNA-induced silencing complex) and accesses the target mRNA by complementary base-pair, and finally, miRNA represses the translation of target mRNA (Wang et al., 2019, Liang, 2015).

Recent reports have mentioned novel miRNAs and their active involvement in nutrient uptake and moving them through the vascular channels of plants (Fischer et al., 2013; Kehr, 2013) under conditions of nutrient stress. Figure 4.2 shows the association of miRNAs with various nutrients in maintaining nutrient homeostasis.

In this chapter, we are focusing on various nutrient-responsive miRNAs exclusively involved in nutrient uptake and mobilization via complex regulatory mechanisms during nutrient scarcity. Understanding the mechanism of miRNA in sensing and signaling the suboptimal availability of essential nutrients might pave the way for enhancing nutrient use efficiency and better crop growth and development.

4.2 MACRONUTRIENT SENSING AND SIGNALING IN PLANTS AS REGULATED BY MIRNAS

4.2.1 MIRNAS RESPONSIVE TO NITROGEN (N) UPTAKE AND TRANSPORT

The indispensable requirement of nitrogen (N) as a macronutrient for plants unable to fix atmospheric N_2 comes by uptaking nitrate and ammonium ions. Thus, the most readily available forms of N taken up by plants are mainly nitrate (NO_3^-) and ammonia (NH_4^+). Nitrate is the most readily

DOI: 10.1201/9781003248453-4

available form and gets dissolved in water and is lost to the soil and environment via leaching, surface runoff, volatilization, and nitrification (Mazahar et al., 2015). Being an important element for the plant, the N fertilizers are used in abundant quantities by farmers for enhancing crop yield (Mazahar et al., 2015, Shahzad et al., 2018). However, the excess application of N harms the environment and hence a balanced N application and improved N utilization efficiency are the essential requirements that can be achieved by maintaining a dynamic equilibrium between nitrogen uptake, translocation, and remobilization (Mazahar et al., 2015).

It is well-established that plants adapt themselves by upregulation and downregulation of proteins involved in the transportation of N in response to soil wherein N has become a limiting factor. Such transporters are differentially regulated by the master controllers belonging to various small miRNA families (Paul et al., 2015). The transport of N depends upon the external acquisition of N by roots from the soil (Paul et al., 2015). Numerous studies on the role of miRNAs in plants under N-starvation have been reported. In *Arabidopsis*, the miRNA responsive to N-starvation was grouped into two depending upon their occurrence through Solexa high throughput sequencing (Paul et al., 2015). Table 4.1 shows the list of N-Starvation Induced (NSI) and Suppressed (NSS) miRNA in *Arabidopsis*.

In major cereal crops such as rice, a global overview of miRNA involved in N response is based on high throughput small RNA sequencing revealing a total of 44 differently expressed miRNA in response to high and low N conditions. Further, the response of miRNA to different N sources such as nitrate and ammonia have also been studied. A study in maize under different N starvation conditions such as chronic and transient nitrate limiting conditions showed unique or overlapping expression in a tissue-dependent manner (Xu et al., 2011). According to Trevisan et al. (2012a), six mature nitrate-responsive miRNAs are involved in maize and the in situ experiment involved in the study explained the importance of the availability of nitrate on the amount and the localization of miRNA which were found to accumulate in the roots of the plant supplied with N, whereas the roots with limited N supply displayed reduced expressions. It was also found that under prolonged N-limited

TABLE 4.1

List of N-Starvation Induced (NSI) and N-Starvation Suppressed (NSS) miRNA in model plant *Arabidopsis*.

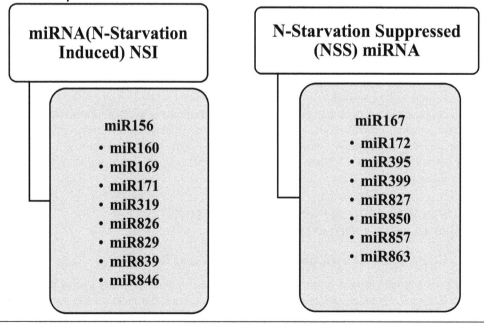

miRNA(N-Starvation Induced) NSI	N-Starvation Suppressed (NSS) miRNA
miR156	miR167
• miR160	• miR172
• miR169	• miR395
• miR171	• miR399
• miR319	• miR827
• miR826	• miR850
• miR829	• miR857
• miR839	• miR863
• miR846	

conditions, the miRNA under study might get suppressed due to activation of the post-transcription event (Trevisan et al., 2012b).

In wheat 38 miRNA (TaMIRs) were found under optimal N and N-depleted conditions. It was observed that most of the TaMIRs were responsive under limited N conditions and showed a distinct pattern of expression where TaMIR399 and TaMIR1133 were expressed as downregulated (Zhao et al., 2015). Another study in two wheat genotypes found a set of the differentially expressed miRNA in plant roots under the condition of normal and limited nitrate supply. The important differences in their expression were dependent on both the genotype and the level of N availability (Sinha et al., 2015).

Wang et al. (2013a) studied soybean the N-responsive miRNA where 16 libraries were generated from two genotypes, that is, low N sensitive and tolerant and explored miRNA in shoot and root, which were expressed differently under stress. Nine potential miRNA was analyzed by RT-PCR and expression of the miR169 family was repressed and the miR172 family was induced in French beans under low nitrate stress (Babu et al., 2014).

There are numerous research reports on the model plant *Arabidopsis* characterizing the involvement and importance of miRNA under nitrogen-limiting conditions. These RNAs are involved in lateral root outgrowth in plants exposed to N as a limiting factor (Gifford et al., 2008). The miR167 in *Arabidopsis* as well as in maize (Zhao et al., 2012; Liang et al., 2012) showing its role in both monocots and dicots plants in response to N-deficiency (Zeng et al., 2013). Primary miR transcript 167a was reported repressed nearly by five times in root pericycle cells and targeting ARF 8 to increase lateral root formation (Pant et al., 2009). NF-YA (nuclear factor) subunit genes that play a pivotal role in balancing N regulated by miR169 (Zhao et al., 2011).

Many miRNAs of N-starved plants are based on the analysis of microarray and in situ hybridization, qRT-PCR, northern blots, etc. Zeng et al. (2013) reported that miRNA responsive to N deficiency belongs to 27 families. The miR156 is abundantly present in *Arabidopsis* which inhibits the negative role of miR172 involved in enhancing the juvenile phase (Paul et al., 2015).

N-Starved condition triggers the induction of miR160 that put breaks on the development of lateral root formation. Although root formation by miR170 is hastened by targeting growth hormones with factors like ARF16/17 and SCL6 (Liang et al., 2012). The root nodule development through the involvement of miR169 and miR172 by regulating AP2 and HAP2 gene expression has been reported in *Medicago tranculata*. Low miR169 during N-limitation upregulates HAP2 gene expression and the nodule primordia (Pant et al., 2009).

The miR169a-h family members (miR169a-h) have been reported to be found in the upregulation of transporter proteins in the model plant *Arabidopsis* and other crops like soybean and corn in response to N-starved condition (Liang et al., 2012, Wang et al., 2013b). Sequencing of NO_3-treated *Arabidopsis* seedlings revealed the nitrate-responsive miRNA at the global level. The miR393 was identified as nitrate-induced small RNAs (Vidal et al., 2013) and was found to be under low availability of N and reduction of its assimilation. The miR393/AFB3 has been reported to be involved in primary and lateral root growth (Vidal et al., 2010, Zeng et al., 2013) respectively and hence regulates nitrate-induced changes in roots via auxin hormone signaling events (Vidal et al., 2010). Studies using technology-based approaches like degradome sequencing detected and validated N stress-responsive miRNAs in *Populus tomentosa* (Chen et al., 2015). In *Chrysanthemum nankingense*, an ornamental plant native to China, a significant proportion of miRNA in roots and leaves has been assessed and reported to be expressed differentially in response to low N levels (Song et al., 2015). These reports cumulatively indicate miRNA as a controlling agent concerning low N level and its differential expression due to differences in the experimental conditions, variation in nitrogen availability, plant tissue under study, and plant species (Shahzad et al., 2018). However deep sequencing technology facilitates the identification of various potential miRNAs under N stress. There is still much more to explore and validate about miRNA and their targets, to discover and reveal miRNA-mediated molecular adaptation in plants under N stress.

4.2.2   MIRNAs IN MAINTAINING PHOSPHORUS (P) HOMEOSTASIS

Phosphorus (P) is another essential macronutrient for the growth and development of plants, involved in vital metabolic events like photosynthesis, energy transfer, and carbohydrate breakdown. P is also a component in nucleic acids like DNA and RNA, sugars, and lipids. Although the concentration of phosphorus in the soil is manifold higher in the soil, plants are often P deficient or the uptake is low. It is due to its fixation in the form of calcium/magnesium or aluminium/iron phosphates, hence making it unavailable for plant uptake (Malhotra et al., 2018). In P-deficient conditions plants modulate their adaptive mechanism by regulating their transcriptional and biochemical pathways in different plant organs and tissues for enhancing P uptake and transport (Shahzad et al., 2018). However, the assimilation and distribution of P mainly depend upon two types of transporters: PHOSPHATE TRANSPORTERS (PHTs) and PHOSPHATE 1 TRANSPORTER (PHO1) (Mlodzinska & Zboinska, 2016).

Recent studies suggest the regulatory role of miRNA during plant response to P starvation. However, the most commonly used method was high throughput sRNA sequencing which helped in the collection of huge data in different plants (Shahzad et al., 2018). During P deficiency the differential regulation of miRNA is based on phosphorus-responsive motifs of the miRNA (Paul et al., 2015). In this perspective *Arabidopsis* and major crops and vegetables have been studied by Pant et al. (2009); Gu et al. (2010); Lundmark et al. (2010); Zhu et al. (2010); Hackenberg et al. (2013); Pei et al. (2013); Xu et al. (2013); Zhao et al. (2013).

Commonly reported miRNA responsive under P-deficiency in different studies are miR159, miR319, miR398, and miR447 (Sun et al., 2012; Sunkar et al., 2012). The role of miR827 is significant in balancing phosphate plants (Kant et al., 2011). Few miRNAs were also to be involved in upregulation under P deficiency. The miR399 reported widely in scientific reports induced under P starvation (Chiou et al., 2006) and P re-addition showed a quick decrease in its expression (Fujii et al., 2005). The abundance of miR778 increases with phosphate starvation in *Arabidopsis*, and with P re-addition, miR2111, and miR778 were decreased by approximately two folds within a time interval of three hours (Paul et al., 2015). However, according to Liang et al. (2015) miR399, miR827, and miR2111 are important in P-starvation response and maintaining homeostasis. Figure 4.1 suggests the recent methods of identifying P-deficient responsive miRNA in various plants.

The scarcity of P in the shoots triggers the synthesis of miR399 in *Arabidopsis* for all six families (miR399A–F) (Paul et al., 2015). However according to the study on the mechanism of PHOSPHATE2 (PHO2) regulation mediated by miR399 in *Arabidopsis*, DRB1 is the major responsive protein for miR399 regulation of phosphate over accumulator2 (PHO2) and DRB2 along with DRB4 play minor roles amongst all four double-stranded RNS binding (DRB) proteins, and they involve mechanisms such as mRNA cleavage and translational repression (Pegler et al., 2019) which is important to maintain P homeostasis in *Arabidopsis*. The miR399 gets attached to five complementary sites of PHO2 transcript at the 5′ untranslated region (UTR) upregulated and involved in mRNA cleavage (Pant et al., 2008). PHO2 is also known as UBC24 which encodes a ubiquitin-conjugating E2 enzyme. The high amount of P in shoots is facilitated due to the reduction of PHO2, being an important phosphate transporter, it aids in P mobilization (Paul et al., 2015). It is reported that the degradation of PHO1 through PHO2 is essential to P homeostasis, and PHO1 is the component of PHO2 dependent pathway (Liu et al., 2012).

Overexpression of miR399 causes enhanced P accumulation in shoots, hence causing loss of chlorophyll (chlorosis) and damage of leaf cells (necrosis) at the tip (Aung et al., 2006; Chiou et al., 2006). In the promoter region this miR (PHR1) binding sites (PIBS) myeloblastosis (MYB) transcription factor is present upstream (Bari et al., 2006; Zeng et al., 2010; Xu et al., 2013). At the *cis* site of the miR399f region, there is a binding of the MYB2 transcription factor that regulates the expression of miR399f (Baek, 2013). Further, the MYB transcription factors are important regulators of the signaling pathway under deficiency (Bustos et al., 2010, Shahzad et al., 2018).

Significantly reduced expression of miR399 was observed in PHR1 mutant of *Arabidopsis* as well as other P-related genes under P starvation; however, under sufficient P availability a cluster of P deficiency responsive genes in PHO2 mutant were found upregulated, hence PHR1 function upstream of miR399 and PHO2 along with many others are needed for miR399 induction in P signaling network (Bari et al., 2006). However, the expression of the PHO2 transporter is assumed to increase by the activity of miR827 and miR399, which targets the *nitrogen-limiting adaptation* (*NLA*) and is active under N-deficient conditions. A further important function of the NLA mutant is to accumulate excess P in low nitrate abundance and maintain P homeostasis (Paul et al., 2015). Likewise, PHO2 mutants were also nitrate dependent, and like the NLA mutants, P toxicity-related phenotype was observed (Shahzad et al., 2018). It was also reported that NLA and PHO2 might operate through the same pathway mediated by miR827 and miR399 respectively, as in comparing NLA or PHO2 single mutant plants with NLA PHO2 double mutants, the excess accumulation of P content was not observed and hence NLA and PHO2 play essential regulatory roles in P homeostasis (Kant et al., 2011). The long-distance traveling of miR399 through systemic repression of PHO2 from shoots to roots maintains P homeostasis in plants, and the numerous functions of miR399 were significantly observed in rice, indicating its positive response to various nutrients (Shahzad et al., 2018).

The analysis of the transgenic rice overexpressing OsmiR399 through Gene Chip technology confirms the switch of several downstream genes associated with various kinds of nutrient stresses, that is, potassium, iron, sodium, and calcium, although the mechanisms of regulation differ for particular nutrients, as shown by OsmiR399 (Hu et al., 2015).

IPS1 (induced by phosphate starvation1) is a non-protein gene in *Arabidopsis* that inhibits the miR399 activity although its sequence was complementary to miR399. This was due to a mismatched loop that significantly interrupted the pairing of *IPS1*-miR399 at the cleaved site and hence known as "target mimicry." As a result, a higher accumulation of PHO2 mRNA and its improved activity were observed along with reduced P levels in shoots (Franco-Zorrilla et al., 2007). Such pseudo interaction represents the regulatory role of its natural target mimicry and hence helps in developing miRNAs based strategies responsible to deal with nutrient stress response in plants. As a result, it knocks down miRNA families (Shahzad et al., 2018, Todesco et al., 2010). Although, in plants, the regulatory component IP S1/miR399/PHO2 is probably found to be conserved (Branscheid et al., 2010; Hu et al., 2011; Huang et al., 2011; Liu et al., 2010; Valdes-Lopez et al., 2008).

The micro-array-based transcriptional analysis in soybean reported an upregulation of 15 miRNAs, and 23 miRNAs were found suppressed under different P levels in the leaves and roots (Zeng et al., 2010). Also, miR159a is upregulated during P starvation, especially in roots; on the contrary, the downregulation of miR319a, miR396a, miR398b, and miR1507a has been proved through qRT-PCR. In P-starved wheat, the molecular analysis has confirmed the upregulation of 9 TaMIRs except for TaMIR408, which is the one repressed under low availability of phosphorous (Zhao et al., 2013). In maize, the sRNA sequencing helped in the characterization of P-responsive miRNA by analyzing the wild-type genotype Qi319 and low P tolerant mutant 99038 and helped in the miRNA expression under normal and limited P-resistant genotypes (Pei et al., 2013). Therefore, the low P-responsive miRNA are significantly regulating the low levels of P in different maize genotypes. There are several P deficiency-responsive miRNAs present in different plant miR169, promoting conserved P signaling pathways (Kuo & Chiou, 2011).

4.2.3 MIRNA RESPONSIVE TO SULPHUR

Sulphur is an important mineral nutrient involved in many physiological processes. It is an important constituent of amino acids and proteins, sulpholipids, vitamins, and sulphated polysaccharides (Leustek et al., 2000). Sulphur is assimilated into methionine, cysteine, glutathione, glucosinolates, and cofactors to promote defence metabolism both at a primary and secondary level under stresses (Rausch & Wachter, 2005). Cysteine is a sulphur-containing amino acid and is involved in

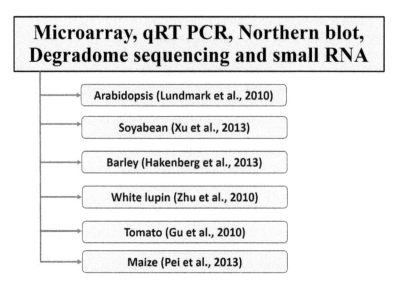

FIGURE 4.1 Recent methods of identifying P-deficient responsive miRNA in various plants.

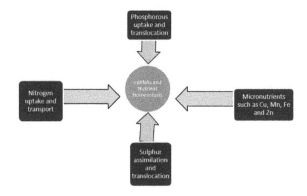

FIGURE 4.2 miRNAs in maintaining nutrient homeostasis.

S-containing compounds and proteins (Anjum et al., 2015). The S uptake in *Arabidopsis* involves a total of 14 putative sulphate transporters that exhibit strong affinity to sulphate-transporters viz. SULTR1;1 and SULTR1;2 and weak affinity towards sulphate-transporters SULTR2;1, SULTR2;2 and SULTR3;5. These transporters promote the S uptake and its translocation in plants (Kataoka et al., 2004).

Sulphate is stored in plant vacuole or it undergoes an array of reactions in plastids wherein it is assimilated with the help of the enzyme ATP sulfurylase. The aforementioned enzyme is the first enzyme of the sulphate assimilation pathway (Saito, 2004). It has been reported that miR395 regulates the expression of sulphate transporter based on the availability of sulphates. Researchers studied different tissue-specific expressions and revealed a positive correlation between target mRNA (SULTR2;1) and miR395 levels in roots and a negative correlation between miR395 and shoots. It is also observed that the two families of genes APS and SULTR2;1 are involved in S metabolism (Liang et al., 2010). In limited S, miR395 is dependent on a transcription factor SLIM1, which is known to promote enhanced sulphate translocation to the shoots and contributes to better S assimilation in aerial parts, particularly in leaves (Kawashima et al., 2011). During S starvation miR395 is upregulated and suppresses its target enzyme genes APS1, APS3, and APS4 (Matthewman et al., 2012).

4.3   MICRONUTRIENTS SENSING AND SIGNALING AS REGULATED BY MIRNAS

Various miRNAs are responsible for regulating and maintaining nutrient homeostasis for micronutrients also, among which are Cu, Mn, Fe, and Zn. Cu is an important constituent of plastocyanin, an electron transport chain, and serves as an important agent in metabolic processes. The superoxide radicals are quenched by superoxide dismutase, which is a zinc/copper-containing enzyme (CSD) involved in maintaining osmotic balance inside the plant cell under oxidative stress (Paul et al., 2015). Under limited Cu, miR398 downregulates various CSDs (Beauclair et al., 2010). Among the members of miRNA, the miR397, miR408, and miR857 are upregulated during limited Cu, and this suppresses the laccase and plastocyanin gene expression (Paul et al., 2015).

Other essential micro-nutrients like iron (Fe), manganese (Mn), and zinc play an important role and satisfy the criteria of essentiality of micro-nutrients. Like other nutrients, their absence affects the growth and development of plants. Several miRNAs have been reported for their role in the uptake and translocation of nutrients within plants. Upregulation of microRNAs, such as miR167, and miR396 under Mn toxicity in *Phaseolus vulgaris* also satisfies the switch-on of various genes and transcription factors (Valdes-Lopez et al., 2010). Zn and Fe are the most important cofactor for several metabolic enzymes and miR163, miR165, and miR172, are extensively studied in Fe-starved *Arabidopsis* (Waters et al., 2012). Studies suggest that under Zn deficiency several miRNAs such as miR166, miR168, miR171, miR319, miR398, and miR528 are downregulated in *Sorghum bicolor* (Li et al., 2013). In *Citrus sinensis* control and B deficient root libraries were elucidated in response to B starvation, 82 miRNAs were found downregulated, and 52 were upregulated suggesting the versatile nature of *C. sinensis* roots during B deficiency tolerance (Lu et al., 2014).

4.4   CONCLUSION

Recently, there have been reports on the role of miRNAs, which are known to be diverse. These molecular controllers are germane and seem quite essential in maintaining nutrient homeostasis in plants subjected to various abiotic stresses. The gene regulation study and the identification of new miRNA provide novel insight into gene expression regulatory mechanisms and also make it more complex. The microRNA-mediated responses under high and low nutrient availability and transducing signals are an interesting area of research. In this chapter, we have discussed the dynamic role of miRNAs in maintaining nutrient homeostasis under the condition of stress. However, the exact mechanisms involved in nutrient sensing and signaling through miRNA are still scarce and thus need to be extrapolated to understand their role and application in agriculture and allied sectors to align food security under changed climatic parameters too.

REFERENCES

Anjum, N. A., Gill, R., Kaushik, M., Hasanuzzaman, M., Pereira, E., Ahmad, I., Tuteja, N., & Gill, S. S. (2015). ATP-sulfurylase, sulfur-compounds and plant stress tolerance. *Frontiers in Plant Science*, 6, 210.
Aung, K., Lin, S. I., Wu, C. C., Huang, Y. T., Su, C. L., & Chiou, T. J. (2006). PHO2, a phosphate overaccumulator, is caused by a nonsense mutation in a microRNA399 target gene. *Plant Physiology*, 141, 1000–1011.
Babu, N., Jyothi, M. N., Shivaram, U., Narayanaswamy, S., Rai, D. V., & Devaraj, V. R. (2014). Identification of miRNAs from French bean (*Phaseolus vulgaris*) under low nitrate stress. *Turkish Journal of Biochemistry*, 39, 1–8.
Baek, D., Kim, M. C., Chun, H. J., Kang, S., Park, H. C., Shin, G., Park, J., Shen, M., Hong, H., Kim, W. Y., Kim, D. H., Lee, S. Y., Bressan, R. A., Bohnert, H. J., & Yun, D. J. (2013). Regulation of miR399f transcription by AtMYB2 affects phosphate starvation responses in *Arabidopsis*. *Plant Physiology*, 161, 362–373.
Bari, R., Pant, B. D., Stitt, M., & Scheible, W. R. (2006). PHO2, microRNA399, and PHR1 define a phosphate-signaling pathway in plants. *Plant Physiology*, 141, 988–999.
Bartel, D. P. (2018). Metazoan microRNAs. *Cell*, 173, 20–51.
Beauclair, L., Yu, A., & Bouche, N. (2010). MicroRNA-directed cleavage and translational repression of the copper chaperone for superoxide dismutase mRNA in *Arabidopsis*. *The Plant Journal*, 62, 454–462.

Branscheid, A., Sieh, D., Pant, B. D., May, P., Devers, E. A., Elkrog, A., Schauser, L., Scheible, W. R., & Krajinski, F. (2010). Expression pattern suggests a role of MiR399 in the regulation of the cellular response to local Pi increase during arbuscular mycorrhizal symbiosis. *Molecular Plant-Microbe Interactions*, *23*, 915–926.

Brant, E. J., & Budak, H. (2018). Plant small non-coding RNAs and their roles in biotic stresses. *Frontiers in Plant Science*, *9*, 1038.

Bustos, R., Castrillo, G., Linhares, F., Puga, M. I., Rubio, V., Perez-Perez, J., Solano, R., Leyva, A., & Paz-Ares, J. (2010). A central regulatory system largely controls transcriptional activation and repression responses to phosphate starvation in *Arabidopsis*. *PLoS Genetics*, *6*, e1001102.

Chen, M., Bao, H., Wu, Q. M., & Wang, Y. W. (2015). Transcriptome-wide identification of miRNA targets under nitrogen deficiency in *Populus tomentosa* using degradome sequencing. *International Journal of Molecular Sciences*, *16*, 13937–13958.

Chiou, T. J., Aung, K., Lin, S. I., Wu, C. C., Chiang, S. F., & Su, C. L. (2006). Regulation of phosphate homeostasis by microRNA in *Arabidopsis*. *Plant Cell*, *18*, 412–421.

Fischer, J. J., Beatty, P. H., Good, A. G., & Muench, D. G. (2013). Manipulation of microRNA expression to improve nitrogen use efficiency. *Plant Science*, *210*, 70–81.

Franco-Zorrilla, J. M., Valli, A., Todesco, M., Mateos, I., Puga, M. I., Rubio-Somoza, I., Leyva, A., Weigel, D., Garcia, J. A., & Paz-Ares, J. (2007). Target mimicry provides a new mechanism for regulation of microRNA activity. *Nature Genetics*, *39*, 1033–1037.

Fujii, H., Chiou, T. J., Lin, S. I., Aung, K., & Zhu, J. K. (2005). A miRNA involved in phosphate-starvation response in *Arabidopsis*. *Current Biology*, *15*, 2038–2043.

Gifford, M. L., Dean, A., Gutierrez, R. A., Coruzzi, G. M., & Birnbaum, K. D. (2008). Cell-specific nitrogen responses mediate developmental plasticity. *Proceedings of the National Academy of Sciences of the United States of America*, *105*, 803–808.

Gu, M., Xu, K., Chen, A., Zhu, Y., Tang, G., & Xu, G. (2010). Expression analysis suggests potential roles of microRNAs for phosphate and arbuscular mycorrhizal signaling in *Solanum lycopersicum*. *Physiologia Plantarum*, *138*, 226–237.

Hackenberg, M., Huang, P. J., Huang, C. Y., Shi, B. J., Gustafson, P., & Langridge, P. (2013). A comprehensive expression profile of microRNAs and other classes of non-coding small RNAs in barley under phosphorous-deficient and -sufficient conditions. *DNA Research*, *20*, 109–125.

Hu, B., Wang, W., Deng, K., Li, H., Zhang, Z., & Chu, C. (2015). MicroRNA399 is involved in multiple nutrient starvation responses in rice. *Frontiers in Plant Science*, *6*, 188.

Hu, B., Zhu, C., Li, F., Tang, J., Wang, Y., Lin, A., Liu, L., Che, R., & Chu, C. (2011). LEAF TIP NECROSIS1 plays a pivotal role in the regulation of multiple phosphate starvation responses in rice. *Plant Physiology*, *156*, 1101–1115.

Huang, C. Y., Shirley, N., Genc, Y., Shi, B., & Langridge, P. (2011). Phosphate utilization efficiency correlates with expression of low-affinity phosphate transporters and noncoding RNA, IPS1, in barley. *Plant Physiology*, *156*, 1217–1229.

Kant, S., Peng, M., & Rothstein, S. J. (2011). Genetic regulation by NLA and microRNA827 for maintaining nitrate-dependent phosphate homeostasis in Arabidopsis. *PLoS Genetics*, *7*, e1002021.

Kataoka, T., Watanabe-Takahashi, A., Hayashi, N., Ohnishi, M., Mimura, T., Buchner, P., Hawkesford, M. J., Yamaya, T., & Takahashi, H. (2004). Vacuolar sulfate transporters are essential determinants controlling internal distribution of sulfate in *Arabidopsis*. *Plant Cell*, *16*, 2693–2704.

Kawashima, C. G., Matthewman, C. A., Huang, S., Lee, B. R., Yoshimoto, N., Koprivova, A., Rubio-Somoza I, Todesco M, Rathjen T, Saito, K., Takahashi, H., Dalmay, T., & Kopriva, S. (2011). Interplay of SLIM1 and miR395 in the regulation of sulfate assimilation in *Arabidopsis*. *The Plant Journal*, *66*, 863–876.

Kehr, J. (2013). Systemic regulation of mineral homeostasis by micro RNAs. *Frontiers in Plant Science*, *4*, 145.

Kuo, H. F., & Chiou, T. J. (2011). The role of microRNAs in phosphorus deficiency signaling. *Plant Physiology*, *156*, 1016–1024.

Leustek, T., Martin, M. N., Bick, J. A., & Davies, J. P. (2000). Pathways and regulation of sulfur metabolism revealed through molecular and genetic studies. *Annual Review of Plant Physiology and Plant Molecular Biology*, *51*, 141–165.

Li, C., & Zhang, B. (2016). MicroRNAs in control of plant development. *Journal of Cellular Physiology*, *231*, 303–313.

Li, Y. L., Zhang, Y., Shi, D. Q., Liu, X. J., Qin, J., Ge, Q., Xu, L. H., Pan, X. L., Li, W., Zhu, Y. Y., & Xu, J. (2013). Spatial–temporal analysis of zinc homeostasis reveals the response mechanisms to acute zinc deficiency in *Sorghum bicolor*. *New Phytologist*, *200*, 1102–1115.

Liang, G., Ai, Q., & Yu, D. (2015). Uncovering miRNAs involved in crosstalk between nutrient deficiencies in *Arabidopsis. Scientific Reports, 5*, 11813.

Liang, G., He, H., & Yu, D. (2012). Identification of nitrogen starvation-responsive microRNAs in *Arabidopsis thaliana. PLoS One, 7*, e48951.

Liang, G., Yang, F., & Yu, D. (2010). MicroRNA395 mediates regulation of sulfate accumulation and allocation in *Arabidopsis thaliana. The Plant Journal, 62*, 1046–1057.

Liu, F., Wang, Z., Ren, H., Shen, C., Li, Y., Ling, H. Q., Wu, C., Lian, X., & Wu, P. (2010). OsSPX1 suppresses the function of OsPHR2 in the regulation of expression of OsPT2 and phosphate homeostasis in shoots of rice. *The Plant Journal, 62*, 508–517.

Liu, T. Y., Huang, T. K., Tseng, C. Y., Lai, Y. S., Lin, S. I., Lin, W. Y., Chen, J. W., & Chiou, T. J. (2012). PHO2-dependent degradation of PHO1 modulates phosphate homeostasis in *Arabidopsis. Plant Cell, 24*, 2168–2183.

Lu, Y. B., Yang, L. T., Qi, Y. P., Li, Y., Li, Z., Chen, Y. B., Huang, Z. R., & Chen, L. S. (2014). Identification of boron-deficiency-responsive microRNAs in *Citrus sinensis* roots by Illumina sequencing. *BMC Plant Biology, 14*, 123.

Lundmark, M., Korner, C. J., & Nielsen, T. H. (2010). Global analysis of microRNA in Arabidopsis in response to phosphate starvation as studied by locked nucleic acid-based microarrays. *Plant Physiology, 140*, 57–68.

Malhotra, H., Vandana., Sharma, S., & Pandey, R. (2018). Phosphorus nutrition: Plant growth in response to deficiency and excess. In M. Hasanuzzaman et al. (Eds.), *Plant Nutrients and Abiotic Stress Tolerance* (pp. 171–190). Springer Nature.

Matthewman, C. A., Kawashima, C. G., Huska, D., Csorba, T., Dalmay, T., & Kopriva, S. (2012). miR395 is a general component of the sulfate assimilation regulatory network in *Arabidopsis. FEBS Letters, 586*, 3242–3248.

Mazahar, S., Sareer, O., Umar, S., & Iqbal, M. (2015). Nitrate accumulation pattern in *Brassica* under nitrogen treatments. *Brazilian Journal of Botany, 38*, 479–486.

Mlodzinska, E., & Zboinska, M. (2016). Phosphate uptake and allocation–a closer look at *Arabidopsis thaliana* L. and *Oryza sativa* L. *Frontiers in Plant Science, 7*, 1198.

Pant, B. D., Buhtz, A., Kehr, J., Scheible, W. R. (2008). MicroRNA399 is a long-distance signal for the regulation of plant phosphate homeostasis. *The Plant Journal, 53*,731–738.

Pant, B. D., Musialak-Lange, M., Nuc, P., May, P., Buhtz, A., Kehr, J., Walther, D., & Scheible, W. R. (2009). Identification of nutrient responsive *Arabidopsis* and rapeseed microRNAs by comprehensive real-time polymerase chain reaction profiling and small RNA sequencing. *Plant Physiology, 150*, 1541–1555.

Paul, S., Datta, S. K., & Datta, K. (2015). miRNA regulation of nutrient homeostasis in plants. *Frontiers in Plant Science, 6*, 232.

Pegler, J. L., Oultram, J. M. J., Grof, C. P. L., & Eamens, A. L. (2019). Profiling the abiotic stress responsive microRNA landscape of *Arabidopsis thaliana. Plants, 8*, 58.

Pei, L., Jin, Z., Li, K., Yin, H., Wang, J., & Yang, A. (2013). Identification and comparative analysis of low phosphate tolerance associated microRNAs in two maize genotypes. *Plant Physiology and Biochemistry, 70*, 221–234.

Rausch, T., & Wachter, A. (2005). Sulfur metabolism: A versatile platform for launching defence operations. *Trends in Plant Science, 10*, 503–509.

Saito, K. (2004). Sulfur assimilatory metabolism. The long and smelling road. *Plant Physiology, 136*, 2443–2450.

Shahzad, R., Harlina, P. W, Ayaad, M., Ewas, M., Nishawy, E., Fahad, S., Subthain, H., & Amar, M. H. (2018). Dynamic roles of microRNAs in nutrient acquisition and plant adaptation under nutrient stress: A review. *Plant Omics Journal, 11*, 58–79, 1836–3644.

Shriram, V., Kumar, V., Devarumath, R. M., Khare, T. S., & Wani, S. H. (2016). MicroRNAs as potential targets for abiotic stress tolerance in plants. *Frontiers in Plant Science, 7*, 817.

Sinha, S. K., Rani, M., Bansal, N., Gayatri, V. K., & Mandal, P. K. (2015). Nitrate starvation induced changes in root system architecture, carbon: Nitrogen metabolism, and miRNA expression in nitrogen-responsive wheat genotypes. *Applied Biochemistry and Biotechnology, 177*, 1299–1312.

Song, A. P., Wang, L. X., Chen, S. M., Jiang, J. F., Guan, Z. Y., Li, P. L., & Chen, F. D. (2015). Identification of nitrogen starvation-responsive microRNAs in *Chrysanthemum nankingense. Plant Physiology and Biochemistry, 91*, 41–48.

Sun, S. B., Gu, M., Cao, Y., Huang, X. P., Zhang, X., Ai, P. H., Zhao, J. N., Fan, X. R., & Xu, G. H. (2012). A constitutive expressed phosphate transporter, OsPht1;1, modulates phosphate uptake and translocation in phosphate-replete rice. *Plant Physiology, 159*, 1571–1581.

Sunkar, R., Li, Y. F., & Jagadeeswaran, G. (2012). Functions of microRNAs in plant stress responses. *Trends in Plant Science, 17,* 196–203.

Todesco, M., Rubio-Somoza, I., Paz-Ares, J., & Weigel, D. (2010). A collection of target mimics for comprehensive analysis of microRNA function in *Arabidopsis thaliana. PLoS Genetics, 6,* e1001031

Trevisan, S., Begheldo, M., Nonis, A., & Quaggiotti, S. (2012b). The miRNA-mediated post-transcriptional regulation of maize response to nitrate. *Plant Signaling & Behavior, 7,* 822–826.

Trevisan, S., Nonis, A., Begheldo, M., Manoli, A., Palme, K., Caporale, G., Ruperti, B., & Quaggiotti, S. (2012a). Expression and tissue-specific localization of nitrate-responsive miRNAs in roots of maize seedlings. *Plant, Cell & Environment, 35,* 1137–1155.

Valdes-Lopez, O., Arenas-Huertero, C., Ramirez, M., Girard, L., Sanchez, F., Vance, C. P., Luis Reyes, J., & Hernandez, G. (2008). Essential role of MYB transcription factor: PvPHR1 and microRNA: PvmiR399 in phosphorus-deficiency signaling in common bean roots. *Plant, Cell & Environment, 3,* 1834–1843.

Valdes-Lopez, O., Yang, S. S., Aparicio-Fabre, R., Graham, P. H., Reyes, J. L., Vance, C. P., & Hernandez, G. (2010). MicroRNA expression profile in common bean (*Phaseolus vulgaris*) under nutrient deficiency stresses and manganese toxicity. *New Phytologist, 187,* 805–818.

Vidal, E. A., Araus, V., Lu, C., Parry, G., Green, P. J., Coruzzi, G. M., Gutierrez, R. (2010). A. nitrate-responsive miR393/AFB3 regulatory module controls root system architecture in *Arabidopsis thaliana. Proceedings of the National Academy of Sciences of the United States of America, 107,* 4477–4482.

Vidal, E. A., Moyano, T. C., Krouk, G., Katari, M. S., Tanurdzic, M., McCombie, W. R., Coruzzi, G. M., & Gutiérrez, R. A. (2013). Integrated RNA-seq and sRNA-seq analysis identifies novel nitrate-responsive genes in Arabidopsis thaliana roots. *BMC Genomics, 14,* 701.

Voinnet, O. (2009). Origin, biogenesis, and activity of plant microRNAs. *Cell, 136,* 669–687.

Wang, J., Mei, J., & Ren, G. (2019). Plant microRNAs: Biogenesis, homeostasis, and degradation. *Frontiers in Plant Science, 10,* 360.

Wang, Y. J., Zhang, C. J., Hao, Q. N., Sha, A. H., Zhou, R., Zhou, X. A., & Yuan, L. P. (2013a). Elucidation of miRNAs-mediated responses to low nitrogen stress by deep sequencing of two soybean genotypes. *PLoS One, 8,* e67423.

Wang, Y., Zhang, C., Hao, Q., Sha, A., Zhou, R., Zhou, X., & Yuan, L. (2013b). Elucidation of miRNAs-mediated responses to low nitrogen stress by deep sequencing of two soybean genotypes. *PLoS One, 8,* e67423. https://doi.org/10.1371

Waters, B. M., Mcinturf, S. A., & Stein, R. J. (2012). Rosette iron deficiency transcript and microRNA profiling reveals links between copper and iron homeostasis in *Arabidopsis thaliana. Journal of Experimental Botany, 63,* 5903–5918.

Xu, F., Liu, Q., Chen, L., Kuang, J., Walk, T., Wang, J., & Liao, H. (2013). Genome-wide identification of soybean microRNAs and their targets reveals their organ-specificity and responses to phosphate starvation. *BMC Genomics, 14,* 66.

Xu, Z., Zhong, S., Li, X., Li, W., Rothstein, S. J., Zhang, S., Bi, Y., & Xie, C. (2011). Genome-wide identification of microRNAs in response to low nitrate availability in maize leaves and roots. *PLoS One, 6,* e28009.

Zeng, H. Q., Zhu, Y. Y., Huang, S. Q., & Yang, Z. M. (2010). Analysis of phosphorus-deficient responsive miRNAs and cis-elements From soybean (*Glycine max* L.). *Journal of Plant Physiology, 167,* 1289–1297.

Zeng, H., Wang, G., Hu, X., Wang, H., Du, L., & Zhu, Y. (2013). Role of microRNAs in plant responses to nutrient stress. *Plant and Soil.* https://doi.org/10.1007/s11104-013-1907-6

Zhao, M., Ding, H., Zhu, J. K., Zhang, F., & Li, W. X. (2011). Involvement of miR169 in the nitrogen-starvation responses in Arabidopsis. *New Phytologist, 190,* 906–915.

Zhao, M., Tai, H., Sun, S., Zhang, F., Xu, Y., & Li, W. X. (2012). Cloning and characterization of maize miRNAs involved in responses to nitrogen deficiency. *PLoS One, 7,* e29669.

Zhao, X., Liu, X., Guo, C., Gu, J., & Xiao, K. (2013). Identification and characterization of microRNAs from wheat (*Triticum aestivum* L.) under phosphorus deprivation. *Journal of Plant Biochemistry and Biotechnology, 22,* 113–123.

Zhao, Y. Y., Guo, L., Lu, W. J., Li, X. J., Chen, H. M, Guo, C. J., & Xiao, K. (2015). Expression pattern analysis of microRNAs in root tissue of wheat (*Triticum aestivum* L.) under normal nitrogen and low nitrogen conditions. *Journal of Plant Biochemistry and Biotechnology, 24,* 143–153.

Zhu, Y. Y., Zeng, H. Q., Dong, C. X., Yin, X. M., Shen, Q. R., & Yang, Z. M. (2010). MicroRNA expression profiles associated with phosphorus deficiency in white lupin (Lupinus albus L.). *Plant Science, 178,* 23–29.

5 Salt-Stress-Responsive Plant miRNAs

Sajad Hussain Shah, Shaistul Islam, Zubair Ahmad Parrey, and Firoz Mohammad

5.1 INTRODUCTION

Salt stress is the accumulation of excessive salt content in the soil which adversely affects the growth, development, and productivity of crop plants. Salt stress induces ion toxicity, ionic imbalance, osmotic imbalance, disorganization of the cell membrane, deleterious reactive oxygen species (ROS) production, and metabolic disorder (Yousuf et al., 2015, 2016a, 2016b; Shah et al., 2021). The ROS production in cells disturbs cell integrity, the structure of proteins, the activity of enzymes, and the disorganization of lipid and nucleic acids leading to a reduction of crop growth and yield (Negrao et al., 2017; Ma et al., 2020; Islam et al., 2021). Plants being sessile organisms employ several mechanisms to resist salt stress, including alteration of different genes expression and synthesis of biomolecules that improve resistance against the adversity of high salinity (Botella et al., 2005; Islam & Mohammad, 2020; Johnson et al., 2021). However, the regulation of gene expression accurately and timely affects different features of plant growth, development, metabolic processes, and responses to stress conditions (Ojolo et al., 2018). Though in response to abiotic stresses, several genes in plants are upregulated or downregulated. Apart from their role in controlling a broad spectrum of essential physio-biochemical and molecular processes, microRNAs (miRNAs) are also involved in plant responses to a different array of abiotic stresses (Liu et al., 2008; Ren et al., 2013). Among the tolerance mechanisms, the role of miRNAs in the modulation of expression of genes at transcriptional and post-transcriptional levels regarding resistance against salt stress is obvious in many plants. The miRNAs consist of a class of small regulatory, non-coding, endogenous RNA molecule that is about 22–25 nucleotides in length (Liu et al., 2017; Wang et al., 2019). The miRNAs are evolutionarily highly conserved. They identify their targets based on the exact complementary sequence. The precursor of miRNA is primary-miRNA (pri-miRNA), transcribed from the genomic DNA template and processed by DCL1 (Dicer-like 1) to form precursor-miRNA (pre-miRNA). Further, the pre-miRNA is methylated by HEN1 (HUA ENHANCER 1), and then transported to the cytosol by HASTY (a homolog of exportin-5 mediates the transport of miRNA from the nucleus to cytosol) and finally embodied into AGO (Argonaut) protein. The AGO-miRNA complex regulates the target genes expression by degradation of mRNA, transcriptional and translational inhibition, remodeling of chromatin, and methylation of DNA (Axtell & Bowman, 2008; Voinnet, 2009; Song et al., 2019). The miRNAs have also been found to be involved in the phase transition in plants from vegetative to reproductive growth (Jover-Gil et al., 2004; Wu, 2013).

Many studies indicated that plants' response to salinity stress could be moderated by miRNAs that guide gene regulation. In *Arabidopsis sp.* and *Zea mays* L. many miRNAs such as miR396, miR319, miR171, miR168, miR167, miR159, and miR156 exhibited differential expression during salinity stress response recognized through microarray analysis (Liu et al., 2008; Ding et al., 2009). The wide-ranging sequencing data produced by NGS (next-generation sequencing) has been used for the observation of salt-stress-responsive miRNAs in many plant species. Nearly 104 differentially expressed miRNAs were analyzed in *Glycine max* L. nodules under salt stress by using this

DOI: 10.1201/9781003248453-5

technology (Dong et al., 2013; Alzahrani et al., 2019). Further, Ren et al. (2013) noticed that among the conserved miRNAs families, seven downregulated and two upregulated miRNAs families were isolated in *Populus tomentosa* L. under salt stress. Li et al. (2013) noticed 132 miRNA families in *Populus euphratica* L. during salt stress. Zhu et al. (2013) revealed that the expression of miRNAs including cin-miR390a, miR396a, cin-miR167b, cin-miR172b, cin-miR157a, cin-miR159a, and cin-miR165a was induced while cin-miR398a was repressed in *Caragana intermedia* L. under salt stress. Moreover, many salt-stress-responsive miRNAs were recognized in some vegetable crops. Under salt stress treatment, 50 miRNAs were isolated to be differentially expressed in *Glycine max* L. (Sun et al., 2015). Likewise, Zhuang et al. (2014) using NGS technology, identified 11 downregulated and three upregulated miRNAs that are differentially expressed in *Solanum linnaeanum* L. after salt stress treatment. Tian et al. (2014) reported 39 new and 42 known miRNAs were differentially expressed in *Brassica oleracea* L. under high salt conditions. Conclusively, many studies inferred that miRNAs-mediated gene regulatory pathways could play important roles in plant adaptive response towards salinity stress. The miRNAs have been discovered as an important modulator in salinity stress conditions via control of the expression of salinity-response genes. Salinity induces miRNAs and downregulates their target miRNAs, which produce negative functional proteins involved in salt stress response (Lotfi et al., 2017; Haldar & Bandyopadhyay, 2021). The miRNAs are also important for maintaining nutrient homeostasis in plants by modulating the expression of transporters that are involved in the uptake and utilization of nutrients. Salt-stress-responsive miRNAs either upregulate or downregulate gene expression to modulate physicochemical and molecular processes and provide tolerance to plants against salinity stress. In this chapter information has been included regarding the role of miRNAs in the growth and development of plants under varied conditions, the regulation of gene expression associated with salt stress response, and their involvement in salt stress mitigation.

5.2 ROLES OF MIRNAS IN GROWTH AND DEVELOPMENT OF PLANTS UNDER DIVERSE CONDITIONS

In many plant species, that is, *Brassica sp.*, *Arabidopsis sp.*, *Triticum aestivum*, *Oryza sativa*, *Hordeum vulgare*, etc., miRNAs have been identified. The miRNAs regulate many plants' developmental processes including growth and development, root initiation, leaf morphogenesis, polarity, floral differentiation, signal transduction, feedback regulation of genes, and responses to environmental stress (Guo et al., 2005; Kim et al., 2005; Lauter et al., 2005; Yang et al., 2007; Meng et al., 2010; Khraiwesh et al., 2012). Most of the miRNAs modulate the expression of TFs that persuade cell fate determination and ultimately affect plant characteristics (Allen et al., 2004; Bartel, 2004).

Under major abiotic stresses (drought, cold, heavy metals, and salinity), plants endure resisting by inducing and modulating various physiochemical and molecular responses. Recently, there have been strong shreds of evidence leading to the proposal that miRNAs play a crucial role at the molecular level to provide resistance against abiotic stresses in *Populus trichocarpa* L. (Lu et al., 2005). Abiotic stresses strongly induced the overexpression of miRNAs such as miR402 in *Arabidopsis thaliana* L. However other miRNAs including miR319 are activated by either cold stress or other stress (Sunkar & Zhu, 2004). Sunkar et al. (2006) reported that miR398 was the foremost miRNA in *Arabidopsis thaliana* L. and had oxidative stress tolerance. Its expression is also transcriptionally downregulated by oxidative stress. Moreover, miR398 targeted two closely related coding genes of copper/zinc superoxide dismutase (CSD): chloroplastic CSD2 and cytosolic CSD1. The low level of miR398 led to enhanced tolerance of transgenic plants as compared to wild-type plants under oxidative stress conditions. Zhao et al. (2007) showed that miR169g was the only member of the miR169 family that was more prominent in root growth and was induced by drought stress in *Oryza sativa* L. Zhang et al. (2005), using expressed sequence tag analyses (EST), showed 25.8% of ESTs containing miRNAs were found in stress-induced plant tissues. Jones-Rhoades and Bartel (2004) revealed that the APS4 (ATP sulfurylases) and the sulphate transporter (AST68) are accumulated at low-sulphur conditions, and both of these genes are controlled by miR395. Lu et al. (2005) recognized a sequence of 48 miRNA from the *Populus trichocarpa* L. genome. Most of these *Populus trichocarpa* L. miRNAs showed their target genes related to the development and stress defence. In another study, Zhang et al. (2008) identified the miRNA involved in submergence tolerance in plants. Fujii et al. (2005) reported that in *Arabidopsis thaliana* L. the overexpression of miR399 resulted in downregulation of the target mRNA transcript, thus interpreting the role of miRNA to deal with mineral nutrition fluctuations. Kantar et al. (2010) identified 28 new miRNAs belonging to 18 families in *Hordeum vulgare* L. under water stress, of which five miRNAs have been experimentally proved for their differential expression. Similarly, Zhang et al. (2010) identified 21 miRNAs in *Zea mays* L. of which 13 are proved to be specific for drought stress. Several studies discovered numerous miRNAs in *Oryza sativa* L. that are involved in different abiotic stresses viz. cold (Lv et al., 2010), heavy metal (Huang et al., 2009), salinity (Zhao et al., 2009), and drought (Jian et al., 2010).

Moreover, miRNAs also modulated the important components of signaling pathways of hormones and further maintained homeostasis of hormones and related developmental processes (Guo et al., 2005). Zhang et al. (2005) noticed that many miRNAs such as miR164, miR167, miR159, and miR160 were characterized from plant tissues induced by gibberellins, jasmonate, abscisic acid, salicylic acid, and other phytohormones. Gray et al. (2001) and Mallory et al. (2005) identified miR393 in *Arabidopsis thaliana* L. that is probably involved in auxin signaling pathways by regulating TIR1, which is a component of an SCF-E3 ubiquitin ligase that degrades Aux/IAA proteins when auxin is present in the cell. Furthermore, Chuck et al. (2007) reported that two miRNAs (miR156 and miR172) are important in plant development and induced transition from juvenile to adult shoots. Besides, Boualem et al. (2008) suggested that miRNAs like miR166 mediated post-transcriptional regulation and were involved in the regulation of legume root architecture. Meng et al. (2010) and Ding et al. (2011) viewed that miRNAs moderated auxin signaling involved in the root development of *Arabidopsis thaliana* L. and *Oryza sativa* L. Breakfield et al. (2012) identified about 133 miRNAs in *Arabidopsis thaliana* L. that are expressed in plant tissue specifically in the root zone. Wang et al. (2014) predicted 88 target genes by using computational analysis for conserved and novel miRNAs, which were expressed in *Prunus mume* L. during flowering time. However, seven target genes, which are encoding SPL (squamosa promoter binding protein-like), ARF (auxin response factor), SCL (scarecrow-like transcription factor), and AP2 (Apetala2-like transcription factors), were verified by 5′3) identified miRNAs and their targets in *Fragaria ananassa* L. during fruit senescence. However, they reported a total of 88 known and 1,224 new miRNAs. Further, they identified 103 target genes cleaved by 19 known miRNAs families and 55 new candidate miRNAs. The targets of these miRNAs were related with metabolism, development, signal transduction, transcriptional regulation, and defence responses. Among these targets, 14 targets including NAC transcription factor, MYB, and ARF were cleaved by six known miRNA families and six predicted candidates. Besides, Zhuang et al. (2014), using customized microarray, identified 106 known miRNAs and 98 potentially novel miRNAs in two *Camellia sinensis* L. cultivars treated with cold stress. They also identified 238 targets which were common under cold stress as well as in control condition, while 455 and 591 genes were identified as cleavage targets of miRNAs detected in cold treatment and in control condition, respectively. Hence, their findings indicated that miRNA target genes were those involved in the regulation of transcription, developmental processes, and stress responses. Sunkar (2010) and Wang et al. (2013) demonstrated that about 40 plant miRNA families are mainly associated with drought and salt stress response. In subsequent study, small RNA and degradome deep sequencing were systematically applied to explore the tissue-specific miRNAs responsible to drought stress in *Solanum lycopersicum* L. Wani et al. (2020) suggested that miRNAs are a dynamic and novel target for enhancement of plant resistance against abiotic factors. In conclusion, the above studies indicate that miRNAs play a diverse role in many aspects of growth, developmental, metabolic processes, and also regulating plants' responses under varying environmental conditions.

5.3   MIRNAS AND MITIGATION OF SALT STRESS IN PLANTS

Salinity, a major abiotic stress, negatively affects growth and developmental processes which cause reduction in productivity of crop plants (Sharif et al., 2019). Plants have developed many regulatory ways to ameliorate salinity stress and maintain their growth and physiological aspects. A number of compounds, biomolecules, and genes have been identified that can modify plants to make them resistant to salinity (Tuteja, 2007; Khan et al., 2020). Recently miRNAs have been explored to mediate adaptation mechanism to cope with salt stress in plants. Plants upregulate genes that play a protective role while they also downregulate their negative expression. Studies revealed that miRNAs regulate salinity stress tolerance responses in plants. For instance, Gao et al. (2011) reported that the overexpression of osa-miR393 resulted in enhanced salinity resistance in *Arabidopsis thaliana* L. Dong et al. (2013) determined 128 novel miRNAs belonging to 64 miRNAs and 110 known miRNAs to 61 miRNA families. Among them, 104 miRNAs were found to be differentially expressed during salt stress. The targets of miRNAs are many diverse functional genes that are involved in biological processes. Moreover, they reported in *Glycine max* L. that the characterization of highly expressed miRNAs in mature nodules showed that miRNAs are more responsive to salinity stress helpful to interpret the molecular mechanisms of controlling nodule development, nitrogen fixation, and salinity resistance to functional nodules. Carnavale Bottino et al. (2013) indicated that 11 miRNAs showed higher expression under salt treatment in *Saccharum officinarum* L. The target genes of miRNAs encode a wide range of proteins, including transcription factors, metabolic enzymes, and genes involved in hormone signaling and probably provide tolerance against salinity. Wu et al. (2016) described a total of 82 conserved miRNAs belonging to 27 miRNAs families, and 17 novel miRNAs were identified and 11 conserved miRNAs families and 4 novel miRNAs exhibited a significant response to salinity stress in *Eutrema salsugineum*. These miRNAs altered the expression of target genes. The expressed genes might control various biological processes responsible for salinity tolerance. Arshad et al. (2017) proposed that the high level (20–50 times) of miR156 resulted in improved biomass, growth, and forage quality of *Medicago sativa* L. under salinity stress. However, the high level of miR156 further reduced the toxic ions uptake, which might contribute to salt tolerance. In addition, miR156 directly or indirectly targeted transcription factors including SPLs (Squamosa-promoter binding protein-like), which in turn might regulate downstream genes in a stress-specific manner leading to salt stress resistance. Ning et al. (2019) identified 21 miRNAs including gma-miR1691–3p, gma-862a, miR5036, and gma-miR398a/b, which were detected to respond to salinity stress. The target genes of these miRNAs are protein phosphatase 2C, ethylene responsive transcription factor 4 and copper/zinc-superoxide dismutase, which had roles in different metabolic processes and in signal transduction in *Glycine max* L. Alzahrani et al. (2019) determined differentially expressed a total of 527 miRNA and 693 miRNAs in *Vicia faba* L. Furthermore, 298 miRNAs upregulated and 395 miRNAs downregulated and 284 upregulated and 243 miRNAs downregulated were obtained in *Vicia faba* L. growing in normal and adverse conditions respectively. Parmar et al. (2020) working on *Oryza sativa* L. reported through target prediction and annotation that miRNAs modulated specific salt-responsive genes, which primarily included genes encoding TFs and superoxide dismutase, laccases, E-box protein, and plantacyanin. However, Gene Ontology (GO) and Kyoto Encyclopedia of Genes and Genomes (KEGG) pathway analyses revealed that the miRNAs were found to be involved in salt-stress-related biological pathways including MAPK signaling pathway, plant hormone signaling transduction, ABC transporter pathway, and phosphatidylinositol signaling system, suggesting that miRNAs play an essential role in salinity stress tolerance. Further, the miRNAs, including conserved and novel ones differentially expressed in shoot and root tissue target TFs like NAC, ARF, MYB, HD-ZIP III, AP2/EREBP domain protein, TCP, SBP, and NF-YA reported to be involved in salt-stress resistance. Wen et al. (2020) identified TAS3a/b/c, tasiARF-1/2/3 (trans-acting small interfering RNAs influencing Auxin Response Factors), miR390, and ARF2/3/4 under salt stress treatment in *Helianthus tuberosus* L. They proposed that ARF2 was likely to play an active role in salt tolerance. Ai et al. (2021) found that the overexpression of conserved miR1861 family increased salt tolerance in *Oryza rufipogon* L. by suppressing the expression of negative regulators. Li et al. (2021) suggested that miRNA mediated regulatory network involved in the process of hydrogen sulphide mitigated alkaline salt stress in *Malus hupehensis* L. roots. Besides, application of miRNAs to develop transgenic plants in which miRNA guided gene regulation made them salt stress tolerant. Zhou et al. (2013) found that the transgenic plants like

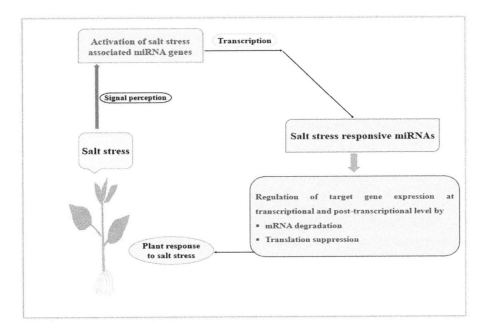

FIGURE 5.1 microRNAs mediated regulation of genes against salt stress: an overview.

Agrostis stolonifera L. that overexpressed the acquired rice miR319 gene (osa-miR319) which exhibited enhanced tolerance to salt stress. Wang et al. (2012) reported that the downregulation of specific bacterial type PEPC (phosphoenolpyruvate carboxylase) gene called Atppc4, by artificial microRNA increased the resistance of *Arabidopsis thaliana* L. against salt stress. Zhou et al. (2013) through the gene expression analysis showed that the improved stress tolerance might be related to a significant downregulation of at least four miR319 target genes, including AsPCF5, AsPCF6, AsPCF8, and AsTCP14 (teosinte branched/Cycloidea/proliferating factors) as well as homolog of rice NAC domain gene AsNAC60. In conclusion, miRNAs play an important role in the mitigation of salt stress by regulating gene expression at transcriptional and post-transcriptional level leading to modulation of various physiological processes in plants. A diagrammatic representation of salinity stress and miRNAs mediated response of plants is shown in **Figure 5.1**.

5.3.1 MIRNAs REGULATE TARGET GENES ASSOCIATED WITH SALT STRESS

MiRNAs target multiple genes that belong to the same gene family in plants. It has been indicated that miRNAs modulate various metabolic processes by targeting the expression of gene associated with them (Sunkar et al., 2012). By sequencing *Oryza sativa* L. miRNAs, data showed that many of these miRNAs target TFs involved in plant developmental processes including organ formation. The transcription factors, MYB and HD-ZIP (homeodomain-leucine zipper protein) were predicted as the targets of zma-miR1661/m, zma-miR164a/b/c/d, and zma-miR159a/b, respectively. Similar results were obtained in *Arabidopsis thaliana* L. (Jones-Rhoades & Bartel, 2004). Yan et al. (2005) and Cheng and Long (2007) suggested that targets of many miRNAs are cytochrome oxidase and NADP-ME (NADP-dependent malic enzyme). Various other TFs were also predicted to be the target of miRNAs including zinc-finger protein and MADS-box proteins which had been reported as salt-stress-responsive factors in *Thellungiella halophila* L. (Fang et al., 2006) and *Oryza sativa* L. (Xu et al., 2008). Achard et al. (2006) reported that miR160 targeted genes that encode B3 DNA-binding domain proteins and ARF in *Arabidopsis thaliana* L. The ARF had effects on various aspects of plant growth, development, and responses to environmental changes. Further, Ding et al. (2009) suggested that sulfurylase and ASP1 genes are regulated by miR395 in salt induced *Glycine max* L.

treated under sulphate deficient conditions. It is also presumed that the role of miR395 is the maintenance of energy supply in non-specific salt stress responding pathways among others. Liu et al. (2009) suggested that miR396 repressed growth-regulating factor (GRF) and had great impacts on plant leaf growth and development of *Oryza sativa* L. De Palo et al. (2012) demonstrated that the targets of cca-miR397 and cca-miR399 were homologous to members of laccase gene family, which are involved in salt stress response. Eren et al. (2015) indicated that three miRNAs (osa-miR44b.2, ppt-miR1074 and hvu-miR5049a) were upregulated in *Triticum aestivum* under salt stress. Further, the salt-stress-responsive miRNAs regulate mostly TFs such as AP2/ERBP, bHLH135-like, and MADS-box. Sun et al. (2016) observed through functional analysis of miR399 that a salt-responsive miRNA plays a role in root growth of *Glycine max* L. by activating genes associated with auxin signaling. Arshad et al. (2017) revealed that miR156 downregulated SPL transcription factors family genes and also modified expression of other important TFs and downstream salt stress responsive genes in *Medicago sativa* L. The miR156 plays a role in moderating physiological and transcriptional responses in the studied plant to salinity stress. Zhao et al. (2017) reported that miRNAs such as sly-miRn23b, sly-miRn50a, sly-miR156e-5p, etc. were found to be involved in salinity stress response of *Solanum pimpinellifolium* L. and *Solanum lycopersicum* L. by regulating gene expression associated with various biological processes. Besides TFs, many miRNAs target genes that encode proteins involved in physiological processes of plant (Samad et al., 2017). Kuang et al. (2019) identified 40 and 51 salt responsive miRNAs in the root and shoot of *Hordeum spontaneum* L. In roots, miR393a, miR319a, miR172b, miR164a, and miR156d targeting NAC079, UGTs, HvFB2/HvTIR1, HvAP2, and TCP4 respectively were probably responsible for salt tolerance. In shoots, miR172b, miR319, miR396e, miR169i, miR159a regulating NHX/LEA7, GRFs, TCP4, and HvAP2 respectively might contribute to salt tolerance. In *Zea mays* L. Shan et al. (2020) reported that zma-miR164 might respond to salinity stress, the expression of zma-miR164 was decreased with salt stress leading to an increase in expression of its target genes: NAM, ATAF, CUC, and NAC TF protein family and electron carrier. Moreover, transcription factor like NAC countered to salinity stress by regulating the expression pattern of their downstream genes. It may be concluded that miRNAs regulate diverse functions in plants in variable conditions by targeting different TFs, genes, and transcripts. The miRNAs and their predicted target genes in different plant species are presented in **Table 5.1**.

TABLE 5.1

List of salt-stress-responsive miRNAs and their predicted target genes in plants.

S. No.	miRNA	Plant species	Target gene	Reference
1	miR156	*Arabidopsis thaliana* L.	Squamosa promoter-binding protein-like 11	Liu et al. (2008)
2	miR159		MYB and TCP transcription factors	
3	miR165		Class III HD-ZIP transcription factors	
4	miR167		Auxin response factor 6 and 8	
5	miR168		ARGONAUTE1	
6	miR169		CAAT binding factor	
7	miR171		Scarecrow-like transcription factor	
8	miR172		APETALA2 transcription factor	
9	miR319		TCP transcription factor	
10	miR393		F-box protein; bHLH transcription factors	
11	miR394		F-box protein	
12	miR396		GRL transcription factors; Rhodenase-like protein; Kinesin-like protein B	
13	miR397		Laccases; β-6 tubulin	

(Continued)

TABLE 5.1 (Continued)
List of salt-stress-responsive miRNAs and their predicted target genes in plants.

S. No.	miRNA	Plant species	Target gene	Reference
14	miR2118	*Phaseolus vulgaris* L.	APS-reductase	Arenas-Huertero et al. (2009)
15	miR162	*Zea mays* L.	Endoribonuclease Dicer	Ding et al. (2009)
16	miR164		Cytochrome P450	
17	miR166		NAC domain protein NAC1	
18	miR167		Homeo domain leucine Zipper protein	
19	miR168		(HD-ZIP)	
20	miR171		Auxin response factor	
21	miR395		PZE40 protein	
			GPI-anchored protein	
			ATP sulfurylase	
22	miR1446	*Populus euphratica* L.	Gibberellin response modulator-like protein	Qin et al. (2011)
23	miR1447		ATP-binding cassette transport protein	
24	miR399	*Thellungiella salsuginea* L.	Transcription factors	Zhang et al. (2013)
25	miR5300	*Solanum linnaeanum* L.	CC-NBS-LRR protein	Zhuang et al. (2014)
26	miR396	*Raphanus sativus* L.	Transcription factors bHLH74	Sun et al. (2015)
27	miR395	*Cucumis sativus* L.	ATP sulfurylase	Zhao et al. (2017)
28	miR172	*Solanum lycopersicum* L.	Ethylene-responsive transcription factor	Zhao et al. (2017)
29	miR408	*Vicia faba* L.	Plantacyanins/uclacyanin-2-like/basic blue-like protein	Alzahrani et al. (2019)
30	miR2111		DNA replication factor CDT1-like protein	
31	miR160	*Setaria viridis* L.	Auxin response factor 1 and 16	Pegler et al. (2020)
32	miR167			

5.4 CONCLUSION AND FUTURE PROSPECTS

Salt stress is an adverse constraint affecting plant physio-biochemical and molecular processes and also speculated to be a major problem in near future. The adverse effects of high salt stress on plants can be noticed at morphological, physiological, and cellular level in terms of reduction in productivity or death of plant. Plants respond to salinity stress by altering their gene expression through post-transcriptional gene regulation. The miRNAs regulate target gene expression at post-transcriptionally involved in salt stress response. Besides its role in growth, development, and maintenance of integrity of genome, miRNAs also play a role in plant stress responses. Hence, understanding how miRNAs regulate gene expression can provide insights to researchers for exploring the role of miRNAs in salt stress responses in different plant species. The experimental and computational approaches proved important to identify many miRNAs and their targets in plants associated with various characteristics. Interestingly as miRNAs are crucial component in gene regulatory network, there is a need of full knowledge about the function of miRNAs in plant resistance to salt stress. Despite the existing information regarding miRNAs, a lot remains to be known concerning the identification and characterization of unknown miRNAs in plants. Several studies indicate that salt-stress-responsive miRNAs and their target gene expression provide tolerance mechanism against salinity stress. However, it still needs to be further explored diverse miRNAs exhibiting alteration of their target gene expression under salinity stress in various crop plants. Recently, there are a few studies showing that application of miRNAs by genetic engineering to guide regulation of gene can help in making salt-stress-tolerant plants. Furthermore, research efforts are needed for identification, characterization, and sequencing of novel salt-stress-responsive miRNAs and their target genes. As our understanding related to the role of miRNAs under salinity stress deepens, the challenges of high salinity may be solved.

REFERENCES

Achard, P., Cheng, H., De Grauwe, L., Decat, J., Schoutteten, H., Moritz, T., & Harberd, N. P. (2006). Integration of plant responses to environmentally activated phytohormonal signals. *Science*, *311*, 91–94.

Ai, B., Chen, Y., Zhao, M., Ding, G., Xie, J., & Zhang, F. (2021). Overexpression of miR1861h increases tolerance to salt stress in rice (*Oryza sativa* L.). *Genetic Resources and Crop Evolution*, *68*, 87–92.

Allen, E., Xie, Z., Gustafson, A. M., Sung, G. H., Spatafora, J. W., & Carrington, J. C. (2004). Evolution of microRNA genes by inverted duplication of target gene sequences in *Arabidopsis thaliana*. *Nature Genetics*, *36*, 1282–1290.

Alzahrani, S. M., Alaraidh, I. A., Khan, M. A., Migdadi, H. M., Alghamdi, S. S., & Alsahli, A. A. (2019). Identification and characterization of salt-responsive microRNAs in *Vicia faba* by high-throughput sequencing. *Genes*, *10*, 303.

Arenas-Huertero, C., Perez, B., Rabanal, F., Blanco-Melo, D., De la Rosa, C., Estrada-Navarrete, G., & Reyes, J. L. (2009). Conserved and novel miRNAs in the legume *Phaseolus vulgaris* in response to stress. *Plant Molecular Biology*, *70*, 385–401.

Arshad, M., Gruber, M. Y., Wall, K., & Hannoufa, A. (2017). An insight into microRNA156 role in salinity stress responses of alfalfa. *Frontiers in Plant Science*, *8*, 356.

Axtell, M. J., & Bowman, J. L. (2008). Evolution of plant microRNAs and their targets. *Trends in Plant Science*, *13*, 343–349.

Bartel, D. P. (2004). MicroRNAs: Genomics, biogenesis, mechanism, and function. *Cell*, *116*, 281–297.

Botella, M. A., Rosado, A., Bressan, R. A., & Hasegawa, P. M. (2005). Plant adaptive responses to salinity stress. *Plant Abiotic Stress*, *21*, 38–70.

Boualem, A., Laporte, P., Jovanovic, M., Laffont, C., Plet, J., Combier, J. P., & Frugier, F. (2008). MicroRNA166 controls root and nodule development in *Medicago truncatula*. *The Plant Journal*, *54*, 876–887.

Breakfield, N. W., Corcoran, D. L., Petricka, J. J., Shen, J., Sae-Seaw, J., Rubio-Somoza, I., & Benfey, P. N. (2012). High-resolution experimental and computational profiling of tissue-specific known and novel miRNAs in Arabidopsis. *Genome Research*, *22*, 163–176.

Candar-Cakir, B., Arican, E., & Zhang, B. (2016). Small RNA and degradome deep sequencing reveal drought- and tissue-specific micrornas and their important roles in drought-sensitive and drought-tolerant tomato genotypes. *Plant Biotechnology Journal*, *14*, 1727–1746.

Carnavale Bottino, M., Rosario, S., Grativol, C., Thiebaut, F., Rojas, C. A., Farrineli, L., & Ferreira, P. C. G. (2013). High-throughput sequencing of small RNA transcriptome reveals salt stress regulated microRNAs in sugarcane. *PLoS One*, *8*, e59423.

Cheng, Y., & Long, M. (2007). A cytosolic NADP-malic enzyme gene from rice (*Oryza sativa* L.) confers salt tolerance in transgenic Arabidopsis. *Biotechnology Letters*, *29*, 1129–1134.

Chuck, G., Cigan, A. M., Saeteurn, K., & Hake, S. (2007). The heterochronic maize mutant Corngrass1 results from overexpression of a tandem microRNA. *Nature Genetics*, *39*, 544–549.

De Paola, D., Cattonaro, F., Pignone, D., & Sonnante, G. (2012). The miRNAome of globe artichoke: Conserved and novel micro RNAs and target analysis. *BMC Genomics*, *13*, 1–14.

Ding, D., Zhang, L., Wang, H., Liu, Z., Zhang, Z., & Zheng, Y. (2009). Differential expression of miRNAs in response to salt stress in maize roots. *Annals of Botany*, *103*, 29–38.

Ding, Y., Chen, Z., & Zhu, C. (2011). Microarray-based analysis of cadmium-responsive microRNAs in rice (*Oryza sativa*). *Journal of Experimental Botany*, *62*, 3563–3573.

Dong, Z., Shi, L., Wang, Y., Chen, L., Cai, Z., Wang, Y., & Li, X. (2013). Identification and dynamic regulation of microRNAs involved in salt stress responses in functional soybean nodules by high-throughput sequencing. *International Journal of Molecular Sciences*, *14*, 2717–2738.

Eren, H., Pekmezci, M. Y., Okay, S. E. Z. E. R., Turktas, M. İ. N. E., Inal, B., Ilhan, E., & Unver, T. (2015). Hexaploid wheat (*Triticum aestivum*) root miRNome analysis in response to salt stress. *Annals of Applied Biology*, *167*, 208–216.

Fang, Q., Xu, Z., & Song, R. (2006). Cloning, characterization and genetic engineering of FLC homolog in *Thellungiella halophila*. *Biochemical and Biophysical Research Communications*, *347*, 707–714.

Fujii, H., Chiou, T. J., Lin, S. I., Aung, K., & Zhu, J. K. (2005). A miRNA involved in phosphate-starvation response in Arabidopsis. *Current Biology*, *15*, 2038–2043.

Gao, P., Bai, X., Yang, L., Lv, D., Pan, X., Li, Y., & Zhu, Y. (2011). osa-MIR393: a salinity-and alkaline stress-related microRNA gene. *Molecular Biology Reports*, *38*, 237–242.

Gray, W. M., Kepinski, S., Rouse, D., Leyser, O., & Estelle, M. (2001). Auxin regulates SCF TIR1-dependent degradation of AUX/IAA proteins. *Nature*, *414*, 271–276.

Guo, H. S., Xie, Q., Fei, J. F., & Chua, N. H. (2005). MicroRNA directs mRNA cleavage of the transcription factor NAC1 to downregulate auxin signals for Arabidopsis lateral root development. *The Plant Cell*, *17*, 1376–1386.

Haldar, S., & Bandyopadhyay, S. (2021). Co-ordinated regulation of miRNA and their target genes by CREs during salt stress in *Oryza sativa* (Rice). *Plant Gene*, 100323.

Huang, S. Q., Peng, J., Qiu, C. X., & Yang, Z. M. (2009). Heavy metal-regulated new microRNAs from rice. *Journal of Inorganic Biochemistry*, *103*, 282–287.

Islam, S., & Mohammad, F. (2020). Triacontanol as a dynamic growth regulator for plants under diverse environmental conditions. *Physiology and Molecular Biology of Plants*, *26*, 871–883.

Islam, S., Zaid, A., & Mohammad, F. (2021). Role of triacontanol in counteracting the ill effects of salinity in plants: A review. *Journal of Plant Growth Regulation*, *40*, 1–10.

Jian, X., Zhang, L., Li, G., Zhang, L., Wang, X., Cao, X., & Chen, F. (2010). Identification of novel stress-regulated microRNAs from *Oryza sativa* L. *Genomics*, *95*, 47–55.

Johnson, R., & Puthur, J. T. (2021). Seed priming as a cost-effective technique for developing plants with cross tolerance to salinity stress. *Plant Physiology and Biochemistry*, *162*, 247–257.

Jones-Rhoades, M. W., & Bartel, D. P. (2004). Computational identification of plant microRNAs and their targets, including a stress-induced miRNA. *Molecular Cell*, *14*, 787–799.

Jover-Gil, S., Candela, H., & Ponce, M. R. (2004). Plant microRNAs and development. *The International Journal of Developmental Biology*, *49*, 733–744.

Kantar, M., Unver, T., & Budak, H. (2010). Regulation of barley miRNAs upon dehydration stress correlated with target gene expression. *Functional & Integrative Genomics*, *10*, 493–507.

Khan, W. U. D., Tanveer, M., Shaukat, R., Ali, M., & Pirdad, F. (2020). An overview of salinity tolerance mechanism in plants. In M. Hasanuzzaman, & M. Tanveer (Eds.), *Salt and drought stress tolerance in plants* (pp. 1–16). Springer.

Khraiwesh, B., Zhu, J. K., & Zhu, J. (2012). Role of miRNAs and siRNAs in biotic and abiotic stress responses of plants. *Biochimica et Biophysica Acta, Gene Regulatory Mechanisms*, *1819*, 137–148.

Kim, J., Jung, J. H., Reyes, J. L., Kim, Y. S., Kim, S. Y., Chung, K. S., & Park, C. M. (2005). MicroRNA-directed cleavage of ATHB15 mRNA regulates vascular development in Arabidopsis inflorescence stems. *The Plant Journal*, *42*, 84–94.

Kuang, L., Shen, Q., Wu, L., Yu, J., Fu, L., Wu, D., & Zhang, G. (2019). Identification of microRNAs responding to salt stress in barley by high-throughput sequencing and degradome analysis. *Environmental and Experimental Botany*, *160*, 59–70.

Lauter, N., Kampani, A., Carlson, S., Goebel, M., & Moose, S. P. (2005). MicroRNA172 down-regulates glossy15 to promote vegetative phase change in maize. *Proceedings of the National Academy of Sciences of the United States of America, 102*, 9412–9417.

Li, B., Duan, H., Li, J., Deng, X. W., Yin, W., & Xia, X. (2013). Global identification of miRNAs and targets in *Populus euphratica* under salt stress. *Plant Molecular Biology*, *81*, 525–539.

Li, H., Yu, T. T., Ning, Y. S., Li, H., Zhang, W., & Yang, H. (2021). Hydrogen sulfide alleviates alkaline salt stress by regulating the expression of microRNAs in *Malus hupehensis* Rehd. Roots. *Frontiers in Plant Science*, *12*, 1020.

Liu, H. H., Tian, X., Li, Y. J., Wu, C. A., & Zheng, C. C. (2008). Microarray-based analysis of stress-regulated microRNAs in *Arabidopsis thaliana*. *RNA*, *14*, 836–843.

Liu, Q., Zhang, Y. C., Wang, C. Y., Luo, Y. C., Huang, Q. J., Chen, S. Y., & Chen, Y. Q. (2009). Expression analysis of phytohormone-regulated microRNAs in rice, implying their regulation roles in plant hormone signaling. *FEBS Letters*, *583*, 723–728.

Liu, W. W., Meng, J., Cui, J., & Luan, Y. S. (2017). Characterization and function of microRNAs in plants. *Frontiers in Plant Science*, *8*, 2200.

Lotfi, A., Pervaiz, T., Jiu, S., Faghihi, F., Jahanbakhshian, Z., Khorzoghi, E. G., & Fang, J. (2017). Role of microRNAs and their target genes in salinity response in plants. *Plant Growth Regulation, 82*, 377–390

Lu, S., Sun, Y. H., Shi, R., Clark, C., Li, L., & Chiang, V. L. (2005). Novel and mechanical stress–responsive microRNAs in *Populus trichocarpa* that are absent from Arabidopsis. *The Plant Cell*, *17*, 2186–2203.

Lv, D. K., Bai, X., Li, Y., Ding, X. D., Ge, Y., Cai, H., & Zhu, Y. M. (2010). Profiling of cold-stress-responsive miRNAs in rice by microarrays. *Gene, 459*, 39–47.

Ma, Y., Dias, M. C., & Freitas, H. (2020). Drought and salinity stress responses and microbe-induced tolerance in plants. *Frontiers in Plant Science, 11*, 1750.

Mallory, A. C., Bartel, D. P., & Bartel, B. (2005). MicroRNA-directed regulation of Arabidopsis AUXIN RESPONSE FACTOR17 is essential for proper development and modulates expression of early auxin response genes. *The Plant Cell, 17*, 1360–1375.

Meng, Y., Ma, X., Chen, D., Wu, P., & Chen, M. (2010). MicroRNA-mediated signaling involved in plant root development. *Biochemical and Biophysical Research Communications, 393*, 345–349.

Negrao, S., Schmöckel, S. M., & Tester, M. (2017). Evaluating physiological responses of plants to salinity stress. *Annals of Botany, 119*, 1–11.

Ning, L. H., Du, W. K., Song, H. N., Shao, H. B., Qi, W. C., Sheteiwy, M. S. A., Yu, D. Y. (2019). Identification of responsive miRNAs involved in combination stresses of phosphate starvation and salt stress in soybean root. *Environmental and Experimental Botany, 167*, 103823.

Ojolo, S. P., Cao, S., Priyadarshani, S. V. G. N., Li, W., Yan, M., Aslam, M., & Qin, Y. (2018). Regulation of plant growth and development: A review from a chromatin remodeling perspective. *Frontiers in Plant Science, 9*, 1232.

Parmar, S., Gharat, S. A., Tagirasa, R., Chandra, T., Behera, L., Dash, S. K., Shaw, B. P. (2020). Identification and expression analysis of miRNAs and elucidation of their role in salt tolerance in rice varieties susceptible and tolerant to salinity. *PLoS One, 15*, e0230958.

Pegler, J. L., Nguyen, D. Q., Grof, C. P., & Eamens, A. L. (2020). Profiling of the salt stress responsive microRNA landscape of C4 genetic model species *Setaria viridis* (L.) Beauv. *Agronomy, 10*, 837.

Qin, Y., Duan, Z., Xia, X., & Yin, W. (2011). Expression profiles of precursor and mature microRNAs under dehydration and high salinity shock in Populus euphratica. *Plant Cell Reports, 30*, 1893–1907.

Ren, Y., Chen, L., Zhang, Y., Kang, X., Zhang, Z., & Wang, Y. (2013). Identification and characterization of salt-responsive microRNAs in Populus tomentosa by high-throughput sequencing. *Biochimie, 95*, 743–750.

Samad, A. F., Sajad, M., Nazaruddin, N., Fauzi, I. A., Murad, A., Zainal, Z., & Ismail, I. (2017). MicroRNA and transcription factor: Key players in plant regulatory network. *Frontiers in Plant Science, 8*, 565.

Shah, S. H., Islam, S., Parrey, Z. A., & Mohammad, F. (2021). Role of exogenously applied plant growth regulators in growth and development of edible oilseed crops under variable environmental conditions: A review. *Journal of Soil Science and Plant Nutrition*, 1–25.

Shan, T., Fu, R., Xie, Y., Chen, Q., Wang, Y., Li, Z., & Wang, B. (2020). Regulatory mechanism of maize (*Zea mays* L.) miR164 in salt stress response. *Russian Journal of Genetics, 56*, 835–842.

Sharif, I., Aleem, S., Farooq, J., Rizwan, M., Younas, A., Sarwar, G., & Chohan, S. M. (2019). Salinity stress in cotton: Effects, mechanism of tolerance and its management strategies. *Physiology and Molecular Biology of Plants, 25*, 807–820.

Song, X., Li, Y., Cao, X., & Qi, Y. (2019). MicroRNAs and their regulatory roles in plant–environment interactions. *Annual Review of Plant Biology, 70*, 489–525.

Sun, X., Xu, L., Wang, Y., Yu, R., Zhu, X., Luo, X., & Liu, L. (2015). Identification of novel and salt-responsive miRNAs to explore miRNA-mediated regulatory network of salt stress response in radish (*Raphanus sativus* L.). *BMC Genomics, 16*, 1–16.

Sun, Z., Wang, Y., Mou, F., Tian, Y., Chen, L., Zhang, S., & Li, X. (2016). Genome-wide small RNA analysis of soybean reveals auxin-responsive microRNAs that are differentially expressed in response to salt stress in root apex. *Frontiers in Plant Science, 6*, 1273.

Sunkar, R. (2010). MicroRNAs with macro-effects on plant stress responses. *Seminars in Cell & Developmental Biology*, 805–811.

Sunkar, R., Kapoor, A., & Zhu, J. K. (2006). Posttranscriptional induction of two Cu/Zn superoxide dismutase genes in Arabidopsis is mediated by downregulation of miR398 and important for oxidative stress tolerance. *The Plant Cell, 18*, 2051–2065.

Sunkar, R., Li, Y. F., & Jagadeeswaran, G. (2012). Functions of microRNAs in plant stress responses. *Trends in Plant Science, 17*, 196–203.

Sunkar, R., & Zhu, J. K. (2004). Novel and stress-regulated microRNAs and other small RNAs from Arabidopsis. *The Plant Cell, 16*, 2001–2019.

Tian, Y., Tian, Y., Luo, X., Zhou, T., Huang, Z., Liu, Y., & Yao, K. (2014). Identification and characterization of microRNAs related to salt stress in broccoli, using high-throughput sequencing and bioinformatics analysis. *BMC Plant Biology, 14*, 1–13.

Tuteja, N. (2007). Mechanisms of high salinity tolerance in plants. *Methods in Enzymology, 428*, 419–438.

Voinnet, O. (2009). Origin, biogenesis, and activity of plant microRNAs. *Cell, 136*, 669–687.

Wang, F., Liu, R., Wu, G., Lang, C., Chen, J., & Shi, C. (2012). Specific downregulation of the bacterial-type PEPC gene by artificial microRNA improves salt tolerance in Arabidopsis. *Plant Molecular Biology Reporter, 30*, 1080–1087.

Wang, J., Mei, J., & Ren, G. (2019). Plant microRNAs: Biogenesis, homeostasis, and degradation. *Frontiers in Plant Science, 10*, 360.

Wang, T., Pan, H., Wang, J., Yang, W., Cheng, T., & Zhang, Q. (2014). Identification and profiling of novel and conserved microRNAs during the flower opening process in *Prunus mume* via deep sequencing. *Molecular Genetics and Genomics, 289*, 169–183.

Wang, Y., Zhang, C., Hao, Q., Sha, A., Zhou, R., Zhou, X., & Yuan, L. (2013). Elucidation of miRNAs-mediated responses to low nitrogen stress by deep sequencing of two soybean genotypes. *PLoS One, 8*, e67423.

Wani, S. H., Kumar, V., Khare, T., Tripathi, P., Shah, T., Ramakrishna, C., & Mangrauthia, S. K. (2020). miRNA applications for engineering abiotic stress tolerance in plants. *Biologia, 75*, 1063–1081.

Wen, F. L., Yue, Y., He, T. F., Gao, X. M., Zhou, Z. S., & Long, X. H. (2020). Identification of miR390-TAS3-ARF pathway in response to salt stress in *Helianthus tuberosus* L. *Gene 738*, 144460.

Wu, G. (2013). Plant microRNAs and development. *Journal of Genetics and Genomics, 40*, 217–230.

Wu, Y., Guo, J., Cai, Y., Gong, X., Xiong, X., Qi, W., & Wang, Y. (2016). Genome-wide identification and characterization of *Eutrema salsugineum* microRNAs for salt tolerance. *Plant Physiology, 157*, 453–468.

Xu, D. Q., Huang, J., Guo, S. Q., Yang, X., Bao, Y. M., Tang, H. J., & Zhang, H. S. (2008). Overexpression of a TFIIIA-type zinc finger protein gene ZFP252 enhances drought and salt tolerance in rice (*Oryza sativa* L.). *FEBS Letters, 582*, 1037–1043.

Xu, X., Yin, L., Ying, Q., Song, H., Xue, D., Lai, T., & Shi, X. (2013). High-throughput sequencing and degradome analysis identify miRNAs and their targets involved in fruit senescence of *Fragaria ananassa*. *PLoS One, 8*, e70959.

Yan, S., Tang, Z., Su, W., & Sun, W. (2005). Proteomic analysis of salt stress-responsive proteins in rice root. *Proteomics, 5*, 235–244.

Yang, T., Xue, L., & An, L. (2007). Functional diversity of miRNA in plants. *Plant Science, 172*, 423–432.

Yousuf, P. Y., Ahmad, A., Aref, I. M., Ozturk, M., Ganie, A. H., & Iqbal, M. (2016a). Salt-stress-responsive chloroplast proteins in Brassica juncea genotypes with contrasting salt tolerance and their quantitative PCR analysis. *Protoplasma, 253*, 1565–1575.

Yousuf, P. Y., Ahmad, A., Ganie, A. H., & Iqbal, M. (2016b). Salt stress-induced modulations in the shoot proteome of *Brassica juncea* genotypes. *Environmental Science and Pollution Research, 23*, 2391–2401.

Yousuf, P. Y., Ahmad, A., Hemant Ganie, A. H., Aref, I. M., & Iqbal, M. (2015). Potassium and calcium application ameliorates growth and oxidative homeostasis in salt-stressed Indian mustard (*Brassica juncea*) plants. *Pakistan Journal of Botany, 47*, 1629–1639.

Zhang, B. H., Pan, X. P., Wang, Q. L., George, P. C., & Anderson, T. A. (2005). Identification and characterization of new plant microRNAs using EST analysis. *Cell Research, 15*, 336–360.

Zhang, Q., Zhao, C., Li, M., Sun, W., Liu, Y., Xia, H., & Zhao, Y. (2013). Genome-wide identification of *Thellungiella salsuginea* microRNAs with putative roles in the salt stress response. *BMC Plant Biology, 13*, 1–13.

Zhang, Z., Wei, L., Zou, X., Tao, Y., Liu, Z., & Zheng, Y. (2008). Submergence-responsive microRNAs are potentially involved in the regulation of morphological and metabolic adaptations in maize root cells. *Annals of Botany, 102*, 509–519.

Zhang, Z., Yu, J., Li, D., Zhang, Z., Liu, F., Zhou, X., & Su, Z. (2010). PMRD: Plant microRNA database. *Nucleic Acids Research, 38*, 806–813.

Zhao, B., Ge, L., Liang, R., Li, W., Ruan, K., Lin, H., & Jin, Y. (2009). Members of miR-169 family are induced by high salinity and transiently inhibit the NF-YA transcription factor. *BMC Molecular Biology, 10*, 1–10.

Zhao, B., Liang, R., Ge, L., Li, W., Xiao, H., Lin, H., & Jin, Y. (2007). Identification of drought-induced microRNAs in rice. *Biochemical and Biophysical Research Communications, 354*, 585–590.

Zhao, G., Yu, H., Liu, M., Lu, Y., & Ouyang, B. (2017). Identification of salt-stress responsive microRNAs from *Solanum lycopersicum* and *Solanum pimpinellifolium*. *Plant Growth Regulation, 83*, 129–140.

Zhou, M., Li, D., Li, Z., Hu, Q., Yang, C., Zhu, L., & Luo, H. (2013). Constitutive expression of a miR319 gene alters plant development and enhances salt and drought tolerance in transgenic creeping bent grass. *Plant Physiology, 161*, 1375–1391.

Zhu, J., Li, W., Yang, W., Qi, L., & Han, S. (2013). Identification of microRNAs in *Caragana intermedia* by high-throughput sequencing and expression analysis of 12 microRNAs and their targets under salt stress. *Plant Cell Reports, 32*, 1339–1349.

Zhuang, Y., Zhou, X. H., & Liu, J. (2014). Conserved miRNAs and their response to salt stress in wild eggplant *Solanum linnaeanum* roots. *International Journal of Molecular Sciences, 15*, 839–849.

6 Heavy Metal-Regulated miRNAs

Zubair Ahmad Parrey, Shaistul Islam, Sajad Hussain Shah, and Firoz Mohammad

6.1 INTRODUCTION

Plants being sessile are exposed to various environmental cues, such as heavy metals (HMs), drought, extreme temperature, salinity, sodicity, etc. (Islam & Mohammad, 2020). The repercussions imposed by these stresses reduce the plant growth and yield parameters. Among abiotic stresses, HM stress is a severe constraint for plants due to their ubiquitous availability in the soil as contamination. The primary sources of HM contamination are excessive usage of inorganic fertilizers, pesticides, industrial effluents, mining operations, and sewage sludge. Heavy metals adversely affect various metabolic processes of plants by disrupting ionic balance and membrane and cellular integrity of plant cells. Heavy metals such as iron (Fe), molybdenum (Mo), manganese (Mn), nickel (Ni), zinc (Zn), and copper (Cu) are required for the proper physio-biochemical processes of plants and are considered essential elements. However, elements like arsenic (As), aluminium (Al), cadmium (Cd), cobalt (Co), chromium (Cr), lead (Pb), and mercury (Hg) are non-essential but toxic metals causing various alterations in physiological processes at the extent of plant death. The HMs toxicity perturbs several aspects of plant growth and development. HMs produce reactive oxygen species, which causes membrane disorganization, lipid peroxidation, and denaturation of proteins, enzymes, and nucleic acids (Mendoza-Soto et al., 2012; Gill & Author, 2014; Ghori et al., 2019). Moreover, they negatively affect cell differentiation and the photosynthetic efficiency of plants (Cheng, 2003). Additionally, HMs cause deleterious effects through the oxidation and cross-linking with protein thiols, inhibition of the H^+-ATPase pump, and variation in the structure and fluidity of lipid membranes in plants (Gill & Author, 2014). Accumulation of Cd in soil interferes with the uptake of essential mineral elements by utilizing Fe and Zn transporter (Zeng et al., 2014). To combat HMs toxicity, plants have many intrinsic counter mechanisms, including modulation of uptake and translocation of HMs, metal chelation, ion homeostasis, osmoregulation, and activation of the antioxidant defence system (Singh et al., 2016; Saini et al., 2021). Apart from these, another counter mechanism is the production of microRNAs (miRNAs) which further strengthen the tolerance mechanisms of plants against HMs stress and is considered an important approach (Ding et al., 2016).

The microRNAs are a class of highly conserved, naturally occurring long non-coding RNA molecules consisting of about 22 nucleotides in most eukaryotes, including plants (Gupta et al., 2014a; Patel et al., 2019; Singh et al., 2021). They regulate stress-responsive genes, proteins, and transcription factors (TFs) (Singh et al., 2021). Many conserved and non-conserved miRNAs are actively involved in plant defence mechanisms. Their expression takes place in a spatiotemporally specific manner depending upon the exposure to different stresses (Shriram et al., 2016). Besides, miRNAs play an essential role in several metabolic and biological pathways such as phytohormones signal transduction, production of secondary metabolites, tissue development, and differentiation processes (Singh et al., 2020). The microRNAs act as essential gene-silencing machinery in eukaryotic organisms, including plants (Achkar et al., 2016). They regulate plant cell homeostasis against HM toxicity by activating specific messenger RNAs (mRNAs) via transcriptional or translational repression

DOI: 10.1201/9781003248453-6

(Achkar et al., 2016; Singh et al., 2020). Besides transcriptional and translational targeted gene control, miRNAs also regulate gene expression by epigenetic modifications like DNA and histone methylation (Wu et al., 2010). The miRNAs in plants are either upregulated or downregulated. This behaviour of miRNAs varies in different plant species and under various types of HM stress (He et al., 2014; Singh et al., 2021). In response to HM toxicities, various miRNAs undergo regulation in many plants such as *Oryza sativa* L., *Platanus acerifolia* L., *Brassica juncea* L., *Medicago truncatula* L., *Brassica napus* L., and *Phaseolus vulgaris* L. (Yang & Chen, 2013). Keeping the importance of miRNAs in the regulation of genes responsible for various physio-biochemical processes in plants, it is highly desirable to study their roles in detail in plants. This chapter describes the biogenesis of miRNAs and the roles of miRNAs in plants, including miRNAs mediated phytohormones signaling under HM stress and miRNA target genes under HM stress.

6.2   BIOGENESIS OF MIRNAS

Plant miRNAs are encoded by independent microRNA genes (MIR) that are transcribed into a long primary miRNA (pri-miRNA) with a hairpin-like structure by RNA pol II (Ding et al., 2020). The RNA binding dawdle (DDL) protein stabilizes pri-miRNA. The DDL protein also converts pri-miRNA to precursor miRNA (pre-miRNA). In some circumstances, introns present downstream of the miRNA precursor can boost its processing when these transcribed regions are spliced down (Bielewicz et al., 2013). The nuclear RNase dicer-like1 (DCL1) and its associated proteins serrate (SE), hyponastic leaves (HYL1), and cap-binding proteins (CBC) process it in two phases, changing it from pri-miRNA to pre-miRNA. The first cut cleaves pri-miRNA into an intermediate distinctive hairpin-like structure, whereas the second cut by DCL1, HYL1, and SE cleaves the nucleotide patch of miRNA/miRNA duplexes, that is, double-stranded RNA and a mature miRNA (Cenik & Zamore, 2011). These miRNA/miRNA duplexes are protected by hua enhancer 1 protein (HEN1) methylation capping at the 3' terminus and then transported into the cytoplasm by an exportin hasty 1 (HST1) protein (Jatan & Lata, 2019). The mature miRNA duplex's guide strand is integrated into an argonaute (AGO), the catalytic protein complex in the cytoplasm, to produce the RNA-induced silencing complex (RISC) (Manavella et al., 2019). Using sequence complementarity, the AGO directs the RISC complex to connect with the target mRNAs, resulting in transcriptional or translational inhibition (Ding et al., 2020). HMs in plants alter the genes involved in miRNA synthesis. DCL1 and AGO1 are required for miRNA biosynthesis in plants and are regulated by HMs (Ding et al., 2011).

6.3   ROLE OF MIRNAS IN PLANTS

The microRNAs play a crucial role in regulating various physio-biochemical and molecular processes of crop plants to cope with adversities caused by HMs stress (Manavella et al., 2019). They regulate transcription by gene silencing and complementary base pairing with their target mRNAs in plants under HMs stress (Yang & Chen, 2013). Moreover, miRNAs regulate the gene *GGT*, which encodes the enzyme glutathione-γ-glutamyl cysteinyl transferase. This enzyme coordinates with the phytochelatin synthase to reduce HMs reactivity in plant cells (Zhang et al., 2013). Fujii et al. (2005) observed that miR399 downregulates ubiquitin-conjugating enzyme (UBC) under less phosphate stress. The downregulation of UBC causes primary root elongation in *Arabidopsis thaliana* L. Laufs et al. (2004) noticed that miRNA164 regulates lateral root development in *Arabidopsis thaliana* L. by controlling the expression of TFs like cup-shaped cotyledon 1 (CUC1) and CUC2. Further, Si-Ammour et al. (2011) observed that miR393 downregulates the expression of transport inhibitor response 1 (*TIR1*), which involves the lateral root and leaf development in *Arabidopsis thaliana* L. Mallory et al. (2005) suggested that miR160 negatively regulates the expression of auxin response factors (*ARFs*) that control the formation of root crown cells in *Arabidopsis thaliana* L.

Lauter et al. (2005) observed that miR172 promotes phase transition from vegetative to flowering by the downregulating *glossy 15* genes in *Zea mays* L. Xu et al. (2013) investigated that miR159 and miR167 regulate genes of HMs transporters, such as natural resistance-associated macrophage protein (*NRAMP*) and ATP- binding cassette transporter in *Brassica napus* L. Furthermore, miR268 inhibits *NRAMP3* (metal transporter encoder) and hamper the seedling growth of *Oryza sativa* L. under Cd stress (Ding et al., 2017). The miRNA5144 mediates the protein disulphide bond formation via targeting protein disulphide isomerase (*PDIL*) gene transcripts. *PDIL* is a key protein-folding catalyst in *Oryza sativa* L. (Xia et al., 2018). The miR160 targets three *ARF* genes, that is, *ARF10, ARF16*, and *ARF17*, which are actively involved in auxin-mediated responses in *Arabidopsis thaliana* L. In addition, miR164 and miR393 also target genes involved in the auxin signaling pathway (Mallory et al., 2005). The miR319 regulates the Teosinte Branched Cycloidea (*TCP*) that controls leaf senescence via JA biosynthesis in *Arabidopsis thaliana* L. The miR838 targets the lipase gene involved in oxylipin synthesis leading to the biosynthesis of JA (Srivastava et al., 2013). Besides regulating HMs and phytohormones, several miRNA families are also engaged in other environmental challenges, such as cold, drought, and salt stress. Gupta et al. (2014a) reported that miR159 and miR393 were downregulated in *Triticum aestivum* L. under cold and salt stress, which upregulate the target MYB and auxin F-box genes, respectively. Both of these target genes have a fundamental role in regulating the growth and development of plants. It may be concluded that miRNAs have a potential role in modulating the growth, root development, leaf senescence, and phase transition from vegetative to flowering and also help maintain the cellular phosphate homeostasis in plants.

6.4   MIRNAS MEDIATED PHYTOHORMONES SIGNALLING UNDER HM STRESS

Under abiotic stress, including HMs stress, miRNAs play an essential role in regulating several phytohormone signaling pathways. The miR167 and miR528 target *IAR3* and *IAR1* transcripts, respectively, to regulate auxin homeostasis by hydrolyzing reaction that converts inactive auxin into the active form (Li et al., 2011; Kinoshita et al., 2012). Moreover, miR167 targets two *ARF*s such as *ARF6* and *ARF8*, which regulate root and ovule development in *Arabidopsis thaliana* L. by auxin homeostasis (Nagpal et al., 2005; Gutierrez et al., 2009). Dubey et al. (2019) reported that miR160 is downregulated under Cr stress in *Oryza sativa* L. and induces the expression of its target gene *ARF*. The *ARF* elevates the expression of auxin required to combat Cr stress. The miR393 targets *TIR1*, an important component of multi-proteins, including F-box containing complex (SCF) E3 ubiquitin ligase. The SCF E3 ubiquitin ligase is responsible for the ubiquitination of AUX/IAA to regulate auxin signaling (Mallory et al., 2005; Yang & Chen, 2013). The miR168 targets the lipoxygenase (LOX1) gene and miR319 targets teosinte branched 1-cycloidea-pcf (*TCP*) and LOX2. The TCP and LOX genes regulate JA biosynthesis (Schommer et al., 2008; Palatnik et al., 2003; Greco et al., 2012). In addition, miR838 targets a lipase-encoding gene that plays a vital role in the biosynthesis of JA (Ding et al., 2020). The miR408 plays an active part in gibberellic acid (GA_3) signaling. The upregulation of miR408 regulates GA biosynthesis, promotes cell elongation, and enhances biomass production in *Arabidopsis thaliana* L. (Song et al., 2018). The miRNAs like miR1535b target the isopentyl transferase (*IPT*) gene involved in cytokinin biosynthesis (Fang et al., 2013). He et al. (2016) reported that miR19 and miR20 target the S-adenosyl-methionine-sterol-C-methyltransferase and serine (Ser)/threonine (Thr) protein phosphatase respectively under Cd stress in *Nicotiana tabacum* L. The S-adenosyl-methionine-sterol-C-methyltransferase is an important protein for ethylene biosynthesis (Ludwików et al., 2009). Auxin and brassinosteroid cellular signaling pathways are regulated by Ser/Thr protein phosphatases (Dai et al., 2012). Dubey et al. (2019) reported the inverse correlation between auxin and MAPK signaling under HMs stress. Further, miR159 acts as an epigenetic regulator of auxin as it suppresses the MAPK cascade and induces the expression of auxin-related *IAA, ARF,* and *PIN* genes.

6.4.1 miRNA Target Genes Under HM Stress

microRNAs improve many physiological and molecular functions in plants, including transcription of target genes, protein folding, chelation of toxic metals, and an antioxidant defence system. Studies revealed that miRNAs regulate many gene expressions associated with tolerance mechanisms in plants under HMs stress [**Figure 6.1**]. The HMs-regulated miRNAs and their target gene are given in **Table 6.1**.

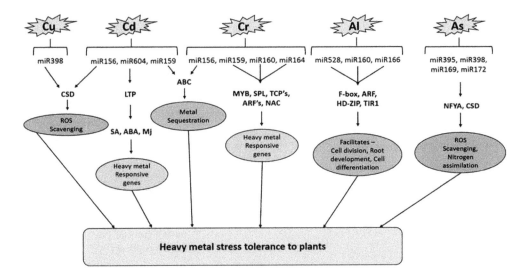

FIGURE 6.1   Heavy metals regulated miRNAs their target genes involved in various metabolic pathways.

TABLE 6.1

Heavy metal regulated miRNAs and their target genes.

miRNAs	Upregulation/ Downregulation	Plant species	Heavy metals	Target genes	Target function	References
miR156	Downregulation	*Medicago truncatula* L.	Hg	*SPL7* protein	Plant development and metal detoxification	Chen et al. (2012)
	Downregulation	*Oryza sativa* L.	Cd			Ding et al. (2011)
	Downregulation	*Brassica juncea* L.	As			Srivastava et al. (2013)
	Downregulation	*Zea mays* L.	Cd			Gao et al. (2019)
miR159	Downregulation	*Brassica napus* L.	Cd	*SPL7* protein	Plant development and metal ion homeostasis	Greco et al. (2012)
	Downregulation	*Brassica juncea* L.	As			Srivastava et al. (2013)
	Downregulation	*Medicago truncatula* L.	Al			Lima et al. (2011)
miR160	Downregulation	*Medicago truncatula* L.	Al	*ARF* transcription factors	Signalling pathway and floral development	Lima et al. (2011)
	Upregulation	*Oryza sativa* L.	Al			Greco et al. (2012)
	Downregulation	*Phaseolus vulgaris* L.	Mn			Valdés-López et al. (2010)

(Continued)

TABLE 6.1 (Continued)
Heavy metal regulated miRNAs and their target genes.

miRNAs	Upregulation/ Downregulation	Plant species	Heavy metals	Target genes	Target function	References
miR167	Upregulation	*Phaseolus vulgaris* L.	Mn	*ARF* TF and *NRAMP* metal transporter	Metal uptake and transport	Valdés-López et al. (2010)
	Upregulation	*Brassica juncea* L.	As			Srivastava et al. (2013)
	Downregulation	*Oryza sativa* L.	As			Liu and Zhang (2012)
	Downregulation	*Zea mays* L.	Cd			Gao et al. (2019)
miR168	Downregulation	*Oryza sativa* L.	Cd	ARGONAUTE, receptor like protein kinase 5, fructose bisphosphate aldolase	Signal transduction, stress response, miRNA processing	Ding et al. (2011)
	Upregulation	*Oryza sativa* L.	Al			Greco et al. (2012)
miR393	Upregulation	*Zea mays* L.	Cd	Transport inhibitor response 1/ auxin F-box	Plant development and response to defence and phosphate starvation, Auxin signalling, root development	Gao et al. (2019)
	Downregulation	*Oryza sativa* L.	As			Greco et al. (2012)
	Upregulation	*Medicago truncatula* L.	Hg			Chen et al. (2012)
miR394	Upregulation	*Brassica napus* L.	Cd	F- box transcription factor	Signal transduction and regulation of cell cycle	Huang et al. (2010)
miR397	Downregulation	*Platanus acerifolia* L.	Pb	Laccase, Laccase-4 precursor	Cu homeostasis and cell wall genesis	Ding et al. (2020)
	Upregulation	*Oryza sativa* L.	As			Liu and Zhang (2012)
	Downregulation	*Phaseolus vulgaris* L.	Mn			Valdés-López et al. (2010)
miR398	Upregulation	*Brassica napus* L.	Cd	*CSD*	ROS detoxification	Greco et al. (2012)
miR528	Upregulation	*Oryza sativa* L.	Cd	Cu^{2+} binding protein, *IAR1* proteins, L- ascorbate oxidase	Metal ion homeostasis, control of cellular free auxin levels and ascorbate metabolism	Ding et al. (2011)

6.4.1.1 Aluminium Stress

Al toxicity is a major limiting factor for plant growth and development by interfering with cellular redox reactions (Lima et al., 2011). Nearly 30–40% of soil is affected by Al accumulation (Gupta et al., 2014b). Plants under Al exposure have short root lengths, impaired water and mineral absorption, disproportionate Ca^{2+} level, and face severe oxidative stress (Silva, 2012). The Al stress regulates various miRNAs, which modulates several developmental processes in plants (Lima

et al., 2011). Lima et al. (2011) revealed that miR528, miR160, and miR166 are upregulated, while miR393 is downregulated in *Oryza sativa* L. under Al stress. The miR160, miR160, and miR166 target F-box, *ARF*, and home domain-leucine zipper (*HD-ZIP*) genes, respectively, and contribute fine control of root development under Al toxicity in *Oryza sativa* L. The miR160 is upregulated and targets *ARF10* and *ARF16* genes in *Arabidopsis thaliana* L. to combat the Al toxicity by inhibiting the root cap growth (Wang et al., 2005). Zeng et al. (2012) revealed that the upregulation of miR160 and miR1514 regulates their target genes *ARF* and *no apical meristem* (*NAM*) gene. Both target genes regulate the inhibition of root development under Al toxicity. Ye et al. (2021) reported that the overexpression of miR160 and miR393 negatively regulates their target gene expression *of auxin-signaling F-box* (*AFB*) and *TIR1*. The loss of function of *ARF10* and *ARF16* genes in *Arabidopsis thaliana* L. results in higher Al tolerance by expressing different protein-encoding genes such as pectin methyl esterase inhibitors (*PMEI*s) and xyloglucan endotransglucosylase/hydrolase (*XTH*s) (Yang et al., 2014). Therefore, auxin signaling associated with these miRNAs is a strategic approach of plants to detoxify Al stress. Silva et al. (2021) reported that the downregulation of miR395 upregulates its target 1-aminocyclopropane-1-carboxylate synthase (*ACS*) gene under Al stress in *Saccharum* spp. The *ACS* gene is linked with the stress regulatory network in plants. Furthermore, the downregulation of miR160 and miR6225 induces the expression of *ARF17* and *KIN12*, respectively. The *ARF17* is associated with auxin signaling and lateral root development in plants (Mallory et al., 2005), whereas *KIN12* plays an active role in cell expansion (Vanstraelen et al., 2006). The downregulation of miR166, miR390, and miR396 upregulate the expression of their target gene *HD-ZIP*, *leucine-rich repeat receptor-like kinase* (*LRR-RLK*) and *growth-regulating factor* (*GRF*) respectively under Al stress in *Glycine max* L. (Al-tolerant BX10). The *HD-ZIP* gene promotes root growth by facilitating cell division in *Arabidopsis thaliana* L. (Singh et al., 2014). *LRR-RLK* is involved in the outer cell layer's specification in root development plants (Song et al., 2008). Zhou et al. (2020) observed that the downregulation of miR160, miR477, and miR3627 upregulates its target genes *ARF*, *ral guanine nucleotide dissociation stimulator-like 1* (RGL1) and H⁺-ATPase gene, respectively under Al stress in *Citrus* spp. The ARF regulates root development, *RGL1* regulates JA and GA signaling cascade in *Citrus* plants under Al toxicity, whereas the H⁺-ATPase is involved to secrete citrate from root tips of *Citrus* spp. and confers higher Al tolerance (Chen et al., 2013).

6.4.1.2 Arsenic Stress

Pandey et al. (2015) revealed that the downregulation of miR395 and miR398 upregulate the target gene expression like cytochrome b5-like heme and copper/zinc superoxide dismutase (*CSD*) in *Oryza sativa* L. under As stress. The cytochrome b5-like heme controls lateral root development and regulates the sulphate assimilation pathway and phytochelatin biosynthesis, whereas *CSD* has an important role in antioxidant defence mechanisms in plants. Liu and Zhang (2012) revealed that the downregulation of miR169, miR172, miR1318, and miR1432 occurs under As stress in *Oryza sativa* L. The miR169 and miR172 upregulate the expression of their target nuclear factor Y subunit A (*NFYA*) and *AP2*, respectively, and miR1318 and miR1432 upregulate the expression of Ca²⁺-ATPase. The *NFYA* enhances nitrogen absorption, thereby improving the development and metabolism of plants. The Ca²⁺-ATPase improves many cellular processes by promoting influx or efflux of Ca²⁺ across biomembranes, while *AP2* plays a crucial role in the phase transition from juvenile to adult stage in plants. Srivastava et al. (2013) suggested that miR319, miR395, miR159, and miR838 are downregulated under As stress in *Brassica juncea* L. The downregulation of miR395 and miR838 upregulate their target genes *ATP sulfurylase/serine acetyltransferase* (*APS/SAT*); miR319 and miR838 upregulate *TCP*, and miR159 upregulates 1-amino-cyclopropane-1-carboxylate synthase (*ACC synthase*). The *APS/SAT* increases sulphate uptake and its assimilation in *Arabidopsis thaliana* L.; *TCP* increases the biosynthesis of jasmonates; *ACC* synthase plays a pivotal role in the activation of defence responses (Liang et al., 2010). Thus, all these target genes play a significant role in increasing the tolerance of plants against As stress.

6.4.1.3 Cadmium Stress

The miRNAs minimize Cd toxicity by improving the defence mechanism of crop plants. Zhou et al. (2012) reported that miR156, miR171, and miR396a downregulation induce the expression of *CSD1/CSD2* transcripts, which modulate the plant mechanism in *Brassica napus* L. to mitigate the damages caused by Cd toxicity. Ding et al. (2018) revealed that the upregulation of miR166 transcript in *Oryza sativa* L. under Cd stress significantly improves chlorophyll content, decreases membrane oxidation, and reduces ROS accumulation and Cd uptake translocation. The miR604 under Cd toxicity is downregulated and, in turn, upregulates the lipid transfer protein (LTP) (Gupta et al., 2014b). The LTP has a vital role in mediating various hormonal signaling pathways under HMs response like salicylic acid (SA), abscisic acid (ABA), ethylene, and methyl jasmonate (MJ) in plants (Kim et al., 2006; Huang et al., 2010). The downregulation of miR159, miR192, and miR167 under Cd stress upregulates the expression of its target genes ABC and NRAMP (Ding et al., 2013; Gupta et al., 2014b). Both *ABC* and *NRAMP* transporters have an important role in the sequestration of Cd to reduce its accumulation in plant cells (Klein et al., 2006). Ding et al. (2011) revealed that the downregulation of miR166 under Cd stress upregulates its target *HD-ZIP* transcription factor in *Oryza sativa* L. The *HD-ZIP* plays a crucial role in leaf polarity and lateral root formation of plants (Hawker & Bowman, 2004; Jones-Rhoades et al., 2006). The downregulation of miR171 under Cd stress upregulates its target Scarecrow TF in *Oryza sativa* L. (Ding et al., 2011). This scarecrow TF controls the floral development in plants (Reinhart et al., 2002). Ding et al. (2011) revealed the downregulated expression of miR444 under Cd stress upregulates its target gene MADS-box TF in *Oryza sativa* L. The TF has many biological functions in plants, including coping with saline and cold stress conditions (Lozano et al., 1998). Greco et al. (2012) reported that the downregulation of miR164 induces the expression of its target gene *monothiol glutaredoxin-S12* (*GRXS*12) under Cd stress in *Brassica napus* L. The *GRXS*12 maintains redox homeostasis in plant cells and prevents them from oxidative stress (Zaffagnini et al., 2012). The miR528 expression is induced under Cd stress in plants (Min Yang & Chen, 2013). The miR528 targets gene code for an ascorbate oxidase to enhance the antioxidant defence system by maintaining an ascorbate redox state (Wu et al., 2009). He et al. (2016) investigated that the downregulation of miR159, miR482, and miR27 under Cd stress upregulates its target genes *TGA* transcription factor, glucose-6-phosphate isomerase (*GPI*) and cytochromes P450 (*CYP*), respectively, in *Nicotiana tabacum* L. The TGA transcription factor regulates redox signaling, GPI enhances the expression of genes involved in antioxidant metabolism, and CYP participates in several biochemical pathways to produce primary and secondary metabolites (alkaloids, cyanogenic glycosides, phenyl propanoids, terpenoids, and glucosinolates) (Kesarwani et al., 2007; Seong et al., 2013; Mizutani & Ohta, 2010). Ding et al. (2016) reported that the downregulation of miR390 expression under Cd stress upregulates its target gene *stress-responsive LRR-like kinase* (*SRKs*) in *Oryza sativa* L. The SRKs are transmembrane proteins that activate a wide array of signaling mechanisms in plants (Tichtinsky et al., 2003; Chae et al., 2009).

6.4.1.4 Chromium Stress

Chromium is continuously accumulated via anthropogenic activities in the environment and is posing a severe threat to overall growth and development patterns by hampering the transportation of water, nutrients, and metabolic activities (such as photosynthesis and respiration) in plants (Singh et al., 2021). For mitigation of oxidative damage due to Cr toxicity, several miRNAs apart from signaling molecules and hormones, like AUX signaling F-box protein, JA, and Ca^{2+}-dependent protein kinase, are involved (Liu et al., 2015). They also revealed the downregulation of several miRNAs like miR159, miR160, and miR164 in *Raphanus sativus* L. under Cr stress. Moreover, downregulation of miRNAs induces the expression of Myb domain protein (*MYBs*), Squamosa promoter-binding-like protein (*SPLs*), TCPs, ARFs, and NAC domain transcription factor (*NACs*), which mediate the expression of HMs responsive genes and transporters to alleviate Cr toxicity in *Raphanus sativus* L. Nie et al. (2021) observed that miR156, miR164, and miR396 under Cr stress induce the expression of target genes *CYP*, sulphate transmembrane transporter and mitogen-activated protein kinase (*MAPK*), respectively, leading to an improvement in physio-biochemical and antioxidant activities in *Miscanthus sinensis* L. Dubey et al. (2019) revealed that the downregulation of miR160, miR169, and miR171 induces the expression of their target genes *ARF*, ATP-binding protein and *TGA*, respectively, under Cr stress in *Oryza sativa* L. This overexpression of these target genes improves the defence and detoxification machinery systems through auxin response, ATP binding cassette transporters, and metal ion transport. Bukhari et al. (2015) investigated that the downregulation of miR156, miR166, and miR167 positively regulates their targets *ABC*, *NRAMP*, and class III-*HD-ZIP* genes respectively under Cr stress in *Nicotiana tabacum* L. The *ABC* and *NRAMP* play a fundamental role in the uptake and translocation of metals in plants (Krämer et al., 2007), whereas class III-*HD-ZIP* regulates lateral root development in plants (Hawker & Bowman., 2004).

6.4.1.5 Copper Stress

The higher concentration of Cu restrains several physiological processes of plants, such as photosynthesis, membrane rigidity, and redox homeostasis of the cell (Noman & Aqeel, 2017). The miR398 targets S*SD* genes. This enzyme is involved in ROS scavenging to improve tolerance in plants under Cu stress (Fang et al., 2013). *CSD* maintains Cu concentration in cells (Yamasaki et al., 2007). The miR398 expression is downregulated under Cu stress, leading to *CSD1* and *CSD2* gene expression (Ding & Zhu, 2009; Noman & Aqeel, 2017). In contrast, miR398, miR397, miR408, and miR857 were found to be upregulated, which mediates the downregulation of the *CSD* gene under low concentrations of Cu (Yamasaki et al., 2007).

6.4.1.6 Lead Stress

Wang et al. (2015) revealed that the downregulation of miR45 and miR108 under Pb stress in *Platanus acerifolia* L. induces the expression of its target gene glutathione S-transferase. The glutathione S-transferase plays a role in ROS detoxification and phytochelatin formation in plants. Wang et al. (2015) reported that miR396 and miR166 are downregulated, whereas miR159 is upregulated under Pb stress in *Raphanus sativus* L. The target genes of miR396, miR166, and miR159, that is, F-box, *HD-ZIP*, and *TCPs*, respectively, are upregulated. These genes play an essential role in the overall growth and physiological functioning of plants under stress conditions.

6.5 CONCLUSION

The survey of the literature showed that miRNAs are key regulators in response to HMs in plants. From miRNA microarrays and high throughput sequencing, the data analysis revealed that the various miRNAs actively improve the metal stress tolerance in plants. The microRNAs are non-coding RNAs that bind with mRNAs and restrict their translation into proteins. They regulate the expression of many protein-coding genes by cleaving complementary base pairs of mRNAs and repressing their translation machinery. These miRNA-associated regulatory mechanisms result in functional improvement against the HM toxicity response in plants. microRNAs have potential roles in regulating their target gene expression and constituting an essential part of the regulatory network that controls different responses of plants against HM stress. In a nutshell, miRNAs promote stress tolerance by regulating various stress-responsive genes, proteins, transcription factors, and phytohormones.

REFERENCES

Achkar, N. P., Cambiagno, D. A., Manavella, P. A. (2016). miRNA biogenesis: A dynamic pathway. *Trends in Plant Science*, *21*, 1034–1044.

Bielewicz, D., Kalak, M., Kalyna, M., Windels, D., Barta, A., Vazquez, F., Szweykowska-Kulinska, Z., & Jarmolowski, A. (2013). Introns of plant pri-miRNAs enhance miRNA biogenesis. *EMBO Reports, 14*, 622–628.

Bukhari, S. A. H., Shang, S., Zhang, M., Zheng, W., Zhang, G., Wang, T. Z., Shamsi, I. H., & Wu, F. (2015). Genome-wide identification of chromium stress-responsive micro RNAs and their target genes in tobacco (Nicotiana tabacum) roots. *Environmental Toxicology and Chemistry, 34*, 2573–2582.

Cenik, E. S., & Zamore, P. D. (2011). Argonaute proteins. *Current Biology, 21*, R446–R449.

Chae, L., Sudat, S., Dudoit, S., Zhu, T., & Luan, S. (2009). Diverse transcriptional programs associated with environmental stress and hormones in the arabidopsis receptor-like kinase gene family. *Molecular Plant, 2*, 84–107.

Chen, L., Wang, T., Zhao, M., Tian, Q., & Zhang, W. H. (2012). Identification of aluminum-responsive microRNAs in Medicago truncatula by genome-wide high-throughput sequencing. *Planta, 235*, 375–386.

Chen, Q., Guo, C. L., Wang, P., Chen, X. Q., Wu, K. H., Li, K. Z., Yu, Y. X., & Chen, L. M. (2013). Up-regulation and interaction of the plasma membrane H+-ATPase and the 14-3-3 protein are involved in the regulation of citrate exudation from the broad bean (Vicia faba L.) under Al stress. *Plant Physiology and Biochemistry, 70*, 504–511.

Cheng, S. (2003). Effects of heavy metals on plants and resistance mechanisms. *Environmental Science and Pollution Research, 6*, 256–264.

Dai, M., Terzaghi, W., & Wang, H. (2012).Multifaceted roles of Arabidopsis PP6 phosphatase in regulating cellular signaling and plant development. *Plant Signaling & Behavior, 8*, 42–46.

Ding, Y., Chen, Z., & Zhu, C. (2011). Microarray-based analysis of cadmium-responsive microRNAs in rice (Oryza sativa). *Journal of Experimental Botany, 62*, 3563–3573.

Ding, Y., Ding, L., Xia, Y., Wang, F., & Zhu, C. (2020). Emerging roles of microRNAs in plant heavy metal tolerance and homeostasis. *Journal of Agricultural and Food Chemistry, 68*, 1958–1965.

Ding, Y., Gong, S., Wang, Y., Wang, F., Bao, H., Sun, J., Cai, C., Yi, K., Chen, Z., & Zhu, C. (2018). MicroRNA166 modulates cadmium tolerance and accumulation in rice. *Plant Physiology, 177*, 1691–1703.

Ding, Y., Qu, A., Gong, S., Huang, S., Lv, B., & Zhu, C. (2013). Molecular Identification and Analysis of Cd-Responsive MicroRNAs. *Journal of Agricultural and Food Chemistry, 61*(47), 11668–11675.

Ding, Y., Wang, Y., Jiang, Z., Wang, F., Jiang, Q., Sun, J., Chen, Z., & Zhu, C. (2017). MicroRNA268 overexpression affects rice seedling growth under cadmium stress. *Journal of Agricultural and Food Chemistry, 65*, 5860–5867.

Ding, Y., Ye, Y., Jiang, Z., Wang, Y., & Zhu, C. (2016). MicroRNA390 is involved in cadmium tolerance and accumulation in rice. *Frontiers in Plant Science, 7*, 1–10.

Ding, Y. F., & Zhu, C. (2009). The role of microRNAs in copper and cadmium homeostasis. *Biochemical and Biophysical Research Communications, 368*, 6–10.

Fang, X., Zhao, Y., Ma, Q., Huang, Y., Wang, P., Zhang, J., Nian, H., & Yang, C. (2013). Identification and comparative analysis of cadmium tolerance-associated miRNAs and their targets in two soybean genotypes. *PLoS One, 8*, 28–34.

Fujii, H., Chiou, T. J., Lin, S. I., Aung, K., & Zhu, J. K. (2005). A miRNA involved in phosphate-starvation response in Arabidopsis. *Current Biology, 15*, 2038–2043.

Gao, J., Luo, M.,; Peng, H., Chen, F., & Li, W. (2019). Characterization of cadmium-responsive MicroRNAs and their target genes in maize (Zea mays) roots. *BMC Molecular Biolog, 20*, 1–9.

Ghori, N. H., Ghori, T., Hayat, M. Q., Imadi, S. R., Gul, A., Altay, V., & Ozturk, M. (2019). Heavy metal stress and responses in plants. *International Journal of Environmental Science and Technology, 16*, 1807–1828.

Gill, M., Author, C. (2014). Heavy metal stress in plants: A review. *International Journal of Advanced Research, 2*, 1043–1055.

Greco, M., Chiappetta, A., Bruno, L., & Bitonti, M. B. (2012). In Posidonia oceanica cadmium induces changes in DNA methylation and chromatin patterning. *Journal of Experimental Botany, 63*, 695–709.

Gupta, O. P., Meena, N. L., Sharma, I., & Sharma, P. (2014a). Differential regulation of microRNAs in response to osmotic, salt and cold stresses in wheat. *Molecular Biology Reports, 41*, 4623–4629.

Gupta, O. P., Sharma, P., Gupta, R. K., & Sharma, I. (2014b). MicroRNA mediated regulation of metal toxicity in plants: Present status and future perspectives. *Plant Molecular Biology, 84*, 1–18.

Gutierrez, L., Bussell, J. D., Păcurar, D. I., Schwambach, J., Păcurar, M., & Bellini, C. (2009). Phenotypic plasticity of adventitious rooting in arabidopsis is controlled by complex regulation of auxin response factor transcripts and microRNA abundance. *Plant Cell, 21*, 3119–3132.

Hawker, N. P., & Bowman, J. L. (2004). Roles for class III HD-zip and Kanadi genes in arabidopsis root development. *Plant Physiology, 135*, 2261–2270.

He, Q., Zhu, S., & Zhang, B. (2014). MicroRNA-target gene responses to lead-induced stress in cotton (Gossypium hirsutum L.). *Functional & Integrative Genomics, 14*, 507–515.

He, X., Zheng, W., Cao, F., & Wu, F. (2016). Identification and comparative analysis of the microRNA transcriptome in roots of two contrasting tobacco genotypes in response to cadmium stress. *Scientific Reports, 6*, 1–14.

Huang, S. Q., Xiang, A. L., Che, L. L., Chen, S., Li, H., Song, J. B., & Yang, Z. M. (2010). A set of miRNAs from Brassica napus in response to sulphate deficiency and cadmium stress. *Plant Biotechnology Journal*, *8*, 887–899.

Hyun Kim, T., Chul Kim, M., Ho Park, J., Sook Han, S., Ryun Kim, B., Yong Moon, B., Chung Suh, M., & Ho Cho, S. (2006). Differential expression of rice lipid transfer protein gene (LTP) classes in response to abscisic acid, salt, salicylic acid, and the fungal pathogen magnaporthe grisea. *Journal of Plant Biology*, *49*, 371–375.

Islam, S., & Mohammad, F. (2020). Triacontanol as a dynamic growth regulator for plants under diverse environmental conditions. Physiology *and Molecular Biology of Plants*, *26*, 871–883.

Jatan, R., & Lata, C. (2019). Role of MicroRNAs in abiotic and biotic stress resistance in plants. *Proceedings of the Indian National Science Academy*, *85*, 553–567.

Jones-Rhoades, M. W., Bartel, D. P., & Bartel, B. (2006). MicroRNAs and their regulatory roles in plants. *Annual Review of Plant Biology*, *57*, 19–53.

Kesarwani, M., Yoo, J., & Dong, X. (2007). Genetic interactions of TGA transcription factors in the regulation of pathogenesis-related genes and disease resistance in Arabidopsis. *Plant Physiology*, *144*, 336–346.

Kinoshita, N., Wang, H., Kasahara, H., Liu, J., MacPherson, C., Machida, Y., Kamiya, Y., Hannah, M. A., & Chuaa, N. H. (2012). IAA-Ala Resistant3, an evolutionarily conserved target of miR167, mediates Arabidopsis root architecture changes during high osmotic stress. *Plant Cell*, *24*, 3590–3602.

Klein, M., Burla, B., & Martinoia, E. (2006). The multidrug resistance-associated protein (MRP/ABCC) subfamily of ATP-binding cassette transporters in plants, *580*, 1112–1122.

Krämer, U., Talke, I. N., & Hanikenne, M. (2007). Transition metal transport. *FEBS Letters*, *581*, 2263–2272.

Laufs, P., Peaucelle, A., Morin, H., & Traas, J. (2004). MicroRNA regulation of the CUC genes is required for boundary size control in Arabidopsis meristems. *Development*, *131*, 4311–4322.

Lauter, N., Kampani, A., Carlson, S., Goebel, M., & Moose, S. P. (2005). microRNA172 down-regulates glossy15 to promote vegetative phase change in maize. *Proceedings of the National Academy of Sciences of the United States of America*, *102*, 9412–9417.

Li, T., Li, H., Zhang, Y. X., & Liu, J. Y. (2011). Identification and analysis of seven H2O2-responsive miRNAs and 32 new miRNAs in the seedlings of rice (Oryza sativa L. ssp. indica). *Nucleic Acids Research*, *39*, 2821–2833.

Liang, G., Yang, F., & Yu, D. (2010). Mediates regulation of sulfate accumulation and allocation in Arabidopsis thaliana. *Plant Journal*, *62*, 1046–1057.

Lima, J. C., Arenhart, R. A., Margis-Pinheiro, M., & Margis, R. (2011). Aluminum triggers broad changes in microRNA expression in rice roots. *Genetics and Molecular Research*, *10*, 2817–2832.

Liu, Q., & Zhang, H. (2012). Molecular Identification and Analysis of Arsenite Stress-Responsive. *Journal of Agricultural and Food Chemistry*, *60*, 6524–6536.

Liu, W., Xu, L., Wang, Y., Shen, H., Zhu, X., Zhang, K., Chen, Y., Yu, R., Limera, C., & Liu, L. (2015). Transcriptome-wide analysis of chromium-stress responsive microRNAs to explore miRNA-mediated regulatory networks in radish (Raphanus sativus L.). *Scientific Reports*, *5*, 1–17.

Lozano, R., Angosto, T., Gómez, P., Payán, C.,; Capel, J., Huijser, P., Salinas, J., Martinez-Zapater, J. M. (1998). Tomato flower abnormalities induced by low temperatures are associated with changes of expression of MADS-box genes. *Plant Physiology*, *117*, 91–100.

Ludwików, A., Kierzek, D., Gallois, P., Zeef, L., & Sadowski, J. (2009). Gene expression profiling of ozone-treated Arabidopsis abi1td insertional mutant: Protein phosphatase 2C ABI1 modulates biosynthesis ratio of ABA and ethylene. *Planta*, *230*, 1003–1017.

Mallory, A. C., Bartel, D. P., & Bartel, B. (2005). MicroRNA-directed regulation of Arabidopsis auxin response factor17 is essential for proper development and modulates expression of early auxin response genes. *Plant Cell*, *17*, 1360–1375.

Manavella, P. A., Yang, S. W., & Palatnik, J. (2019). Keep calm and carry on: miRNA biogenesis under stress. *Plant Journal*, *99*, 832–843.

Mendoza-Soto, A. B., Sánchez, F., & Hernández, G. (2012). MicroRNAs as regulators in plant metal toxicity response. *Frontiers in Plant Science*, *3*, 1–7.

Min Yang, Z., & Chen, J. (2013). A potential role of microRNAs in plant response to metal toxicity. *Metallomics*, *5*, 1184–1190.

Mizutani, M., & Ohta, D. (2010). Diversification of P450 genes during land plant evolution. *Annual Review of Plant Biology*, *61*, 197–204.

Nagpal, P., Ellis, C. M., Weber, H., Ploense, S. E., Barkawi, L. S., Guilfoyle, T. J., Hagen, G., Alonso, J. M., Cohen, J. D., Farmer, E. E., Ecker, J. R., & Reed, J. W. (2005). Auxin response factors ARF6 and ARF8 promote jasmonic acid production and flower maturation. *Development*, *132*, 4107–4118.

Nie, G., Liao, Z., Zhong, M., Zhou, J., Cai, J., Liu, A., Wang, X., & Zhang, X. (2021). MicroRNA-mediated responses to chromium stress provide insight into tolerance characteristics of miscanthus sinensis. *Frontiers in Plant Science, 12*, 666117

Noman, A., & Aqeel, M. (2017). miRNA-based heavy metal homeostasis and plant growth. *Environmental Science and Pollution Research, 24*, 10068–10082.

Palatnik, J. F., Allen, E., Wu, X., Schommer, C., Schwab, R., Carrington, J. C., & Weigel, D. (2003).Control of leaf morphogenesis by microRNAs. *Nature, 425*, 257–263.

Pandey, C., Raghuram, B., Sinha, A. K., & Gupta, M. (2015). MiRNA plays a role in the antagonistic effect of selenium on arsenic stress in rice seedlings. *Metallomics, 7*, 857–866.

Patel, P., Yadav, K., Ganapathi, T. R., & Penna, S. (2019). *Plant miRNAome: Cross talk in abiotic stressful times: Vol. I.* Springer International Publishing.

Reinhart, B. J., Weinstein, E. G., Rhoades, M. W., Bartel, B., & Bartel, D. P. (2002). MicroRNAs in plants. *Genes & Development, 16*, 1616–1626.

Saini, S., Kaur, N., & Pati, P. K. (2021). Phytohormones: Key players in the modulation of heavy metal stress tolerance in plants. *Ecotoxicology and Environmental Safety, 223*, 112578.

Seong, E. S., Yoo, J. H., Lee, J. G., Kim, H. Y., Hwang, I. S., Heo, K., Lim, J. D., Lee, D. K., Sacks, E. J., & Yu, C. Y. (2013). Transient overexpression of the Miscanthus sinensis glucose-6-phosphate isomerase gene (MsGPI) in Nicotiana benthamiana enhances expression of genes related to antioxidant metabolism. *Plant OMICS, 6*, 408–414.

Shriram, V., Kumar, V., Devarumath, R. M., Khare, T. S., & Wani, S. H. (2016). Micrornas as potential targets for abiotic stress tolerance in plants. *Frontiers in Plant Science, 7*, 1–18.

Si-Ammour, A., Windels, D., Arn-Bouldoires, E., Kutter, C., Ailhas, J., Meins, F., & Vazquez, F. (2011). miR393 and secondary siRNAs regulate expression of the TIR1/AFB2 auxin receptor clade and auxin-related development of Arabidopsis leaves. *Plant Physiology, 157*, 683–691.

Silva, J. D. O. L., Silva, R. G. da., Nogueira, L. D. F., & Zingaretti, S. M. (2021). MicroRNAs regulate tolerance mechanisms in sugarcane (Saccharum spp.) under aluminum stress. *Crop Breeding and Applied Biotechnology, 21*, 1–8.

Silva, S. (2012). Aluminium toxicity targets in plants. *Journal of Botany, 5*, 1–8.

Singh, A., Singh, S., Panigrahi, K. C. S., Reski, R., & Sarkar, A. K. (2014). Balanced activity of microRNA166/165 and its target transcripts from the class III homeodomain-leucine zipper family regulates root growth in Arabidopsis thaliana. *Plant Cell Reports, 33*, 945–953.

Singh, D. K., Mehra, S., Chatterjee, S., & Purty, R. S. (2020). In silico identification and validation of miRNA and their DIR specific targets in Oryza sativa Indica under abiotic stress. *Non-Coding RNA Research, 5*, 167–177.

Singh, P., Dutta, P., & Chakrabarty, D. (2021). miRNAs play critical roles in response to abiotic stress by modulating cross-talk of phytohormone signaling. *Plant Cell Reports, 40*, 1617–1630.

Singh, S., Parihar, P., Singh, R., Singh, V. P., & Prasad, S. M. (2016). Heavy metal tolerance in plants: Role of transcriptomics, proteomics, metabolomics, and ionomics. *Frontiers in Plant Science, 6*, 1–36.

Song, D., Li, G., Song, F., & Zheng, Z. (2008). Molecular characterization and expression analysis of OsBISERK1, a gene encoding a leucine-rich repeat receptor-like kinase, during disease resistance responses in rice. *Molecular Biology Reports, 35*, 275–283.

Song, Z., Zhang, L., Wang, Y., Li, H., Li, S., Zhao, H., & Zhang, H. (2018). Constitutive expression of mir408 improves biomass and seed yield in arabidopsis. *Frontiers in Plant Science, 8*, 1–14.

Srivastava, S., Chiappetta, A., & Beatrice, M. (2013). Identification and profiling of arsenic In Posidonia oceanica cadmium induces changes in DNA microRNAs in Brassica methylation and patterning. *Journal of Experimental Botany, 64*, 303–315.

Tichtinsky, G., Vanoosthuyse, V., Cock, J. M., & Gaude, T. (2003). Making inroads into plant receptor kinase signalling pathways. *Trends in Plant Science, 8*, 231–237.

Valdés-López, O., Yang, S. S., Aparicio-Fabre, R., Graham, P. H., Reyes, J. L., Vance, C. P.& Hernández, G. (2010). GMicroRNA expression profile in common bean (Phaseolus vulgaris) under nutrient deficiency stresses and manganese toxicity. *New Phytologist, 187*, 805–818.

Vanstraelen, M., Inzé, D., & Geelen, D. (2006). Mitosis-specific kinesins in Arabidopsis. *Trends in Plant Science, 11*, 167–175.

Wang, J. W., Wang, L. J., Mao, Y. B., Cai, W. J., Xue, H. W., & Chen, X. Y. (2005).Control of root cap formation by MicroRNA-targeted auxin response factors in Arabidopsis. *Plant Cell, 17*, 2204–2216.

Wang, Y., Liu, W., Shen, H., Zhu, X., Zhai, L., Xu, L., Wang, R., Gong, Y., Limera, C., & Liu, L. (2015). Identification of radish (Raphanus sativus L.) miRNAs and their target genes to explore miRNA-mediated regulatory networks in lead (Pb) stress responses by high-throughput sequencing and degradome analysis. *Plant Molecular Biology Reporter, 33*, 358–376.

Wu, L., Zhang, Q., Zhou, H., Ni, F., Wu, X., & Qi, Y. (2009). Rice microrna effector complexes and targets. *Plant Cell*, *21*, 3421–3435.

Wu, L., Zhou, H., Zhang, Q., Zhang, J., Ni, F., Liu, C., & Qi, Y. (2010). DNA methylation mediated by a microRNA pathway. *Molecular Cell*, *38*, 465–475.

Xia, K., Zeng, X., Jiao, Z., Li, M., Xu, W., Nong, Q., Mo, H., Cheng, T., & Zhang, M. (2018). Formation of protein disulfide bonds catalyzed by OsPDIL1;1 is mediated by microRNA5144–3p in rice. *Plant and Cell Physiology*, *59*, 331–342.

Xu, L., Wang, Y., Zhai, L., Xu, Y., Wang, L., Zhu, X., Gong, Y., Yu, R., Limera, C., & Liu, L. (2013). Genome-wide identification and characterization of cadmium-responsive microRNAs and their target genes in radish (Raphanus sativus L.) roots. *Journal of Experimental Botany*, *64*, 4271–4287.

Yamasaki, H., Abdel-Ghany, S. E., Cohu, C. M., Kobayashi, Y., Shikanai, T., & Pilon, M. (2007). Regulation of copper homeostasis by micro-RNA in Arabidopsis. *Journal of Biological Chemistry*, *282*, 16369–16378.

Yang, Z. B., Geng, X., He, C., Zhang, F., Wang, R., Horst, W. J., & Ding, Z. (2014). TAA1-regulated local auxin biosynthesis in the root-apex transition zone mediates the aluminum-induced inhibition of root growth in Arabidopsis. *Plant Cell*, *26*, 2889–2904.

Ye, Z., Zeng, J., Long, L., Ye, L., & Zhang, G. (2021). Identification of microRNAs in response to low potassium stress in the shoots of Tibetan wild barley and cultivated. *Current Plant Biology*, *25*, 1–14.

Zaffagnini, M., Bedhomme, M., Marchand, C. H., Morisse, S., Trost, P., & Lemaire, S. D. (2012). Redox regulation in photosynthetic organisms: Focus on glutathionylation. *Antioxidants and Redox Signaling*, *16*, 567–586.

Zeng, H., Wang, G., Hu, X., Wang, H., Du, L., & Zhu, Y. (2014). Role of microRNAs in plant responses to nutrient stress. *Plant and Soil*, *374*, 1005–1021.

Zeng, Q. Y., Yang, C. Y., Ma, Q. Bin, Li, X. P., Dong, W. W., & Nian, H. (2012). Identification of wild soybean miRNAs and their target genes responsive to aluminum stress. *BMC Plant Biology*, *12*, 210–219.

Zhang, L. W., Song, J. B., Shu, X. X., Zhang, Y., & Yang, Z. M. (2013). MiR395 is involved in detoxification of cadmium in Brassica napus. *Journal of Hazardous Materials*, *250–251*, 204–211.

Zhou, H., Liu, Q., Li, J., Jiang, D., Zhou, L., Wu, P., Lu, S., Li, F., Zhu, L., Liu, Z., Chen, L., Liu, Y. G., & Zhuang, C. (2012). Photoperiod- and thermo-sensitive genic male sterility in rice are caused by a point mutation in a novel non-coding RNA that produces a small RNA. *Cell Research*, *22*, 649–660.

Zhou, Y. F., Wang, Y. Y., Chen, W. W., Chen, L. S., & Yang, L. T. (2020). Illumina sequencing revealed roles of microRNAs in different aluminum tolerance of two citrus species. *Physiology and Molecular Biology of Plants*, *26*, 2173–2187.

7 Tolerance to Radiation Stress in Plants With Reference to microRNAs

Sumira Malik, Shilpa Prasad, Rahul Kumar,
Shristi Kishore, and Nitesh Singh

7.1 INTRODUCTION

Radiation is the form of energy that emerges from a particular source and then proceeds by traveling through space with the speed of light, that is, 3×10^8 m/s (Percuoco, 2014). Electric as well as magnetic field radiation has wave-like characteristics. Radiation can also infer as "electromagnetic waves." The electromagnetic radiation obtained from solar radiation consists of various ranges of wavelength that emerge from the sun. The different ranges are visible light, ultraviolet radiation (UV), x-rays, and gamma (y) rays. Among them, x-rays and gamma rays have energy that is capable to interact with atoms and can remove electrons leading to ionized atoms (Percuoco, 2014). The intensity and quality of radiation play an important role in affecting the environment around us, and that also includes the growth and development of the whole plant. If the quality of radiation is inappropriate in any aspect of quality or quantity, then it acts as stress or radiation stress for the plant system. Solar radiation is an indispensable part of sustainable life on earth. Our biosphere receives solar radiations in various wavelengths that range between 290 nm–3,000 nm that reach the earth's surface and are used as a major source of energy for photoautotrophs. The ultraviolet spectrum can be further divided into three regions respectively, UV-A (320–400 nm), UV-B (280–320 nm), and UV-C (200–280 nm) (Singh et al., 2020).

It has been observed in the last few decades that environmental issues such as deforestation and continuous use of fossil fuels have increased the accumulation of various greenhouse gases, that is, carbon dioxide, methane, chlorofluorocarbons, nitrous oxide, and tropospheric ozone. As a result of the increased concentration of greenhouse gases, our environment experiences an increment in solar radiation quantity that is received by the earth's surface due to the greenhouse effect, thus causing the serious issue of global warming (Singh et al., 2020). To cope with the issue of global warming and increased radiation, plant responses are mainly controlled by different molecular processes. Different biotic and abiotic stresses affect plant growth, but to control their deleterious effects, plants express trailblazing and synchronized mechanisms by interconnected defence networks and various physiological, molecular, and cellular adaptations also such as altered gene expression, changes in growth rates, and crop yields (Singh et al., 2020). Studies of ultraviolet radiation stress have shown the involvement of leaves as sites for incoming radiation. UV-induced radiation stress is a complex oxidative stress that is capable of triggering the production of various reactive oxygen species (ROS) in plants, which are sufficient to induce different morphological and physiological effects (Shriram et al., 2016; Singh et al., 2020).

The microRNAs (miRNAs) are used for various applications in molecular biology. They are endogenous, single-stranded non-coding RNAs with an average length of 20–22 nucleotides that paramount roles in the regulation of gene expression. The miRNAs are mainly formed after transcription from DNA sequences through a series of steps. These steps include the formation of primary

 DOI: 10.1201/9781003248453-7

miRNAs that further get processed into precursor miRNAs and then finally mature miRNAs (Ha & Kim, 2014). The miRNAs have also been reported to activate translation or regulate transcription; miRNAs can interact with target genes, and this mechanism is dependent on different factors, that is, their subcellular location, abundance, and the affinity of miRNA-mRNA interactions. Intracellular miRNA is regulated under various levels of control that include transcription, processing, RNA modification, RNA-induced silencing complex (RISC) assembly, miRNA-target interaction, and turnover. Extracellular miRNAs can also be generated that can work as chemical messengers. The discovery of the first microRNA (miRNA) was of *lin-4*, in the year 1993 by the Ambros and Ruvkun groups in *Caenorhabditis elegans*, and this has transformed the field of molecular biology in different aspects (Kar & Raichaudhuri, 2021). The miRNAs that are present in plants play an important role in their developmental and metabolic processes, pattern formation, hormone regulation, regulation of biotic and abiotic stress response, and also self-regulation of the miRNA biogenesis (Jones-Rhoades et al., 2006; Bej & Basak, 2014). Plant miRNAs are differentially regulated under different types of stresses.

The miRNAs reported in plants are a type of small RNAs encoded by MIR genes that are transcribed by the role of RNA polymerase II (poly II) and produce the primary miRNA transcripts, known as pri-miRNA (Voinnet, 2009; Lee et al., 2004). The MIR genes are of several kilobases and composed of a hairpin structure that form precursor of miRNA after processing that is known as pre-miRNA and is of stem-loop structure (Kim, 2005). The DCL1, in plants, is responsible for converting the pri-miRNA to pre-miRNA (Fang & Spector, 2007; Kar & Raichaudhuri, 2021). There are different stress conditions in which miRNAs have been reported to be engaged, such as soil salinity (Gao et al., 2011), heat (Goswami et al., 2014), UV-B radiation-stress (Casadevall et al., 2013), and metal stress (Gupta et al., 2014). The present chapter will focus on explaining the functional role of miRNAs in regulating plants during radiation stress. The miRNAs play an efficient role in mediating stress responses, and they can be explored as novel targets for engineering radiation stress-tolerant crop varieties.

7.2 DIFFERENT TYPES OF PLANT MIRNAS INVOLVED IN MANAGING RADIATION STRESS

Various studies have identified different miRNAs that are playing functional roles in managing abiotic stresses, among which radiation stress is one. Plant miRNAs have been reported to regulate gene expression at the post-transcriptional level only by suppressing mRNA translation and cleaving target mRNAs. Different recent approaches that comprise of traditional cloning method together with computational prediction using bioinformatics tools and high throughput sequencing of small RNA libraries have been used to identify several classes of small RNAs with functions in plants. These classes of small RNAs include miRNAs, natural antisense transcript-derived small interfering RNAs (nat-siRNAs), repeat-associated small interfering RNAs (ra-siRNAs), heterochromatic small interfering RNAs (hc-siRNAs), trans-acting small interfering RNAs (ta-siRNAs), primary siRNAs, secondary transitive siRNA, and long small interfering RNAs (lsiRNAs) (Khraiwesh et al., 2012).

Among various ranges of electromagnetic radiation, UV-B fraction is harmful and deteriorating radiation that is capable to affect plants, directly as well as indirectly. UV-B radiation can induce molecular-level damage to DNA, RNA, and proteins and also alter the morphology and physiology of plants (Guleria et al., 2011). As per studies based on computational and statistical approaches, it has been reported that UV-B responsive miRNAs possess the same array of proximal promoter motifs as present in UV-B responsive protein-encoding genes, and the inferred expression of miRNAs should be negatively correlated with the expression of target genes. It has been identified that 21 miRNAs from 11 miRNA families are upregulated at the time of radiation stress. UV-B responsive miRNA families are miR156/157, miR159/319, miR160, miR165/166, miR167, miR169, miR170/171, miR172, miR393, miR398, and miR401. These miRNAs, except for miR393, miR398, and miR401, have been observed to target genes that encode transcription factors that further affected the expression of related genes (Zhou et al., 2007; Guleria et al., 2011). Differential expression of approximately 1,062 miRNAs in 41 plant species under 35 different abiotic stress conditions have been observed (Zhang et al., 2013). It has been reported by Wang et al. that Tae-miRNA6000 in wheat is expressed differentially as a response to exposure to UV-B. It has also been observed that miRNA171, miRNA159, and miRNA167a were upregulated, while miRNA156, miRNA167a, miRNA159, and miRNA164 were downregulated as a result of exposure of the plant to UV-B, indicating their anticipated role in stress tolerance to UV-B radiation (Wang et al., 2013). Among radiation stress-responsive miRNAs, miRNA398 (miR398) has been pronounced to be directly linked with the plant stress regulatory network. UV radiation stress leads to an oxidative burst in plant defence from the production of ROS (McKenzie et al., 2003; Zhu et al., 2011a). Zhou et al. (2007) reported that by utilizing computational approaches to identify miRNA genes induced by UV-B radiation in *Arabidopsis*, miR398 was 1 of the 11 putative UV-B-responsive miRNAs (Zhu et al., 2011a).

Few miRNA families that were predicted to be upregulated by UV-B radiation in *Arabidopsis* (miR156, miR160, miR165/166, miR167, miR398, and miR168) were reported to be upregulated by UV-B radiation in *Populus tremula* (Jia et al., 2009). Interestingly, three families, that is, miR159, miR169, and miR393, that were predicted to be upregulated in *Arabidopsis* were downregulated in *P. tremula*. These studies suggest that responses to UV-B radiation stress may be species-specific and vary accordingly (Khraiwesh et al., 2012).

Tradescantia (BNL clone 3340) has been reported as an important bio-indicator that could be used to measure the effect of radiation. Using *Tradescantia* after radiation stress, it has been reported that 37 miRNAs belonged to 36 different miRNA families, and among them, five randomly selected miRNAs were confirmed for their responsiveness to γ-IR (gamma-IR) stress. Target prediction has revealed that 37 miRNAs targeted 149 genes that were involved in stress tolerance regulation, light response, redox systems, signaling pathways, DNA repair, and transcription factors (Subburaj et al., 2017).

A dose and time-dependent study for analyzing expression profiles of three miRNAs osa-miR414, osa-miR164e, and osa-miR408, and also their targeted helicase genes, *OsABP*, *OsDBH*, and *OsDSHCT*, were done in response to different doses of γ-rays delivered, but the irradiated rice seeds were reported to show DNA damage and reactive species accumulation, but no dose- or time-dependent expressions were observed (Macovei & Tuteja, 2013). However, a study in rice seedlings has identified seven H_2O_2-responsive miRNAs: miR169, miR397, miR528, miR827, miR1425, miR319a.2, and miR408–5p (Bej & Basak, 2014). Oxidative stress caused by UV radiation has been reported to modulate two Cu-Zn superoxide dismutase (CSD) genes by upregulating them depending on miR398 levels. The cytosolic CSD1 and plastidic CSD2 genes were targeted by miR398. Under radiation stress, the expression of miR398 is downregulated, which led to increased accumulation of CSD1 and CSD2 mRNAs and further accumulation of the highly toxic superoxide free radicals. In *Arabidopsis*, it has been reported that the miR398 family is represented by two members with three loci: miR398a, miR398b, and miR398c (Bej & Basak, 2014).

7.3   BIOGENESIS OF MIRNAS

The miRNA biogenesis starts inside the nucleus. The transcription of plant miRNAs takes place from non-coding genes known as microRNA (MIR) genes that are present usually within intronic regions of other genes in a sense or anti-sense directions (Jones-Rhoades et al., 2006). Similar to other transcribing genes, MIR also contains a promotor, a transcribing part, and a terminator (Lee et al., 2004). The MIR genes are transcribed by DNA-dependent RNA polymerase II (RNA pol II) to form a primary transcript known as primary-miRNA or pri-miRNA, which is then capped at 5' end, tailed at 3' end, and folded into a stem-loop-like structure that is further stabilized by a DAWDLE (DDL) factor (Jones-Rhoades et al., 2006; Khraiwesh et al., 2012). This stem loop-like structure is now known as precursor-miRNA or pre-miRNA. One arm of the pre-miRNA represents the mature miRNA sequences while the other one represents miRNA*. In plants, stem-loop-like pre-miRNA is processed mainly by three components: SERRATE (SE), HYPONSTIC LEAVES 1 (HYL1), and DICER-LIKE-1 (DCL1) endonuclease (Fukudome & Fukuhara, 2017). The endonuclease activity

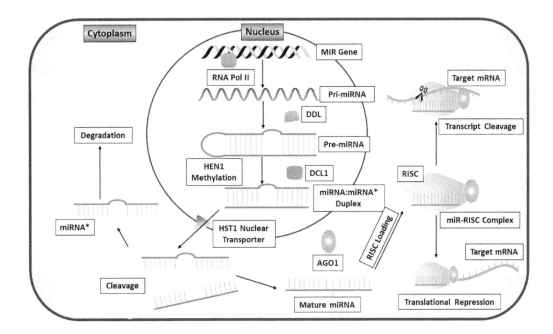

FIGURE 7.1 Biogenesis of miRNAs in plants.

of DCL1 helps to cleave the pre-miRNA at the loop (16–17 base pairs from the single strand-double strand junction) to form miRNA:miRNA* duplex (Zhu et al., 2013). DCL1 needs the presence of HYL1 and SE to function properly. The newly formed miRNA:miRNA* duplex can be degraded inside the nucleus. To ensure its stability, HUA ENHANCER 1 (HEN1) adds a methyl group at the 3' end of each strand, and the duplex is now transported via HASTY1 (HST1) plant nuclear transport protein to the cytoplasm for further processing (Xie et al., 2014).

In the cytoplasm, the duplex is opened up/cleaved to form a mature miRNA strand and a miRNA*. The mature miRNA is complexed to RNA-induced gene silencing complex (RISC) to regulate gene expressions of target mRNA. ARGONAUATE 1 (AGO1) factor helps in the stabilization of this miRNA-RISC complex (Arribas-Hernández et al., 2016). The mature miRNA either leads to direct cleavage of the target mRNA or blocks its expression by translational repression. Earlier it was thought that mRNA expression was only blocked by the mature miRNA, but recent reports have confirmed the regulation of gene expression by the miRNA* strand too (Zhang et al., 2018). The mechanism of miRNA biogenesis is summarized in Figure 7.1.

7.4   MECHANISM OF ACTION OF PLANT MIRNAS IN RESPONSE TO RADIATION STRESS

Mature miRNAs mainly act at the post-transcriptional level. Transcript cleavage was thought to be the primary mode of action of mature miRNAs due to the presence of an almost perfect complementarity between miRNAs and their target mRNAs (Chen, 2009; Voinnet, 2009). However, translational repression is not refracted by a high complementation ratio (Yang et al., 2012). Thus, the degree of complementation does not play an important role in recognizing the mechanism of action of miRNAs. Plant miRNAs principally act during the level of post-transcriptional gene silencing and help in the cleavage of target mRNA molecules (Zhu et al., 2011b). In plants, transcript cleavage is observed more often than translational repression in which either read-through of the ribosomes or ribosomal movement is blocked by miRNAs, resulting in the cessation of the translational process (Wang et al., 2008; Iwakawa & Tomari, 2013; Li et al., 2013).

The role of miRNAs in managing abiotic stress in plants is well-established through various pieces of literature. After the biogenesis of miRNAs from pre-miRNA precursors, they are found to regulate various transcription factors (TFs) associated with the development of the plant (Chávez Montel et al., 2014). Recent studies have reported that radiations not merely direct the bioaccumulation of miRNAs but guide their biogenesis and processing as well. Moreover, the functioning of the RNA-induced silencing complex (RISC) is affected by the radiation stress in plants (Sánchez-Retuerta et al., 2018).

It is well-known that light plays an important role in plants' growth and development. Genome-wide analysis of *Arabidopsis* has shown that nearly one-fifth of the total genes present are responsive to light (Wang et al., 2014). It has been reported that miRNAs mediate several light-responsive processes such as photoperiod signaling, pigment synthesis, and photomorphogenesis (Sánchez-Retuerta et al., 2018). It has also been found that miRNAs may regulate the expression of transcription factors to get involved in the phytochrome-b-mediated light transduction pathway. The basis of this finding is the results of high-throughput sequencing (HTS) and degradome sequencing (DS) through comparative analysis. The use of HTS and DS helped to identify expectant miRNAs engaged in phytochrome-b-mediated light transduction pathways with their respective targets. A total of 135 differentially expressible miRNAs (38 upregulated and 97 downregulated) were identified as expectant light-sensitive miRNAs involved in the pathway (Sun et al., 2015). In a study involved in identifying miRNAs responsive to γ-irradiated stress tolerance in *Trasectania*, abiotic-stress sensitive *cis*-elements as well as light-responsive motifs were found to be present in the promotor sites of miRNA target genes (Subburaj et al., 2017). These findings suggest the role of miRNAs in improved plant growth under varying radiation stress.

Identification of miRNAs has become easier with the help of techniques such as next-generation sequencing (NGS). In *Arabidopsis*, RNA sequencing helped in the identification of various light-responsive miRNA genes expressed differentially when exposed to different radiations (Hernando et al., 2017). Radiations of different wavelengths affect the expression of miRNAs in different ways. For example, exposure to visible light during the night can promote the expressions of miR157c, miR163, miR398b, and miR398c but downregulates the expression of miR256c, miR408, and miR824a (Shikata et al., 2014). On the other side, the levels of miR167, miR160, miR939, and miR939 are decreased during UV irradiation (Zhou et al., 2007). In a study, alteration in miRNA expression with changes in light was discovered. In a long period of light distress, HYL1, a protein required for miRNA processing, was found to be degraded, which led to gene silencing; nonetheless, exposure to light led to rapid dephosphorylation of HYL1, resuming the miRNA biogenesis, and maximizing the uptake (Achkar et al., 2018). In a similar study, a light-responsive miR163 was discovered with its respective target paraxanthine-methyltransferase 1. However, it was found that the miR163 expresses only in the preliminary growth phase in roots and is suppressed by light in later stages (Chung et al., 2016). This suggests the role of light stress in affecting the miRNA expression and ultimately the plant's growth. In the peach plant, UV-B-responsive miRNAs were identified and characterized. From the identified miRNAs, miRNAs including miR398a-5p, miR398a-3p, miR6263, miR6260, and miR319a were found to be upregulated, and miR1511, miR171c, and miR3627–3p were found to be downregulated upon UV-B irradiation (Li et al., 2017). The miR1771c was predicted to be involved in the chlorophyll metabolism, growth of meristematic tissues, and other growth-related activities by targeting SCL27, SCL22, and SCL6 TFs. Further analysis showed that carbohydrate metabolism is also regulated by miRNAs, particularly miR3627 under the UV-stress (Li et al., 2017). A similar investigation was performed in juvenile maize plants by exposing them to UV-B for eight hours. The findings of the experiment showed that miRNAs, including miR398, miR166, and miR165 were upregulated, whereas miR529, miR396, miR172, miR171, and miR156 were downregulated in exposure to UV-B. Additionally, some miRNAs such as miR529 and miR156 displayed contrary expression profiles to their respective targets, thus hinting about some post-transcriptional pathways (Casati, 2013).

Electromagnetic γ-rays are also considered harmful to living tissues because of their penetrance capability. A study was performed to analyze the effects of γ-irradiation on plant miRNAs. After treating *Arabidopsis* plantlets for four hours with γ-rays, they were left to grow in a normal environment for a few days. After analysis, it was found that miR850, miR840, miR827, miR403, miR172b, and miR164a were responsive to γ-irradiation stress (Kim et al., 2016). Apart from UV and γ-rays, high light also induces radiation stress in plants. In a study, *Arabidopsis thaliana* roots were put under high-light stress for two hours. The findings showed that high-light stress led to the induction of transcriptional changes in the root cells and also affected the expressions of miRNAs. Significant changes were found in upregulated miRNAs including miR160b, miR394a, and miR8175, and downregulated miR169f. Predicted targets of miR160b and miR169f-3p were found to be auxin-response factors (ARFs) and probable phosphoinositide phosphatase respectively (Anna et al., 2019). Therefore, it is concluded that miRNAs play a vital role in mitigating the radiation stress in plants via regulating various TFs and signaling pathways. Figure 7.2 summarizes some radiation-responsive miRNAs and their respective targets.

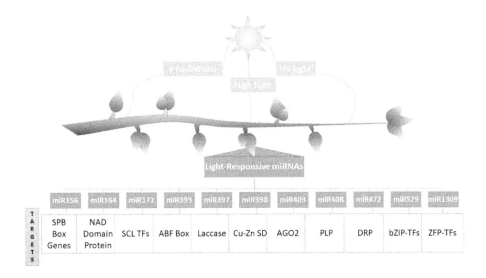

FIGURE 7.2 Radiation-responsive miRNAs and their respective targets (SPB: Squamosa Promoter Binding Protein, SCL: Scarecrow-Like, TFs: Transcription Factors, ABF: Auxin-Binding F Box, SD: Superoxide Dismutase, AGO: Argonaute, bZIP: basic leucine Zipper, ZFP: Zinc Finger Proteins).

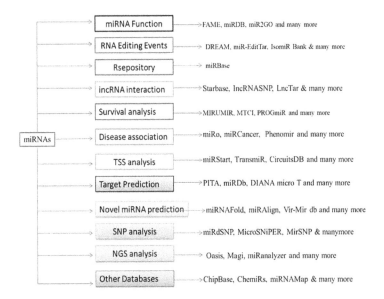

FIGURE 7.3 An overview of various computational tools for miRNA study.

7.5   INVOLVEMENT OF COMPUTATIONAL TOOLS IN THIS FIELD

The introduction of next-generation sequencing (NGS) technology has opened up a wide range of possibilities for the biological analysis of sequencing data. With modest bio-informatics skills, here we have outlined the tools which could be used to do an in-depth interpretation of micro RNA (miRNAs) sequencing information. An overview of various bioinformatics methods for the studies of miRNA is summarized in Figure 7.3.

7.6   SUMMARY AND PERSPECTIVES

The miRNAs are one of the major regulators in causing plant stress in response to radiation stress. Therefore, through the identification of specific miRNA/s and their target gene, the mechanisms mediating the tolerance level of plants against radiation stress can be moderated through various approaches. In the future, a mechanistic concept involving molecular, functional, and computational approaches-based characterization may aid in attaining the deeper knowledge and scope to gain an insight into effective targets relevant to the application of miRNAs in enhancing and managing plant tolerance against radiation stress with their respective targets in plants. Further advances in these studies will elucidate the molecular mechanism in the deployment of miRNA-mediated crop improvement strategies in tolerance to radiation stress.

REFERENCES

Achkar, N. P., Cho, S. K., Poulsen, C., Arce, A. L., Re, D. A., Giudicatti, A. J., Karayekov, E., Ryu, M. Y., Choi, S. W., Harholt, J., & Casal, J. J. (2018). A quick HYL1-dependent reactivation of microRNA production is required for a proper developmental response after extended periods of light deprivation. *Developmental Cell*, *46*, 236–247. https://doi.org/10.1016/j.devcel.2018.06.014

Anna, B. B., Grzegorz, B., Marek, K., Piotr, G., & Marcin, F. (2019). Exposure to high-intensity light systemically induces micro-transcriptomic changes in arabidopsis thaliana roots. *International Journal of Molecular Sciences*, *20*, 5131. https://doi.org/10.3390/ijms20205131

Arribas-Hernández, L., Marchais, A., Poulsen, C., Haase, B., Hauptmann, J., Benes, V., Meister, G., & Brodersen, P. (2016). The slicer activity of argonaute1 is required specifically for the phasing, not production, of trans-acting short interfering RNAs in Arabidopsis. *Plant Cell*, *28*, 1563–1580. https://doi.org/10.1105/tpc.16.00121.

Bej, S., & Basak, J. (2014). MicroRNAs: The Potential biomarkers in plant stress response. *American Journal of Plant Sciences, 5*,748–759. http://dx.doi.org/10.4236/ajps.2014.55089

Casadevall, R., Rodriguez, R. E., Debernardi, J. M., Palatnik, J. F., & Casati, P. (2013). Repression of growth regulating factors by the microRNA396 inhibits cell proliferation by UV-B radiation in Arabidopsis leaves. *Plant Cell*, *25*, 3570–3583. https://doi.org/10.1105/tpc.113.117473

Casati, P. (2013). Analysis of UV-B regulated miRNAs and their targets in maize leaves. *Plant Signaling & Behavior*, *8*, e26758. https://doi.org/10.4161/psb.26758

Chávez Montes, R. A., de Fátima Rosas-Cárdenas, F., De Paoli, E., Accerbi, M., Rymarquis, L. A., Mahalingam, G., Marsch-Martínez, N., Meyers, B. C., Green, P. J., & de Folter, S. (2014). Sample sequencing of vascular plants demonstrates widespread conservation and divergence of microRNAs. *Nature Communications*, *5*, 3722. https://doi.org/10.1038/ncomms4722.

Chen, X. (2009). Small RNAs and their roles in plant development. *Annual Review of Cell and Developmental Biology*, *25*, 21–44.

Chung, P. J., Park, B., Wang, H., Liu, J., Jang, I. C., & Chua, N. H. (2016). Light-inducible miR163 targets PXMT1 transcripts to promote seed germination and primary root elongation in Arabidopsis. *Plant Physiology*, *170*, 1772–1782. https://doi.org/10.1104/pp.15.01188

Fang, Y., & Spector, D. L. (2007). Identification of nuclear dicing bodies containing proteins for microRNA biogenesis in living Arabidopsis plants. *Current Biology, 17*, 818–823. https://doi.org/10.1016/j.cub.2007.04.005

Fukudome, A., & Fukuhara, T. (2017). Plant dicer-like proteins: Double-stranded RNA cleaving enzymes for small RNA biogenesis. *Journal of Plant Research*, *130*, 33–44.

Gao, P., Bai, X., Yang, L., Lv, D., Pan, X., Li, Y., Cai, H., Ji, W., Chen, Q., & Zhu, Y. (2011). Osa-MIR393: A salinity- and alkaline stress-related microRNA gene. *Molecular Biology Reports*, *38*, 237–242. https://doi.org/10.1007/s11033-010-0100-8

Goswami, S., Kumar, R., & Rai, R. (2014). Heat-responsive microRNAs regulate the transcription factors and heat shock proteins in modulating thermo-stability of starch biosynthesis enzymes in wheat (Triticum aestivum L.) under the heat stress. *Australian Journal of Crop Science*, 8, 697–705. kumar_8_5_2014_697_705. pdf. cropj.com

Guleria, P., Mahajan, M., Bhardwaj, J., & Yadav, S. K. (2011). Plant small RNAs: Biogenesis, mode of action and their roles in abiotic stresses. *Genomics Proteomics & Bioinformatics*, 9(9), 183–199. https://doi. org/10.1016/S1672–0229(11)60022–3

Gupta, O. P., Sharma, P., Gupta, R. K., & Sharma, I. (2014). MicroRNA mediated regulation of metal toxicity in plants: Present status and future perspectives. *Plant Molecular Biology, 84*, 1–18. https://doi.org/10.1007/s11103-013-0120-6

Ha, M., & Kim, V. N. (2014). Regulation of microRNA biogenesis. *Nature Reviews Molecular Cell Biology, 15*, 509–524. https://doi.org/10.1038/nrm3838.

Hernando, C. E., Garcia, C., & Mateos, J. L. (2017). Casting away the shadows: Elucidating the role of light-mediated posttranscriptional control in plants. *Photochemistry and Photobiology, 93*, 656–665. https://doi.org/10.1111/php.12762

Iwakawa, H. O., & Tomari, Y. (2013). Molecular insights into microRNA-mediated translational repression in plants. *Molecular Cell*, 52, 591–601.

Jia, X., Ren, L., Chen, Q.-J., Li, R., & Tang, G. (2009). UV-B-responsive microRNAs in populus tremula. *Journal of Plant Physiology, 166*, 2046–2057.

Jones-Rhoades, M. W., Bartel, D. P., & Bartel, B. (2006). MicroRNAs and their regulatory roles in plants. *Annual Review of Plant Biology*, 57, 19–53. http://dx.doi.org/10.1146/annurev.arplant.57.032905.105218

Kar, M. M., & Raichaudhuri, A. (2021). Role of microRNAs in mediating biotic and abiotic stress in plants. *Plant Gene, 26*, 100277. https://doi.org/10.1016/j.plgene.2021.100277

Khraiwesh, B., Zhu, J.-K., & Zhu, J. (2012). Role of miRNAs and siRNAs in biotic and abiotic stress responses of plants. *Biochimica et Biophysica Acta, 1819*, 137–148. https://doi.org/10.1016/j.bbagrm.2011.05.001

Kim, J. H., Go, Y. S., Kim, J. K., & Chung, B. Y. (2016). Characterization of microRNAs and their target genes associated with transcriptomic changes in gamma-irradiated Arabidopsis. *Genetics and Molecular Research, 15*, gmr.15038386.

Kim, V. N. (2005). MicroRNA biogenesis: Coordinated cropping and dicing. *Nature Reviews Molecular Cell Biology, 6*, 376–385. https://doi.org/10.1038/nrm1644

Lee, Y., Kim, M., Han, J., Yeom, K. H., Lee, S., Baek, S. H., & Kim, V. N. (2004). MicroRNA genes are transcribed by RNA polymerase II. *The EMBO Journal, 23*, 4051–4060. https://doi.org/10.1038/sj.emboj.7600385.

Li, B., Duan, H., Li, J., Deng, X. W., Yin, W., & Xia, X. (2013). Global identification of miRNAs and targets in populus euphratica under salt stress. *Plant Molecular Biology, 81*, 525–539

Li, S., Shao, Z., Fu, X., Xiao, W., Li, L., Chen, M., Sun, M., Li, D., & Gao, D. (2017). Identification and characterization of prunus persica MiRNAs in response to UVB radiation in greenhouse through high-throughput sequencing. *BMC Genomics, 18*, 938.

Macovei, A., & Tuteja, N. (2013). Different expression of miRNAs targeting helicases in rice in response to low and high dose rate γ-ray treatments. Plant *Signaling & Behavior, 8*,8, e25128. https://doi.org/10.4161/psb.2512

McKenzie, R. L., Bjorn, L. O., Bais, A., & Ilyasad, M. (2003). Changes in biologically active ultraviolet radiation reaching the Earth' s surface. *Photochemical & Photobiological Sciences, 2*(1), 5–15. https://doi.org/10.1039/b211155c

Percuoco, R. (2014). Chapter 1-plain radiographic imaging. In D. M. Marchiori (Ed.), *Clinical imaging* (3rd ed., pp. 1–43). Mosby. https://doi.org/10.1016/B978-0-323-08495-6.00001-4.

Sánchez-Retuerta, C., Suaréz-López, P., & Henriques, R. (2018). Under a new light: Regulation of light-dependent pathways by non-coding RNAs. *Frontiers in Plant Sciences, 9*, 962. https://doi.org/10.3389/fpls.2018.00962

Shikata, H., Hanada, K., Ushijima, T., Nakashima, M., Suzuki, Y., & Matsushita, T. (2014). Phytochrome controls alternative splicing to mediate light responses in Arabidopsis. *Proceeding of the Natural Acadamy of Sciences USA, 111*, 18781–18786.

Shriram, V., Kumar, V., Devarumath, R. M., Khare, T. S., & Wani, S. H. (2016). MicroRNAs as potential targets for abiotic stress tolerance in plants. *Frontiers in Plant Science, 7*, 817. https://doi.org/10.3389/fpls.2016.00817

Singh, S., Fatima, A., Tiwari, S., & Prasad, S. M. (2020). Chapter 5-plant responses to radiation stress and its adaptive mechanisms. In D. M. Tripathi, V. P. Singh, D. K. Chauhan, S. Sharma, S. M. Prasad, N. K. Dubey, & N. Ramawat (Eds.), *Plant life under changing environment* (pp. 105–122). Academic Press. https://doi.org/10.1016/B978-0-12-818204-8.00006-0

Subburaj, S., Ha, H. J., Jin, Y. T., Jeon, Y., Tu, L., Kim, J. B., Kang, S. Y., & Lee, G. J. (2017). Identification of
γ-radiation-responsive microRNAs and their target genes in tradescantia (BNL clone 4430). *Journal of
Plant Biology*, *60*, 116–128. https://doi.org/10.1007/s12374-016-0433-5

Sun, W., Xu, X. H., Wu, X., Wang, Y., Lu, X., Sun, H., & Xie, X. (2015). Genome-wide identification of
microRNAs and their targets in wild type and phyb mutant provides a key link between microRNAs
and the phyn-mediated light signaling pathway in rice. *Frontiers in Plant Science*, *6*, 372. https://doi.
org/10.3389/fpls.2015.00372

Voinnet, O. (2009). Origin, biogenesis, and activity of plant microRNAs. *Cell*, *136*, 669–687. https://doi.
org/10.1016/j.cell.2009.01.046D

Wang, B., Malik, R., Nigg, E. A., & Korner, R. (2008). Evaluation of the low-specificity protease elastase for
large-scale phosphoproteome analysis. *Analytical Chemistry*, *80*, 9526–9533.

Wang, B., Sun, Y. F., Song, N., Wang, X. J., Feng, H., Huang, L. L., & Kang, Z. S. (2013). Identification
of UV-B-induced microRNAs in wheat. *Genetics and Molecular Research,* *12*, 4213–4221. https://doi.
org/10.4238/2013.October.7.7

Wang, H., Chung, P. J., Liu, J., Jang, I. C., Kean, M. J., Xu, J., & Chua, N. H. (2014). Genome-wide identification of long noncoding natural antisense transcripts and their responses to light in Arabidopsis. *Genome
Research, 24, 444–453. https://doi.org/10.1101/gr.165555.113*

Xie, M., Zhang, S., & Yu, B. (2014). MicroRNA biogenesis, degradation and activity in plants. *Cellular and
Molecular Life Sciences*, *72*, 87–99. https://doi.org/10.1007/s00018-014-1728-7.

Yang, L., Wu, G., & Poethig, R. S. (2012). Mutations in the GW-repeat protein SUO reveal a developmental
Function for microRNA-mediated translational repression in Arabidopsis. *Proceedings of the Natural
Acadamy of Sciences USA*, *109*, 315–320.

Zhang, B., & Unver, T. (2018). A critical and speculative review on microRNA technology in crop improvement: Current challenges and future directions. *Plant Science*, *274*, 193–200. https://doi.org/10.1016/j.
plantsci.2018.05.031

Zhang, S., Yue, Y., Sheng, L., Wu, Y., Fan, G., Li, A., Hu, X., ShangGuan, M., & Wei, C. (2013). PASmiR:
A literature-curated database for miRNA molecular regulation in plant response to abiotic stress. *BMC
Plant Biology, 13*, 33. https://doi.org/10.1186/1471-2229-13-33

Zhu, C., Ding, Y., & Liu, H. (2011a). MiR398 and plant stress responses. *Physiologia Plantarum*, *143*, 1–9.
https://doi.org/10.1111/j.1399-3054.2011.01477.x

Zhu, H., Hu, F., Wang, R., Zhou, X., Sze, S. H., Liou, L. W., Barefoot, A., Dickman, M., & Zhang, X. (2011b).
Arabidopsis argonaute10 specifically sequesters miR166/165 to regulate shoot apical meristem sevelop-
ment. *Cell*, *145*, 242–256.

Zhu, H., Zhou, Y., Castillo-González, C., Lu, A., Ge, C., Zhao, Y. T., Duan, L., Li, Z., Axtell, M. J., Wang,
X. J., & Zhang, X. (2013). Bidirectional processing of pri-miRNAs with branched terminal loops by
Arabidopsis dicer-like1. *Nature Structural & Molecular Biology*, *20*, 1106–1115. https://doi.org/10.1038/
nsmb.2646

8 miRNAs and Plant-Pathogen Interactions

Bipin Maurya[$], Vishnu Mishra[$], and Shashi Pandey Rai

[$]These authors contributed equally

8.1 INTRODUCTION

To fulfil the demand of world populations, we are highly dependent on agriculture. However, agriculture augmentation promotes plant stress *via* various abiotic and biotic environmental factors such as low temperature, high salt conditions, drought, herbivores, variety of pathogens, respectively (Isaac-Renton et al., 2018; Lesk et al., 2017; Neumann et al., 2017; Porter et al., 2020; Reyer et al., 2013; Seidl et al., 2017; Zhang et al., 2018a). Both plants and animals are continuously attacked by various pests, pathogens, and parasites. Recently, it was observed that approximately 16% of global loss of crop productivity is directly by pathogens (Savary et al., 2019). Besides certain diseases, the yield losses and disease severity are also dependent on environmental conditions and the pathogen's aggressiveness (Ficke et al., 2018). Due to continuous exposure to infectious microbes, pests, and pesticides, gradually plants have developed a variety of advanced immune systems, and these acquired immune systems have various common characteristics with animal systems recognizing pathogen-associated molecular patterns (PAMPs) (Huang et al., 2019; Ingle et al., 2006; Jones et al., 2016). Plants recognize PAMP with the help of plasma membrane-associated pattern recognition receptors protein (PRRs) activates PAMP-induced trigger immunity (PTI) (Chisholm et al., 2006; Jones & Dangl, 2006). However, pathogens suppress PTI for successful colonization and lead to disease development by suppressing the host immune system (Jones & Dangl, 2006). Previously some plants have developed effector-triggered immunity (ETI), which involves various nucleotide-binding domains, leucine-rich repeat (LRR), and nucleotide-binding leucine-rich repeat (NLR) genes. The suppression of one or more components of PTI or ETI leads to the downregulation of the plant immune system, which promotes pathogen colonization in host cells (Jones & Dangl, 2006; Nguyen et al., 2021). Currently, miRNAs have emerged to be an important modulator of gene function and regulate plant immune responses under pathogen attack, thereby regulating plant-pathogen interaction processes (Huang et al., 2019). The miRNAs are endogenously produced non-coding RNAs having pivotal functions in a variety of growth and developmental processes by inversely regulating their target gene(s). The miRNA biogenesis is a complex process (Gautam et al., 2017). These miRNAs are generally present between the coding regions of two proteins. They are first transcribed by RNA polymerase II to give rise to primary miRNA (**Figure 8.1**). This initially synthesized pri-miRNA has been further converted into precursor miRNAs (pre-miRNAs) by multiple protein complexes such as DAWDLE, DICER-LIKE1 protein (DCL1), SERRATE protein (SE), and HYPONASTIC LEAVES1 protein (HYL1) (Chen, 2009; Gautam et al., 2017). Further, this pre-miRNA is converted into miRNA/miRNA* duplex by the activity of multiple protein complexes such as DCL1, HYL1, and SE complex. Further, this miRNA strand interacts with ARGONAUTE (AGO1) protein to give complementary target gene mRNA either by degradation/cleavage or by inhibition of the translational process. The RNA-induced silencing complex (RISC) binds to the target gene and negatively regulates the expression of the target gene either through cleavage or translational inhibition (Chen, 2009; Gautam et al., 2017). DCL1, SE, and HYL1 are the central components of miRNA biogenesis

DOI: 10.1201/9781003248453-8

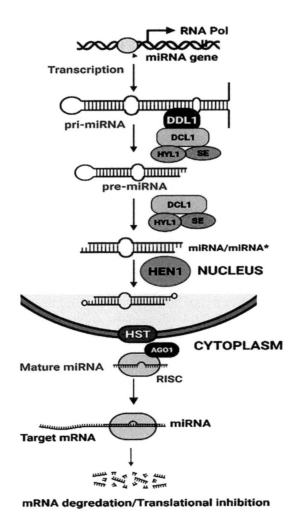

FIGURE 8.1 The miRNA synthesis and mechanism of target gene regulation. The miRNA gene is first transcribed into a capped and polyadenylated pri-miRNA by the action of RNA pol II. This pri-miRNA is then processed into pre-miRNA, which is further converted into a miRNA/miRNA* duplex by its downstream processing. After methylation by HEN1, it is loaded into the AGO1 complex. The activated RISC-bearing mature miRNA negatively regulates the expression of target genes.

(**Figure 8.1**). However, recently the elongator complex is an important factor for RNA Polymerase II (RNAPII) occupancy at miRNA loci, which associate with DCL1 and SE and play a crucial role in miRNA biogenesis (Achkar et al., 2016).

The miRNAs have been known to be involved in various plant growth and developmental processes, such as miR172, which has been shown in shoot development and vegetative phase transition; miR166 is associated with leaf development; miR399 in stress response; miR396, miR165/66, and miR394 are involved in root apical meristem and shoot apical meristem development; miR166 in primary root development (Chiou et al., 2005; Gautam et al., 2017; Millar, 2020; Rodriguez et al., 2015; Singh et al., 2014). Besides their regulatory role in differentiation, growth, and developmental processes; normal growth; and development, the miRNAs have been also reported in plant defence mechanisms and actively participate in host immunity against pathogen attack by modulating their target genes (Islam et al., 2017). The differential regulation of these miRNAs inversely affects their target gene, which ultimately regulates plant immunity against various pathogens such as bacteria, viruses, fungi, nematodes, and pests. Some miRNAs such as miR393 are induced by PAMPs and activate plant immune responses against these biotic factors. Furthermore, miR160, miR167, and miR825 are modulated by bacterial infection and might be involved in plant defence response (Ruiz-Ferrer & Voinnet, 2009). Besides these, some other miRNAs such as miR156, miR164, miR171, miR172, miR393, miR396, miR399, miR398, miR408, miR482, miR444, miR825, miR827, and miR1885 have also been documented having their regulatory role in defence responses (Islam et al., 2017; Khraiwesh et al., 2012; Sunkar et al., 2012; Zhu et al., 2013). Here, we summarize the recent update related to miRNAs during plant-pathogen interactions. Additionally, we also discuss the possible role of novel strategies such as miPEPs and genome editing in miRNA-mediated disease resistance and crop protection to improve quality and yield.

8.2   ROLE OF MIRNAS IN PLANT-PATHOGEN INTERACTIONS

To attain successful life organisms are highly dependent on plants for food and energy (Islam et al., 2017; Lin et al., 2017). A wide variety of plant pathogens are key biotic elements majorly affecting the growth, productivity, and yield of crop plants (Islam et al., 2017). These pathogens either directly or indirectly activate plant defence response by modulating plant cell machinery via miRNAs **(Figure 8.2)** (Adnan et al., 2019; Islam et al.,2017). To counter these pathogens, plants have developed various mechanisms like physical barriers such as bark and waxy cuticles, stomata, and chemical defence such as secondary metabolites. One of these enlightened mechanisms is the involvement of miRNAs in defence mechanisms against pathogens. These miRNAs are particularly involved in the functional regulation of various defence-related signals and metabolic pathways as defence response, for example, regulation of *NUCLEOTIDE-BINDING SITE LEUCINE-RICH REPEAT* (NBS-LRR) gene expression, hormonal response, generation of ROS, and another gene silencing, which is further involved in plants and pathogen interaction process. In the following, we will discuss the involvement of miRNAs in various crops and non-crop plants during pathogen interaction.

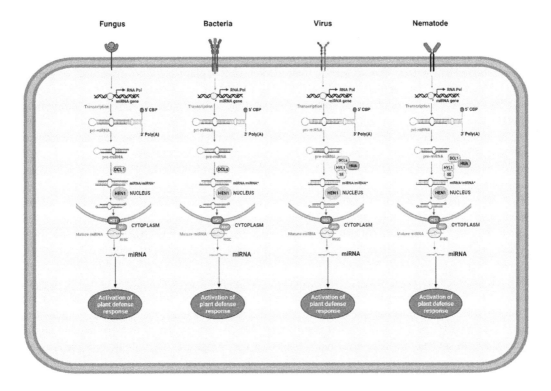

FIGURE 8.2   Picture showing the process of plant defence response *via* generation of miRNA against a variety of pathogen attacks.

8.2.1   ROLE OF MIRNAS IN *ARABIDOPSIS* DURING PATHOGEN INTERACTION

Arabidopsis, a weed, non-crop plant, belongs to the Cruciferae/Brassicaceae family. Several ecotypes of *Arabidopsis thaliana*, such as Wassilewskija (Ws), Columbia (Col-0), Landsberg erecta (L*er*), and C24, etc. are naturally available (Feuerstein et al., 2010; Miao et al., 2018). These ecotypes are varying in size, shape, and morphological and physiological aspects. Several molecular regulators such as transcription factors, epigenetic modifiers, signal receptors, and various biosynthetic enzymes are involved significantly in various plant-pathogenic interaction processes (Alonso et al., 2019; Falak et al., 2021; Gullner et al., 2018; Satapathy et al., 2017; Tang et al., 2017; Zhu et al., 2016). Apart from genes, TFs, and epigenetic modifiers, plant microRNAs are known to play a significant role against pathogens in *Arabidopsis* (Islam et al., 2017). The various roles of miRNAs, which actively participate in *Arabidopsis* during plant-pathogen interaction, is explained as follows **(Table 8.1)**. Recently, miR156 has been shown in plant immune systems response in *Arabidopsis* against its target *SQUAMOSA PROMOTER BINDING PROTEIN-LIKE* (*SPL*) transcription factor (TF). The detailed studies showed the overexpression of miR156 *Arabidopsis* seedlings accumulated lower ROS and salicylic acid (SA) signaling genes compared to *MIM156* and *ProSPL9:rSPL9* (Yin et al., 2019). This lower accumulation of ROS and downregulation of SA signaling genes which are involved in overexpression of miR156 induced susceptibility against *Pseudomonas syringae pv. Tomato DC3000*. Further investigation showed miR156/*SPL* module is involved in *Arabidopsis* biotic factor-induced immune response by modulating ROS production and activation of SA signaling (Yin et al., 2019). Some miRNAs such as miR159, miR160, miR163, miR167, miR393, miR393*b, miR398, miR400, miR408, miR472, miR773, miR825, miR863, and miR863–3p have been modulated by *Pst. DC3000*. In *Arabidopsis*, miR159 showed to be downregulated by *Pst DC3000* (EV and avrRpt2) at six hours of post-inoculation (hpi) but upregulated at 14-hpi by *Pst DC3000* (avrRpt2). The miR159 negatively regulates TFs *MYELOBLASTOSIS* (*MYB*)/*ISOTRICHODERMIN C-15* HYDROXYLASE (*HIC-15*), *MYB33*, *MYB65*, and *MASTER REGULATOR OF CELL CYCLE ENTRY* and *PROLIFERATIVE METABOLISM* (*MYC101*) and activates gibberellic acid (GA) signaling (Islam et al., 2017; Millar & Gubler, 2005; Reyes & Chua, 2007; Zhang et al., 2011). Similarly, miR393 is upregulated by *Pst DC3000* (avrRpt2) but downregulated by *PstDC3000* (EV), and miR162 is upregulated by *Pst DC300* (EV). However, miRNAs such as miR408 and miR822 are downregulated by *Pst DC3000*, suggesting their involvement in plant defence response (Zhang et al., 2011). The miR398 is upregulated by *Pst DC3000* (EV) at 6dpi, however, a detectable difference in miR398 expression was not observed (Jagadeeswaran et al., 2009; Zhang et al., 2011). Additionally, miR319 and miR852 are upregulated by *Pst DC3000* (hrcC) and *Pst DC3000* (avrRpt2) at 14hpi. The deep sequencing data analysis, however, failed to detect the differential accumulation of these miRNAs (Zhang et al., 2011). In most cases, expression of miRNAs is highly upregulated by infection of avirulent strain *Pst DC3000* (avrRpt2) compared to virulent strains *Pst DC3000* (EV), and non-pathogenic *Pst DC3000* (hrcC) such as miR158, miR159, miR166, and miR319 (Zhang et al., 2011). The miR159 and miR319 have been known to be involved in abscisic acid (ABA) and jasmonic acid (JA) signaling and promote SA-mediated defence response by negatively regulating their targets *MYB/HIC-15*, *MYB33*, *MYB65*, *MYC101*, and *TEOSINTE BRANCHED(TCP)* respectively (Zhang et al., 2011). The miR160 has shown its role in root growth, embryo, and seed development by targeting several *AUXIN RESPONSIVE FACTORs* (*ARFs*) (Mallory et al., 2005; Wang et al., 2005). However, the role of miR160 is well documented in plant-pathogen interaction. The miR160 induced by flg22 and non-pathogenic *hrc*C mutant *P. syringae* which inversely regulate their target *ARF10/16/17* and thus provide a very crucial role in plant defence mechanism by modulating auxin signaling (Li et al., 2011). It has been found that the expression of miR160 is unregulated in response to soybean mosaic virus and fungal infection suggesting the role of miR160 in plant defence response to viral and fungal infection (Yin et al., 2013; Zhao et al., 2012). Another miRNA, miR163 has been shown in plant defence and disease protection. The miR160 is a non-conserved miRNA having its targets *PARAXANTHINE METHYLTRANSFERASE 1 (PXMT*1) and *FARNESOIC ACID METHYLTRANSFERASE (FAMT)*, which are induced by *PstDC3000* infection in *Arabidopsis*, suggesting their involvement in pathogen-induced disease resistance. It is also reported that the miR163 mutant was more resistant compared to the overexpression of miR163 indicating their contributions to plant defence response (Chow & Ng, 2017). Recently, miR167 showed involvement in SA-mediated defence response against *P. syringae* by negatively regulating target genes *ARF6* and *ARF8*. Further, miR167 overexpression promotes resistance against *P. syringae* (Caruana et al., 2020). Some other miRNAs including miR390, miR393, miR393*b, miR408, miR472, miR773, miR863, and miR863–3p have been implicated in plant-pathogen interaction by regulating their target gene(s). Previously shown *Arabidopsis* miR164 targets NAM, ATAF, and CUC (*NAC*) TFs and has a pivotal role against biotic responses against such as *P. syringae*, *Alternaria brassicicola*, *Botrytis cinerea*, and powdery mildew fungus (Wang et al., 2018). However, a recent investigation on miR164 showed rice (*Oryza sativa*) miRNA, Osa-miR164a negatively modulates rice immunity against *Magnaporthe oryzae* (*M. oryzae*) (Li et al., 2019; Wang et al., 2018). Furthermore, miR396 has been previously studied in PAMP-triggered immune responses in the Solanaceae family, *Medicago*, soybean, citrus, tomato, and *Arabidopsis* by targeting *GROWTH RESPONSE FACTOR* (GRFs) in response to *Plectosphaerellacucumerina*, *B. cinerea*, *Colletotrichum higginsianum*, *Phytophthora infestans*, and *Phytophthora nicotianae* (Chen et al., 2015; Islam et al., 2017; Song et al., 2019; Soto-Suárez et al., 2017). Recently, the miR169/*SUBUNIT A OF NUCLEAR FACTOR Y (NF-YA)* module was also shown against *R. solanacearum* in *Arabidopsis*. Further, it has been observed that Osa-miR169/NF-YA negatively regulates immunity against blast fungus *M. oryzae* in rice (Hanemian et al., 2016; Li et al., 2017b; Song et al., 2019). Some other miRNAs such as miR825 and miR825* involved in *Bacillus cereus* AR156 triggered systemic resistance against *Pst DC3000* in *A. thaliana* (Nie et al., 2017, 2019). However, it has recently been shown that miR825 and miR825* downregulated in Col-0 upon pretreatment with *B. cereus* AR156 compared to *B. cinerea* B1301 infection. The attenuated expression of miR825 and miR825* showed more tolerance to *B. cinerea* B1301 compared to the overexpression of miR825 and miR825* (more susceptible) to the pathogen *B. cinerea* B1301 (Nie et al., 2017, 2019). Besides these, some other miRNAs such as miR166, miR398, and miR400 participate in plant-pathogen interactions by regulating their target gene(s) respectively.

8.2.2 ROLE OF miRNAs IN THE SOLANACEAE FAMILY DURING PATHOGEN INTERACTION

In the plant genome, there are large numbers of leucine-rich repeat (LRR) and intracellular LRR receptors that recognized specific pathogens and trigger the effective resistance response. Earlier, it has been found that unregulated expression of *NB-LRR* genes results in autoimmunity in crop plants by inhibiting growth and developmental processes. In the family of Solanaceae plants like potato, tobacco, and tomato, miRNAs like miR6019 and miR6020 target the *R* gene and *TIR-NBS-LRR*TFs, which have been shown to play a major role against the viral pathogen (**Table 8.1**) (Huang et al., 2016b; Li et al., 2012). Further investigation showed overexpression of miR6019 and miR6020 attenuating *N*-gene-induced resistance to tobacco mosaic virus (TMV). Additionally, miR6021, miR6022, and miR6023 have been shown to target the gene Homologs *of CLADOSPORIUM FULVUM RESISTANCE 9 (Hcr9)*, which have a protective role in providing immunity against TMV (Huang et al., 2019; Li et al., 2012).

8.2.3 ROLE OF miRNAs IN RICE DURING PATHOGENESIS

The miRNAs play specific roles in biochemical and physiological processes in plants by regulating their respective target gene(s). The miR164 targets *NAC4* and *NAC60*TFs, which induce pathogenicity in rice. In rice, the other miRNA, miR7695 expression is regulated by *M. oryzae* (Baldrich & San

TABLE 8.1

Role of different miRNAs in plant pathogenesis.

S. No	miRNAs	Target	Regulatory role in plant-pathogen interactions	Plants	Pathogen	Nature of pathogen	References
1	miR156	SPL	Regulates ROS accumulation, immune response, and SA signaling	Arabidopsis	P. syringae pv. tomato (Pst) DC3000	Bacteria	(Yin et al., 2019)
2	miR159	MYB/HIC-15, MYB33, MYB65, MYC101	Regulate GA and ABA signaling	Arabidopsis	Pst. DC3000	Bacteria	(Islam et al., 2017; Zhang et al., 2011)
3	miR160	ARF10, ARF16, ARF17	Increase PAMP-induced cellulose deposition	Arabidopsis	Pst. DC3000	Bacteria	(Islam et al., 2017; Li et al., 2011)
4	miR163	FAMT	Pathogen immune response	Arabidopsis	Pst. DC3000	Bacteria	(Song et al., 2019)
5	miR164	NAC4/NAC60	Pathogen immune response	Arabidopsis, O. sativa	M. oryzae	Fungus	(Lee et al., 2017; Wang et al., 2018)
6	miR166	HD-ZIPIII/CLP-1/RDD1	Pathogen immune response	Arabidopsis, Cotton	Tomato yellow leaf curl China virus (TYLCCNV), Verticillium dahlia	Virus	(Song et al., 2019; Yang et al., 2008; Zhang et al., 2016)
7	miR167	ARF8, ARF6	Regulate Auxin signaling and plant defence response	Arabidopsis	Pst. DC3000	Bacteria	(Fahlgren et al., 2007; Islam et al., 2017; Zhang et al., 2011)
8	miR169	NFYA/HAP2	Defence Responses	Arabidopsis, O. sativa	Ralstonia solanacearum, M. oryzae	Bacteria and fungus	(Hanemian et al., 2016; Li et al., 2017a; Song et al., 2019)
9	miR390	TAS3	Pathogen immunity response triggers the synthesis of tasiRNAs, thereby regulating ARF3 and ARF4 expression	Arabidopsis	Phytophthora syringae	Fungus-like microorganism	(Islam et al., 2017; Zhang et al., 2011)
10	miR393	TIR1, AFB2, AFB3	Auxin signaling and plant defence response	Arabidopsis	Pst. DC3000	Bacteria	(Islam et al., 2017)

(Continued)

TABLE 8.1 (Continued)
Role of different miRNAs in plant pathogenesis.

S. No	miRNAs	Target	Regulatory role in plant-pathogen interactions	Plants	Pathogen	Nature of pathogen	References
11	miR393*b	MEBM12	Regulate synthesis of pathogenesis-related protein PR1	Arabidopsis, N. benthamiana	Pst. DC3000	Bacteria	(Islam et al., 2017; Zhang et al., 2011)
12	miR396	GRF	triggered immune responses	Arabidopsis, Solanaceae, Medicago, soybean, citrus, tomato	Plectosphaerella cucumerina, B. cinerea, Colletotrichum higginsianum, Phytophthora infestans, Phytophthora nicotianae	Fungus and fungus-like microorganisms	(Chen et al., 2015; Islam et al., 2017; Song et al., 2019; Soto-Suárez et al., 2017)
13	miR398	CSD1/CSD2/CCS1 and COX5b.1	Inhibit callose deposition, downregulate auxin signaling, suppression detoxification of ROS	Arabidopsis, Barley	Pst. DC3000, Blumeriagraminis f. sp. Hordei	Bacteria and fungus	(Islam et al., 2017; Jagadeeswaran et al., 2009; Li et al., 2011; Xu et al., 2014)
14	miR400	PPR	Defence response to diverse pathogens	Arabidopsis	Pst. DC3000, B. cinerea	Bacteria and fungus	(Islam et al., 2017; Park et al., 2014)
15	miR408	LACCASE COPPER PROTEIN, COPPER and COPPER ION BINDING PROTEIN GENES PROTEIN PLANTACYANIN GENE	Unknown function	Arabidopsis	Pst. DC3000	Bacteria	(Islam et al., 2017)
16	miR472	CC-NBS-LRR	Pathogen immunity response and overexpression of miR472 decreases plant resistance to bacteria	Arabidopsis	Pst. DC3000	Bacteria	(Boccara et al., 2014; Islam et al, 2017)
17	miR773	MET2	Negatively regulate callose deposition and resistance to bacteria	Arabidopsis	Pst. DC3000	Bacteria	(Salvador-Guirao et al., 2018)

(Continued)

TABLE 8.1 (Continued)
Role of different miRNAs in plant pathogenesis.

S. No	miRNAs	Target	Regulatory role in plant-pathogen interactions	Plants	Pathogen	Nature of pathogen	References
18	miR825	REMORIN, ZINC FINGER HOMEOBOX FAMILY, FRATAXIN-RELATED	Cellular defence responses (hydrogen peroxide production and callose deposition)	Arabidopsis	Pst. DC3000, Bacillus cereus AR156 (Rhizobacterium), B. cinerea	Bacteria	(Nie et al., 2017, 2019)
19	miR863	ARLPK1, ARLPK2, SE	Regulate disease resistance and plant immunity	Arabidopsis	Pst. DC3000	Bacteria and fungus	(Nie et al., 2017)
20	miR863–3p	RLPK1/2	Regulate disease resistance and plant immunity	Arabidopsis	Pst. DC3000	Bacteria	(Nie et al., 2017)
21	miR32	ADP-RIBOSYLATION FACTOR 1-LIKE, BRASSINOSTEROID INSENSITIVE 1-ASSOCIATED RECEPTOR KINASE 1 and SERINE/THREONINE-PROTEIN PHOSPHATASE	Provide indicators of phytoplasma infection	P. tomentosa × P. fortunei	Phytoplasma	Bacteria	(Fan et al., 2015)
22	miR34	NBS-LRR RESISTANCE PROTEIN, AUXIN-INDUCED PROTEIN PCNT1 and LECTIN-RECEPTOR LIKE PROTEIN KINASE 3	Provide indicators of phytoplasma infection	P. tomentosa × P. fortunei	Phytoplasma	Bacteria	(Fan et al., 2015)
23	miR41	CALCIUM-TRANSPORTING ATPASE 4, GLUTATHIONE S-TRANSFERASE T1-LIKE and ELONGATION FACTOR TUA	Provide indicators of phytoplasma infection	P. tomentosa × P. fortunei	Phytoplasma	Bacteria	(Fan et al., 2015)
24	miR46b	UNCHARACTERIZED PROTEIN	Provide indicators of phytoplasma infection	P. tomentosa × P. fortunei	Phytoplasma	Bacteria	(Fan et al., 2015)
25	miR90	PENTATRICOPEPTIDEREPEAT-CONTAINING PROTEIN	Provide indicators of phytoplasma infection	P. tomentosa × P. fortunei	Phytoplasma	Bacteria	(Fan et al., 2015)

(Continued)

TABLE 8.1 (Continued)
Role of different miRNAs in plant pathogenesis.

S. No	miRNAs	Target	Regulatory role in plant-pathogen interactions	Plants	Pathogen	Nature of pathogen	References
26	miR4414	SUBTILASE	Provide indicators of phytoplasma infection	P. tomentosa × P. fortunei	Phytoplasma	Bacteria	(Fan et al., 2015)
27	miR403	AGO2, AGO3, abkC, ANKYRIN REPEAT FAMILY PROTEIN and MALATE DEHYDROGENASE	Hormone homeostasis, cell proliferation/death and other metabolic pathways	P. tomentosa, Soybean	Phytoplasma, Phytophthora sojae	Bacteria and fungus-like microorganism	(Fan et al., 2015; Guo et al., 2011; Litholdo Junior et al., 2017)
28	miR6019	N (R gene), TIR-NBS-LRR	Overexpression of miR6019 attenuates N-gene-mediated resistance to viruses	N. tabacum	Tobacco mosaic virus	Virus	(Li et al., 2012)
29	miR6020	N (R gene), TIR-NBS-LRR	Overexpression of miR6020 attenuates N-gene-mediated resistance to viruses	N. tabacum	Tobacco mosaic virus	Virus	(Li et al., 2012)
30	miR6021	Homologs of Cladosporium fulvum resistance 9 (Hcr9)	Pathogen immunity response	N. tabacum	Tobacco mosaic virus	Virus	(Li et al., 2012)
31	miR6022	Hcr9	Pathogen immunity response	N. tabacum	Tobacco mosaic virus	Virus	(Li et al., 2012)
32	miR6023	Hcr9	Pathogen immunity response	N. tabacum	Tobacco mosaic virus	Virus	(Li et al., 2012)
33	miR827	NLA/SPX-MFS	Controls NLA for defence	N. benthamiana	Heteroderaschachtii	Nematode	(Hewezi et al., 2016)
34	miR2111	F-BOX/KELCH-REPEAT PROTEIN, TOO MUCH LOVE	Pathogen immunity response, regulatory plant growth, and development, rhizobium-legume interaction	N. benthamiana, L. japonicus.	A. rhizogenes	Bacteria	(Huang et al., 2019; Huen et al., 2018; Tsikou et al., 2018)
35	miR168	AGO1	A virulence mechanism triggered by the fungus that affects the plant siRNA machinery and AGO1 mutant shows compromised innate immunity and disease resistance	O. sativa, B. napus	Verticillium longisporum, F. oxysporum	Fungus	(Baldrich & San Segundo, 2016; Li et al., 2011; Litholdo Junior et al., 2017; Staiger et al., 2013)

(Continued)

TABLE 8.1 (Continued)
Role of different miRNAs in plant pathogenesis.

S. No	miRNAs	Target	Regulatory role in plant-pathogen interactions	Plants	Pathogen	Nature of pathogen	References
36	miR319	*TCP*	Pathogen immunity response by suppression of jasmonic acid	*O. sativa*	RRSV	Virus	(Song et al., 2019)
37	miR444	*MADS*	Pathogen immunity response	*O. sativa*	RSV	Virus	(Song et al., 2019)
38	miR528	*AO/LAC*	Pathogen immunity response	*O. sativa*	RBSDV and RSV	Virus	(Wu et al., 2017)
39	miR535	*RESISTANCE-LIKE PROTEIN*	Pathogen immunity response	*O. sativa, Cassava*	RSV	Virus	(Li et al., 2013; Pérez-Quintero et al., 2012; Song et al., 2019)
40	miR7695	*NRAMP6*	Disease resistance and plant immunity	*O. sativa, N. tabacum*	*M. oryzae*	Fungus	(Campo et al., 2013)
41	miR172	*AP2*	Symbiotic relationship and nitrogen fixation	*S. lycopersicum*	Potato spindle tuber viroid (PSTVd-D)	Viroid	(Tsushima et al., 2015)
42	miR5300	R genes, *Solyc05g008650, tm-2*	Fungus infection downregulates the accumulation of miR5300	*S. lycopersicum*	*F. oxysporum*	Fungus	(Litholdo Junior et al., 2017; Ouyang et al., 2014)
43	miR482	R genes, *NBS-LRR*	Defence against pathogen attack and downregulates the expression of miR482	*S. tuberosum, S. lycopersicum, Cotton*	*Pst. DC3000, Verticillium dahliae,* ToMV, CMV and TRV	Bacteria and virus	(Islam et al., 2017; Litholdo Junior et al., 2017; Shivaprasad et al., 2012; Yang et al., 2015; Zhu et al., 2013)
44	miR2118	*NBS-LRR*	Promote bacterial colonization for the growth of the N-fixing nodule	*S. lycopersicum* and *M. truncatula*	*Sinorhizobium (Ensifer) meliloti 1021*	Bacteria	(Sós-Hegedüs et al., 2020; Tiwari et al., 2021)

(Continued)

TABLE 8.1 (Continued)
Role of different miRNAs in plant pathogenesis.

S. No	miRNAs	Target	Regulatory role in plant-pathogen interactions	Plants	Pathogen	Nature of pathogen	References
45	miR171h	*NSP2*	Inhabit the over-colonization of roots by arbuscular mycorrhizal fungi *via* miRNA-mediated regulation of NSP2	*M. truncatula*	*Rhizophagusirregularis*	Fungus	(Hofferek et al., 2014)
46	miR1507	*NBS-LRR*	Formation of the nitrogen-fixing nodule by bacterial colonization	*M. truncatula*	*Sinorhizobium* (*Ensifer*) *meliloti* 1021	Bacteria	(Sós-Hegedüs et al., 2020)
47	miR2109	*NBS-LRR*	Facilitates bacterial colonization and the development of an N-fixing nodule	*M. truncatula*	*Sinorhizobium* (*Ensifer*) *meliloti* 1021	Bacteria	(Sós-Hegedüs et al., 2020)
48	miR394	*F-BOX DOMAIN* and *CYP450*	Regulates resistance against *F. oxysporum*	*Allium sativum L*	*F. oxysporum*	Fungus	(Chand et al., 2016)
49	miR395	*WRKY26*	Regulates leaf spot disease resistance in apple	Apple	*Alternaria alternaria*	Fungus	(Zhang et al., 2017)
50	miR397	*LAC*	Involves in pathogen resistance through cell wall protection	*M. hupehensis*	*Botryosphaeria dothidea*	Fungus	(Yu et al., 2020)
51	miR399	*PHO2*	Involve in phosphorus homeostasis	Citrus	*Candidatus Liberibacter asiaticus*	Bacteria	(Islam et al., 2017; Zhao et al., 2013)
52	miR477	*PHENYLALANINE AMMONIA-LYASE* and *CBP60A*	Overexpressing Csn-miR477	*C. sinensis*	*Pseudopestalotiopsis* species	Fungus	(Wang et al., 2020)

(Continued)

TABLE 8.1 (Continued)
Role of different miRNAs in plant pathogenesis.

S. No	miRNAs	Target	Regulatory role in plant-pathogen interactions	Plants	Pathogen	Nature of pathogen	References
53	miR530	*UNCHARACTERIZED PROTEIN*	Pathogen immunity response	*C. arietinum*	*F. oxysporum*	Fungus	(Kohli et al., 2014)
54	miR1023	*FGSG_03101*	Suppresses invasion of *Fusarium* species	Wheat	*F. graminearum*	Fungus	(Jiao & Peng, 2018)
55	miR1885	*TNL1, CP24*	Innate immunity and plant growth and response to viral infection	*B. napus*	Turnip mosaic virus (TuMV)	Virus	(Cui et al, 2020)
56	miR9863	*MLA1(R* gene)	Cell death signaling and fungal resistance	*H. vulgare*, *T. aestivum*	*B. graminis*	Fungus	(Liu et al., 2014)

Segundo, 2016; Campo et al., 2013). The miR7695 downregulates the expression level of *NATURAL RESISTANCE-ASSOCIATED MACROPHAGE PROTEIN 6*, (OsNRAMP6) transcript (Sánchez-Sanuy et al., 2019). In rice, OsNRAMP6 protein is involved in iron and manganese transportation. The loss of function mutant plant OsNRAMP6 showed increased resistance against *M. oryzae*, indicating OsNRAMP6 negatively modulates rice immunity (Peris-Peris et al., 2017). Furthermore, miR169 targets *NFYA/HAPLESS 2* (*HAP2*) transcription factor and promotes immunity against the biotic factors causing disease in rice crops. Thus, miR169 has a negative regulatory function in plant immunity. In rice, miR319 targets TCPTFs and induces an immunity response (Zhang et al., 2018b). OsTCP21 is shown to play a positive role against the blast disease in rice. It has been observed that *LIPOXYGENASE2* (*LOX2*) and *LOX5* were involved in downregulating the expression of miR319 (Zhang et al., 2018b). LOXs are the crucial enzyme for jasmonic acid involved in the synthesis of hydroperoxy-octadecadienoic acid. Detailed investigation showed that enhanced OsLOX2/5 transcript also provides resistance against blast disease (Zhang et al., 2018b). Additionally, some other miRNAs such as miR528, miR535, and miR164 have been shown to plant immunity. The miR528 targets *L-ASCORBATE OXIDASE* (*AO*) gene and modulate immunity against rice stripe virus (RSV) and rice black-streaked dwarf virus (RBSDV) (Wu et al., 2017). The miR535 targets resistance-like protein, which provides immunity against the viral pathogen RSV in *O. sativa* and *Cassava* (Li et al., 2013; Pérez-Quintero et al., 2012; Song et al., 2019). The miR164 targets NAC TFs. In wheat, miR164 provides resistance to stripe rust by targeting TaNAC21/22 against *Puccinia striiformi* (Feng et al., 2015).

8.2.4 INVOLVEMENT OF MIRNAs IN *PULWONIA SPECIES* DURING PATHOGENESIS

The two species *P. tomentosa* and *P. fortunei* belonging to the family Paulowniaceae are deciduous trees, and it is native to central and western China, commonly known as dragon tree. These plants are economically very important for timber, and these timbers of *Pulwonia* are extensible and affected by a disease caused by phytoplasma, widely known as Paulownia witches' broom (PaWB) (Fan et al., 2015). It has been found that several miRNAs like miR32, miR34, miR41, miR46b, miR90, and miR4414 are involved in plant hormone signaling and pathogen interactions by targeting the pathogen-related genes (**Table 8.1**) (Fan et al., 2015). So these findings support that miRNAs such as miR32, miR34, miR41, miR46b, miR90, and miR4414 play an important role against Paulownia witches' broom, caused by a phytoplasma (Fan et al., 2015).

8.3 SOME OTHER MIRNAS DURING PLANT-PATHOGEN INTERACTION

The plant miRNAs have a differential regulatory role in response to a variety of pathogen attacks (**Figure 8.3**). These miRNAs play a critical role in inducing immunity and have various isoforms of the AGO protein in the functioning of miRNA (Li et al., 2022). In *Arabidopsis*, AGO1/AGO2 previously have been shown to provide resistance to specific viruses (Litholdo Junior et al., 2017). The miR168 is negatively involved in the regulation of *ARGONAUTE1* and provides innate immunity against the fungal pathogen *Verticillium longisporum* and *F. oxysporum* in *O. sativa* and *B. napus* (Litholdo Junior et al., 2017). Many legumes have been demonstrated to form symbiotic relationships with arbuscular mycorrhizal fungi (AMF) and rhizobia (Scheublin & van der Heijden, 2006). This type of endosymbiosis is regulated by *NODULATION SIGNALING PATHWAY 2* (*NSP2*). It has been found that miR171h, which targets *NSP2*, induced *Medicago* root cells during the colonization of *Rhizophagus irregularis* (Lauressergues et al., 2012). Furthermore, miR171h reduces fungal colonization by mimicking the root morphology mutant NSP2 (Lauressergues et al., 2012). The miR394 was significantly upregulated in garlic seedlings in response to *Fusarium oxysporum* infection (Chand et al., 2016). When the garlic cultivar was exposed to FOC, miR394 was induced, and its target was inversely affected in both varieties (CBT-As153 and CBT-As21) (Chand et al., 2016). However, in the resistant genotypes, the expression pattern of miR394 was delayed. This observation

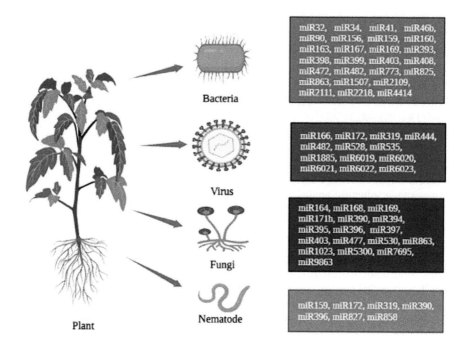

Bacteria

miR32, miR34, miR41, miR46b, miR90, miR156, miR159, miR160, miR163, miR167, miR169, miR393, miR398, miR399, miR403, miR408, miR472, miR482, miR773, miR825, miR863, miR1507, miR2109, miR2111, miR2218, miR4414

Virus

miR166, miR172, miR319, miR444, miR482, miR528, miR535, miR1885, miR6019, miR6020, miR6021, miR6022, miR6023,

Fungi

miR164, miR168, miR169, miR171h, miR390, miR394, miR395, miR396, miR397, miR403, miR477, miR530, miR863, miR1023, miR5300, miR7695, miR9863

Nematode

miR159, miR172, miR319, miR390, miR396, miR827, miR858

Plant

FIGURE 8.3   Differential function of miRNAs in pathogen attacks against bacteria, viruses, fungi, and nematodes (A general illustration).

demonstrated that miR394 has a specific role in inhibiting FOC resistance (Chand et al., 2016). Recently, in apple (*Malus × domestica*), Md-miR156ab and Md-miR395 has been identified to target a novel WRKY TF, *MdWRKYN1*, and *MdWRKY26* respectively, upon infection with *Alternaria alternata f. sp. mali* (ALT1), an apple leaf spot fungus (Zhang et al., 2017). Furthermore, it was shown that both overexpression of *MdWRKYN1* and *MdWRKYN26* or knocked down of Md-miR156ab induced disease resistance by upregulating WRKY-regulated pathogenesis-related (PR) protein-encoding genes (Zhang et al., 2017). Contrastingly, overexpression of Md-miR156ab and Md-miR395 promotes susceptibility to *Alternaria alternata* (ALT1) (Zhang et al., 2017). Currently, it was reported that transient overexpression of miR397b activates sensitivity in *Malus hupehensis* after infection with *Botryosphaeria dothidea* by lowering *LAC7* expression and lignin concentration (Yu et al., 2020). Previously, miR399 is induced by phosphorus starvation (Pant et al., 2008). Further research showed that miR399 is induced by *Candidatus Liberibacter asiaticus* (Las), the bacteria, citrus stubborn disease with symptoms similar to Huanglongbing (Litholdo Junior et al., 2017; Zhao et al., 2013). In *Paulownia tomentosa*, miR403 has a positive role against phytoplasma (Fan et al., 2015). Furthermore, when soybean roots and stems are infected with *Phytophthora sojae*, miR403 has been shown to downregulate, a miRNA that targets *AGO2* and modulates plant immunity (Huang et al., 2016a). The monocot-specific miR444 is upregulated and inversely regulates the MIKC(C)-type MADS-box proteins (Wang et al., 2016). The induced expression of miR444 tail off the inhibitory effect of *OsMADS23*, *OsMADS27a*, and *OsMADS57* on *OsRDR1* transcription and activate *the OsRDR1*-dependent defence pathway (Wang et al., 2016). The miR477 have a PHENYLALANINE AMMONIA LYASE gene target and were found to be downregulated in tea plants after infection with *Pseudo pestalotiopsis species* (Wang et al., 2020). Furthermore, when plants overexpressing Csn-miR477 were infected with a pathogen, the damage was more pronounced in wild-type plants (Wang et al., 2020). The overexpression of a miR482 family member increased potato sensitivity to fungal infection by targeting *the NBS-LRR* genes family (Islam et al., 2017). Further research showed that when cotton leaves and roots were infected with *Verticillium dahlia* (*V. dahlia*), three miR482 variants were downregulated, whereas 10 of the 11 predicted miR482 target genes were upregulated (Islam et al., 2017). Cotton *NBS-LRRs* have conserved sequences that miR482 recognizes, similar to tomato *NBS-LRRs*. As a result, miR482-dependent regulation of NBS-LRR in cotton is a master regulator for induced immune responses (Islam et al., 2017). In *Cicer arietinum*, miR530 was highly upregulated under FOC and regulates the synthesis of microtubule-associated proteins, suggesting miR530 has a crucial role in plant defence (Kohli et al., 2014). In *Arabidopsis*, miR827 is highly upregulated in response to *Heterodera schachti* (Hewezi et al., 2016). Further analysis shows upregulation of miR827 induces susceptibility to the strain *H. schachtii* (Hewezi et al., 2016). More than 60% of the NB-LRR genes in the *Medicago truncatula* genome could be targeted by the members of the miR482 superfamily or phasiRNAs family (Sós-Hegedűs et al., 2020). It has been shown that when *M. truncatula* was infected with the alfalfa mosaic virus, the various miRNA such as miR2118a, miR1507, and miR2109 have been downregulated in the cotyledons at four days post-inoculation (Sós-Hegedűs et al., 2020). After inoculating with *S. melilotiexo* Y mutant, miR2118a, miR1507, and miR2109 were shown to be induced in *Medicago* root that was similar to WT rhizobia (Sós-Hegedűs et al., 2020). Thus, by regulating NB-LRR genes, these miRNAs (miR2118a, miR1507, and miR2109) have an important role in the development of symbiotic interactions with rhizobia (Sós-Hegedűs et al., 2020). Both the immune receptor gene and plant developmental gene have been demonstrated as a target for miR1885 in *Brassica* (Cui et al., 2020). The miR1885 has been demonstrated to directly target the TIR-NBS-LRR class of the R gene *BraTNL1* (Cui et al., 2020). On the other side, miR1885 represses *BraCP24* by targeting the gene *BraTIR1* (Cui et al., 2020). During early infection with *Plasmodiophora brassicae*, miR1885 was shown to regulate disease resistance in *Brassica rapa* by modulating TIR-NBS genes. Recently, in *Lotus japonicus* root, miR2111 has been found to regulate *TOO MUCH LOVE* (*TML*) non-cell autonomously known to nodulation processes (Huang et al., 2019; Tsikou et al., 2018). The Mla alleles of Barley (*Hordeum vulgare* L.) encode NB-LRR receptors, for the powdery mildew fungus *Blumeria graminis* f. sp. *hordei* (Liu et al., 2014). Further miR9863 family differentially involve in Mla alleles and reduces the accumulation of MLA1 protein in Barley and *N. benthamiana* (Liu et al., 2014). Furthermore, overexpression of miR9863 reduces cell death and disease resistance signaling mediated by MLA1 (Liu et al., 2014). Thus, using miRNAs and their targets is one of the most effective methods for understanding the precise roles of individual pathogen stress. However, understanding the specific function of miRNA towards pathogen interaction required more functional and expression studies.

8.4 USE OF IN SILICO APPROACHES IN ADVANCING PLANT MIRNA RESEARCH

In the past few years, the mythology for the isolation and structural and functional characterization of miRNAs has been developed to a great extent and emerged novel bioinformatics and experimental approaches. Besides traditional techniques used for genetic modifications such as loss of function, a gain of function, cloning, RNA blot, and microarray, next-generation sequencing (NGS) technology has been implicated in various studies such as prediction, validation, and identification of miRNAs. The basic strategy for miRNAs analysis required miRNAs datasets, background knowledge, target identification, prediction through alignment score, and downstream pathway of target genes. Additionally, the detection and validation of miRNAs through molecular approaches were further supported by using in silico approaches (Bonnet et al., 2004). These approaches also provide complete knowledge of identifying miRNAs which is difficult to clone due to their lower expression such as miR395 and miR399 (Bellato et al., 2019). By using in silico approaches, several strategies have been developed for miRNAs detection and identification in various plants such as *Arabidopsis*, maize, foxtail millet, tomato, soybean, *Brassica napus*, grape, apple, and some other economically important plants (Bellato et al., 2019). The advancing of in silico approaches thus provides the correct identification of precursor and mature miRNAs by using machine learning–based applications such as miRBase, miRPara, MiPred, and miREval (Gao et al., 2013; Wu et al., 2011; Yadav et al., 2020). Further, miR-BAG and miRDeep servers are used to analyze NGS data and, recently, miR-Deep-P2, which is procured from miRDeep-P with new strategies and overhauled algorithm for accurate and fast analysis (Kuang et al., 2019; Yadav et al., 2020). The advancement of sequencing technologies thus provides complete and accurate knowledge of miRNA profiling that could differentiate a single base pair between two fragments. With the advancement of in silico approaches and the availability of genomic sequences, now it is possible to provide accurate information on miRNA annotation. Even huge plant miRNA datasets can be obtained from NGS, yet these annotations are still not clear due to false annotation of siRNAs as miRNAs (Axtell & Meyers, 2018; Yadav et al., 2020). Hence, the "era of big data" needs a reliable and user-friendly, trustworthy web-based server that highlights the criteria for minimizing false positives (Yadav et al., 2020). Short Stack is the most customizable all-in-one contesting software that can effectively provide comprehensive annotation by analyzing highly diverse datasets and providing the required information for novel classification of small RNAs including miRNAs and hairpin-association, small RNA size distribution, strandedness, phasing, repetitiveness, and quantification (Yadav et al., 2020).

8.5 BIOTECHNOLOGICAL APPLICATIONS OF NOVEL TECHNOLOGIES

The miRNAs have been known for their implication in various developmental processes in both plants and animals, miRNAs identification, detection, quantification, target prediction, and validation are exciting areas of research. Currently, a multiplexed northern technique developed for the detection of multiple miRNAs in a single run (Yadav et al., 2020). Nowadays, molecular breeding, antisense RNA technology, and genome editing technologies such as clustered regularly interspersed short palindrome repeats (CRISPR/Cas9), zinc-finger nucleases (ZFN), transcription activator-like effector nucleases, and (TALENs) have been used for the improvement of plant fitness against various diseases (Li et al., 2020; Yadav et al., 2020). Recently, micro peptides (miPEPs) have been shown to regulate plant growth and developmental processes *by* modulating the rhizospheric microbiota (Middleton et al., 2021; Yadav et al., 2020). Moreover, the application of miPEPs could be utilized to improve plant immunity against various pathogens to enhance the productivity of plants. Thus, miPEPs technology could be a promising tool for holobiont engineering (Middleton et al., 2021). Alternatively, the applications of mutations on cis-regulatory elements of pathogens and disease-related *miRNA* genes have also been reported for the modulation and expression of miRNAs (Yadav et al., 2020). Undoubtedly, the applications of novel techniques such as CRISPR/Cas9, ZFN, TALENs, and miPEPs along with miRNA-target modules have interaction between plants and pathogens (Yadav et al., 2020).

8.6 CONCLUSION AND FUTURE PERSPECTIVES

With the increasing world population, the demand for food consumption is increasing day by day. However, the unpredictable environmental conditions facilitate both abiotic and biotic stresses which further decreased natural resources and required a continuous supply of nutrients and fertilizers (such as nitrogen, phosphorus, and potassium) to improve plant fitness against invasive pathogens. The miRNAs have been already demonstrated to cope with various pathogen responses and modulate plant defence mechanisms against pathogens. Therefore, research on miRNA-target provides valuable information related to the various molecular pathways during plant-pathogen interaction. Recent evidence largely showed that miRNA has regulatory action of gene expression and emerged as a powerful technique and become a regulatory hub and versatile toolbox against invading pathogens. Moreover, miRNAs have emerged as a useful and prognostic tool as biomarkers for disease-resistance traits in designing different breeding programs for biotic resistance. Although various miRNAs have been already characterized for their possible involvement in plant-pathogen interaction and disease resistance. However, the proper function and molecular mechanisms for miRNA-mediated gene regulation in plants need more exploration. It has been recently reported that cross-kingdom communication of miRNAs played a crucial role in plant-microbe interactions, especially in rhizospheric microbial communities. However, the more fundamental questions and deciphering the molecular mechanism behind it are required to uncover the interaction of plant-pathogen and the involvement of miRNAs in their regulation.

8.6.1 ACKNOWLEDGEMENTS

We acknowledge the DST-FIST and CAS (Botany) facilities of the Department of Botany, Banaras Hindu University, Varanasi, India. VM is grateful to DBT-eLibrary Consortium (DeLCON) for e-resources and the Department of Biotechnology (DBT), NIPGR, New Delhi, India (DBT/JRF/15/AL/223) for fellowship.

8.6.2 AUTHOR CONTRIBUTIONS

BM and VM conceptualized the outline of the manuscript. SP provided scientific feedback and critical suggestions.

8.6.3 CONFLICT OF INTEREST

All authors declare that they have no conflict of interest.

REFERENCES

Achkar, N. P., Cambiagno, D. A., & Manavella, P. A. (2016). miRNA biogenesis: A dynamic pathway. *Trends in Plant Science, 21,* 1034–1044.

Adnan, M., Islam, W., Shabbir, A., Khan, K. A., Ghramh, H. A., Huang, Z., Chen, H. Y. H., & Lu, G.-d. (2019). Plant defense against fungal pathogens by antagonistic fungi with Trichoderma in focus. *Microbial Pathogenesis, 129,* 7–18.

Alonso, C., Ramos-Cruz, D., & Becker, C. (2019). The role of plant epigenetics in biotic interactions. *The New Phytologist, 221,* 731–737.

Axtell, M. J., & Meyers, B. C. (2018). Revisiting criteria for plant microRNA annotation in the era of big data. *Plant Cell, 30,* 272–284.

Baldrich, P., & San Segundo, B. (2016). MicroRNAs in rice innate immunity. *Rice (N Y), 9,* 6.

Bellato, M., De Marchi, D., Gualtieri, C., Sauta, E., Magni, P., Macovei, A., & Pasotti, L. (2019). A bioinformatics approach to explore microRNAs as tools to bridge pathways between plants and animals. Is DNA damage response (DDR) a potential target process? *Frontiers in Plant Science, 10,* 1535.

Boccara, M., Sarazin, A., Thiébeauld, O., Jay, F., Voinnet, O., Navarro, L., & Colot, V. (2014). The Arabidopsis miR472-RDR6 silencing pathway modulates PAMP- and effector-triggered immunity through the post-transcriptional control of disease-resistance genes. *PLoS Pathogens, 10,* e1003883.

Bonnet, E., Wuyts, J., Rouzé, P., & Van de Peer, Y. (2004). Detection of 91 potential conserved plant microRNAs in Arabidopsis thaliana and Oryza sativa identifies important target genes. *Proceedings of the National Academy of Sciences of the United States of America, 101,* 11511–11516.

Campo, S., Peris-Peris, C., Siré, C., Moreno, A. B., Donaire, L., Zytnicki, M., Notredame, C., Llave, C., & San Segundo, B. (2013). Identification of a novel microRNA (miRNA) from rice that targets an alternatively spliced transcript of the Nramp6 (Natural resistance-associated macrophage protein 6) gene involved in pathogen resistance. *The New Phytologist, 199,* 212–227.

Caruana, J. C., Dhar, N., & Raina, R. (2020). Overexpression of Arabidopsis microRNA167 induces salicylic acid-dependent defense against Pseudomonas syringae through the regulation of its targets ARF6 and ARF8. *Plant Direct, 4,* e00270.

Chand, S. K., Nanda, S., & Joshi, R. K. (2016). Regulation of miR394 in response to fusarium oxysporum f. Sp. Cepae (FOC) infection in garlic (Allium sativum L). *Frontiers in Plant Science, 7,* 258.

Chen, L., Luan, Y., & Zhai, J. (2015). Sp-miR396a-5p acts as a stress-responsive genes regulator by conferring tolerance to abiotic stresses and susceptibility to phytophthora nicotianae infection in transgenic tobacco. *Plant Cell Reports, 34,* 2013–2025.

Chen, X. (2009). Small RNAs and their roles in plant development. *Annual Review of Cell and Developmental Biology, 25,* 21–44.

Chiou, T.-J., Aung, K., Lin, S.-I., Wu, C.-C., Chiang, S.-F., & Su, C.-l. (2005). Regulation of phosphate homeo-stasis by microRNA in Arabidopsis. *The Plant Cell, 18*, 412–421.

Chisholm, S. T., Coaker, G., Day, B., & Staskawicz, B. J. (2006). Host-microbe interactions: Shaping the evolu-tion of the plant immune response. *Cell, 124*, 803–814.

Chow, H. T., & Ng, D. W. K. (2017). Regulation of miR163 and its targets in defense against Pseudomonas syringae in Arabidopsis thaliana. *Scientific Reports, 7*, 46433.

Cui, C., Wang, J. J., Zhao, J. H., Fang, Y. Y., He, X. F., Guo, H. S., & Duan, C. G. (2020). A brassica miRNA regulates plant growth and immunity through distinct modes of action. *Molecular Plant, 13*, 231–245.

Fahlgren, N., Howell, M. D., Kasschau, K. D., Chapman, E. J., Sullivan, C. M., Cumbie, J. S., Givan, S. A., Law, T. F., Grant, S. R., Dangl, J. L., & Carrington C. J. (2007). High-throughput sequencing of Arabidopsis microRNAs: Evidence for frequent birth and death of miRNA genes. *PLoS One, 2*, e219.

Falak, N., Imran, Q. M., Hussain, A., & Yun, B.-W. (2021). Transcription factors as the "blitzkrieg" of plant defense: A pragmatic view of nitric oxide's role in gene regulation. *International Journal of Molecular Sciences, 22*, 522.

Fan, G., Niu, S., Xu, T., Deng, M., Zhao, Z., Wang, Y., Cao, L., & Wang, Z. (2015). Plant-pathogen interaction-related microRNAs and their targets provide indicators of phytoplasma infection in paulownia tomentosa × paulownia fortunei. *PloS One, 10*, e0140590.

Feng, H., Wang, B., Zhang, Q., Fu, Y., Huang, L., Wang, X., & Kang, Z. (2015). Exploration of microRNAs and their targets engaging in the resistance interaction between wheat and stripe rust. *Frontiers in Plant Science, 6*, 469–469.

Feuerstein, A., Niedermeier, M., Bauer, K., Engelmann, S., Hoth, S., Stadler, R., & Sauer, N. (2010). Expression of the AtSUC1 gene in the female gametophyte, and ecotype-specific expression differences in male reproductive organs. *Plant Biology (Stuttgart, Germany), 12*(Suppl 1), 105–114.

Ficke, A., Cowger, C., Bergstrom, G., & Brodal, G. (2018). Understanding yield loss and pathogen biology to improve disease management: Septoria nodorum blotch-a case study in wheat. *Plant Disease, 102*, 696–707.

Gao, D., Middleton, R., Rasko, J. E. J., & Ritchie, W. (2013). miREval 2.0: A web tool for simple microRNA prediction in genome sequences. *Bioinformatics, 29*, 3225–3226.

Gautam, V., Singh, A., Verma, S., Kumar, A., Kumar, P., Mahima, Singh, S., Mishra, V., & Sarkar, A. K. (2017). Role of miRNAs in root development of model plant Arabidopsis thaliana. *Indian Journal of Plant Physiology, 22*, 382–392.

Gullner, G., Komives, T., Király, L., & Schröder, P. (2018). Glutathione s-transferase enzymes in plant-pathogen interactions. *Frontiers in Plant Science, 9*, 1836.

Guo, N., Ye, W. W., Wu, X. L., Shen, D. Y., Wang, Y. C., Xing, H., & Dou, D. L. (2011). Microarray profiling reveals microRNAs involving soybean resistance to Phytophthora sojae. *Genome, 54*, 954–958.

Hanemian, M., Barlet, X., Sorin, C., Yadeta, K. A., Keller, H., Favery, B., Simon, R., Thomma, B. P., Hartmann, C., Crespi, M., Marco, Y., Tremousaygue, D., & Deslandes, L. (2016). Arabidopsis CLAVATA1 and CLAVATA2 receptors contribute to Ralstonia solanacearum pathogenicity through a miR169-dependent pathway. *The New phytologist, 211*, 502–515.

Hewezi, T., Piya, S., Qi, M., Balasubramaniam, M., Rice, J. H., & Baum, T. J. (2016). Arabidopsis miR827 mediates post-transcriptional gene silencing of its ubiquitin E3 ligase target gene in the syncytium of the cyst nematode Heterodera schachtii to enhance susceptibility. *The Plant journal: For Cell and Molecular Biology, 88*, 179–192.

Hofferek, V., Mendrinna, A., Gaude, N., Krajinski, F., & Devers, E. A. (2014). MiR171h restricts root symbio-ses and shows like its target NSP2 a complex transcriptional regulation in Medicago truncatula. *BMC Plant Biology, 14*, 199.

Huang, C. Y., Wang, H., Hu, P., Hamby, R., & Jin, H. (2019). Small RNAs-big players in plant-microbe interac-tions. *Cell Host Microbe, 26*, 173–182.

Huang, J., Yang, M., Lu, L., & Zhang, X. (2016a). Diverse functions of small RNAs in different plant-pathogen communications. *Frontiers in Microbiology, 7*, 1552.

Huang, J., Yang, M., & Zhang, X. (2016b). The function of small RNAs in plant biotic stress response. *Journal of Integrative Plant Biology, 58*, 312–327.

Huen, A., Bally, J., & Smith, P. (2018). Identification and characterisation of microRNAs and their target genes in phosphate-starved Nicotiana benthamiana by small RNA deep sequencing and 5'RACE analysis. *BMC Genomics, 19*, 940.

Ingle, R. A., Carstens, M., & Denby, K. J. (2006). PAMP recognition and the plant-pathogen arms race. *BioEssays: News and Reviews in Molecular, Cellular and Developmental Biology, 28*, 880–889.

Isaac-Renton, M., Montwé, D., Hamann, A., Spiecker, H., Cherubini, P., & Treydte, K. (2018). Northern forest tree populations are physiologically maladapted to drought. *Nature Communications, 9*, 5254.

Islam, W., Islam, S. U., Qasim, M., & Wang, L. (2017). Host-pathogen interactions modulated by small RNAs. *RNA Biology, 14*, 891–904.

Jagadeeswaran, G., Saini, A., & Sunkar, R. (2009). Biotic and abiotic stress down-regulate miR398 expression in Arabidopsis. *Planta, 229*, 1009–1014.

Jiao, J., & Peng, D. (2018). Wheat microRNA1023 suppresses invasion of Fusarium graminearum via targeting and silencing FGSG_03101. *Journal of Plant Interactions, 13*, 514–521.

Jones, H. R., Robb, C. T., Perretti, M., & Rossi, A. G. (2016). The role of neutrophils in inflammation resolution. *Seminars in Immunology, 28*, 137–145.

Jones, J. D. G., & Dangl, J. L. (2006). The plant immune system. *Nature, 444*, 323–329.

Khraiwesh, B., Zhu, J.-K., & Zhu, J. (2012). Role of miRNAs and siRNAs in biotic and abiotic stress responses of plants. *Biochim Biophys Acta, 1819*, 137–148.

Kohli, D., Joshi, G., Deokar, A. A., Bhardwaj, A. R., Agarwal, M., Katiyar-Agarwal, S., Srinivasan, R., & Jain, P. K. (2014). Identification and characterization of Wilt and salt stress-responsive microRNAs in chickpea through high-throughput sequencing. *PLoS One, 9*, e108851.

Kuang, Z., Wang, Y., Li, L., & Yang, X. (2019). miRDeep-P2: Accurate and fast analysis of the microRNA transcriptome in plants. *Bioinformatics, 35*, 2521–2522.

Lauressergues, D., Delaux, P. M., Formey, D., Lelandais-Brière, C., Fort, S., Cottaz, S., Bécard, G., Niebel, A., Roux, C., & Combier, J. P. (2012). The microRNA miR171h modulates arbuscular mycorrhizal colonization of Medicago truncatula by targeting NSP2. *The Plant Journal: For Cell and Molecular Biology, 72*, 512–522.

Lee, M. H., Jeon, H. S., Kim, H. G., & Park, O. K. (2017). An Arabidopsis NAC transcription factor NAC4 promotes pathogen-induced cell death under negative regulation by microRNA164. *The New Phytologist, 214*, 343–360.

Lesk, C., Coffel, E., D'Amato, A. W., Dodds, K., & Horton, R. (2017). Threats to North American forests from southern pine beetle with warming winters. *Nature Climate Change, 7*, 713–717.

Li, F., Pignatta, D., Bendix, C., Brunkard, J. O., Cohn, M. M., Tung, J., Sun, H., Kumar, P., & Baker, B. (2012). MicroRNA regulation of plant innate immune receptors. *Proceedings of the National Academy of Sciences of the United States of America, 109*, 1790–1795.

Li, J., Li, H., Chen, J., Yan, L., & Xia, L. (2020). Toward precision genome editing in crop plants. *Molecular Plant, 13*, 811–813.

Li, S., Liu, K., Zhang, S., Wang, X., Rogers, K., Ren, G., Zhang, C., & Yu, B. (2017a). STV1, a ribosomal protein, binds primary microRNA transcripts to promote their interaction with the processing complex in Arabidopsis. *Proceedings of the National Academy of Sciences of the United States of America, 114*, 1424–1429.

Li, Y., Jeyakumar, J. M. J., Feng, Q., Zhao, Z.-X., Fan, J., Khaskheli, M. I., & Wang, W.-M. (2019). The roles of rice microRNAs in rice-Magnaporthe oryzae interaction. *Phytopathology Research, 1*, 33.

Li, Y., Lu, Y.-G., Shi, Y., Wu, L., Xu, Y.-J., Huang, F., Guo, X.-Y., Zhang, Y., Fan, J., Zhao, J.-Q., Wu, X. J., Wang, P. R., & Wang, W. M. (2013). Multiple rice microRNAs are involved in immunity against the blast fungus magnaporthe oryzae *Plant Physiology, 164*, 1077–1092.

Li, Y., Wang, W., & Zhou, J.-M. (2011). Role of small RNAs in the interaction between Arabidopsis and Pseudomonas syringae. *Frontiers in Biology, 6*, 462–467.

Li, Y., Zhao, S.-L., Li, J.-L., Hu, X.-H., Wang, H., Cao, X.-L., Xu, Y.-J., Zhao, Z.-X., Xiao, Z.-Y., Yang, N., Huang, F. F., & Wang, W. M. (2017b). Osa-miR169 negatively regulates rice immunity against the blast fungus magnaporthe oryzae. *Frontiers in Plant Science, 8*, 2–2.

Li, Z., Li, W., Guo, M., Liu, S., Liu, L., Yu, Y., Mo, B., Chen, X., & Gao, L. (2022). Origin, evolution and diversification of plant argonaute proteins. *The Plant Journal: For Cell and Molecular Biology, 109*, 1086–1097.

Lin, Y., Chen, M., Lin, H., Hung, Y.-C., Lin, Y., Chen, Y., Wang, H., & Shi, J. (2017). DNP and ATP induced alteration in disease development of Phomopsis longanae Chi-inoculated longan fruit by acting on energy status and reactive oxygen species production-scavenging system. *Food Chemistry, 228*, 497–505.

Litholdo Junior, C., Schwedersky, R., Hemerly, A., & Ferreira, P. (2017). The role of microRNAs in plant-pathogen interactions. *RAPP-Revisão Anual de Patologia de Plantas, 25*, 41–58.

Liu, J., Cheng, X., Liu, D., Xu, W., Wise, R., & Shen, Q. H. (2014). The miR9863 family regulates distinct Mla alleles in barley to attenuate NLR receptor-triggered disease resistance and cell-death signaling. *PLoS Genetics, 10*, e1004755.

Mallory, A. C., Bartel, D. P., & Bartel, B. (2005). MicroRNA-directed regulation of Arabidopsis auxin response factor17 is essential for proper development and modulates expression of early auxin response genes. *Plant Cell, 17*, 1360–1375.

Miao, R., Wang, M., Yuan, W., Ren, Y., Li, Y., Zhang, N., Zhang, J., Kronzucker, H. J., & Xu, W. (2018). Comparative analysis of Arabidopsis ecotypes reveals a role for brassinosteroids in root Hydrotropism. *Plant Physiology, 176*, 2720–2736.

Middleton, H., Yergeau, É., Monard, C., Combier, J. P., & El Amrani, A. (2021). Rhizospheric Plant-microbe interactions: miRNAs as a key mediator. *Trends in Plant Science, 26*, 132–141.

Millar, A. A. (2020). The Function of miRNAs in Plants. *Plants (Basel, Switzerland), 9*, 198.

Millar, A. A., & Gubler, F. (2005). The Arabidopsis GAMYB-like genes, MYB33 and MYB65, are microRNA-regulated genes that redundantly facilitate anther development. *The Plant Cell, 17*, 705–721.

Neumann, M., Mues, V., Moreno, A., Hasenauer, H., & Seidl, R. J. G. c. b. (2017). Climate variability drives recent tree mortality in Europe. *Global Change Biology, 23*, 4788–4797.

Nguyen, Q. M., Iswanto, A. B. B., Son, G. H., & Kim, S. H. (2021). Recent advances in effector-triggered immunity in plants: New pieces in the puzzle create a different paradigm. *International Journal of Molecular Sciences, 22*.

Nie, P., Chen, C., Yin, Q., Jiang, C., Guo, J., Zhao, H., & Niu, D. (2019). Function of miR825 and miR825* as negative regulators in bacillus cereus AR156-elicited Systemic resistance to botrytis cinerea in Arabidopsis thaliana. *International Journal of Molecular Sciences, 20*.

Nie, P., Li, X., Wang, S., Guo, J., Zhao, H., & Niu, D. (2017). Induced systemic resistance against botrytis cinerea by bacillus cereus AR156 through a JA/ET- and NPR1-Dependent signaling pathway and activates PAMP-triggered immunity in Arabidopsis. *Frontiers in Plant Science, 8*, 238.

Ouyang, S., Park, G., Atamian, H. S., Han, C. S., Stajich, J. E., Kaloshian, I., & Borkovich, K. A. (2014). MicroRNAs suppress NB domain genes in tomato that confer resistance to fusarium oxysporum. *PLoS Pathogens, 10*, e1004464–e1004464.

Pant, B. D., Buhtz, A., Kehr, J., & Scheible, W. R. (2008). MicroRNA399 is a long-distance signal for the regulation of plant phosphate homeostasis. *The Plant Journal: For Cell and Molecular Biology, 53*, 731–738.

Park, Y. J., Lee, H. J., Kwak, K. J., Lee, K., Hong, S. W., & Kang, H. (2014). MicroRNA400-guided cleavage of pentatricopeptide repeat protein mRNAs renders Arabidopsis thaliana more susceptible to pathogenic bacteria and fungi. *Plant and Cell Physiology, 55*, 1660–1668.

Pérez-Quintero, Á. L., Quintero, A., Urrego, O., Vanegas, P., & López, C. (2012). Bioinformatic identification of cassava miRNAs differentially expressed in response to infection by xanthomonas axonopodis pv. Manihotis. *BMC Plant Biology, 12*, 29.

Peris-Peris, C., Serra-Cardona, A., Sánchez-Sanuy, F., Campo, S., Ariño, J., & San Segundo, B. (2017). Two NRAMP6 isoforms function as iron and manganese transporters and contribute to disease resistance in Rice. *Molecular Plant-Microbe Interactions, 30*, 385–398.

Porter, S. S., Bantay, R., Friel, C. A., Garoutte, A., Gdanetz, K., Ibarreta, K., Moore, B. M., Shetty, P., Siler, E., & Friesen, M. L. J. F. E. (2020). Beneficial microbes ameliorate abiotic and biotic sources of stress on plants. *Functional Ecology, 34*, 2075–2086.

Reyer, C. P., Leuzinger, S., Rammig, A., Wolf, A., Bartholomeus, R. P., Bonfante, A., de Lorenzi, F., Dury, M., Gloning, P., Abou Jaoudé, R., Klein, T., Kuster, T. M., Martins, M., Niedrist, G., Riccardi, M., Wohlfahrt, G., de Angelis, P., de Dato, G., François, L., Menzel, A., & Pereira, M. (2013). A plant's perspective of extremes: Terrestrial plant responses to changing climatic variability. *Global Change Biology, 19*, 75–89.

Reyes, J. L., & Chua, N. H. (2007). ABA induction of miR159 controls transcript levels of two MYB factors during Arabidopsis seed germination. *The Plant Journal: For Cell and Molecular Biology, 49*, 592–606.

Rodriguez, R. E., Ercoli, M. F., Debernardi, J. M., Breakfield, N. W., Mecchia, M. A., Sabatini, M., Cools, T., De Veylder, L., Benfey, P. N., & Palatnik, J. F. (2015). MicroRNA miR396 regulates the switch between stem cells and transit-amplifying cells in Arabidopsis roots. *The Plant Cell, 27*, 3354–3366.

Ruiz-Ferrer, V., & Voinnet, O. (2009). Roles of plant small RNAs in biotic stress responses. *Annual Review of Plant Biology, 60*, 485–510.

Salvador-Guirao, R., Baldrich, P., Weigel, D., Rubio-Somoza, I., & San Segundo, B. (2018). The microRNA miR773 is involved in the Arabidopsis immune response to fungal pathogens. *Molecular Plant-Microbe Interactions, 31*, 249–259.

Sánchez-Sanuy, F., Peris-Peris, C., Tomiyama, S., Okada, K., Hsing, Y. I., San Segundo, B., & Campo, S. (2019). Osa-miR7695 enhances transcriptional priming in defense responses against the rice blast fungus. *BMC Plant Biology, 19*, 563.

Satapathy, L., Kumar, D., & Mukhopadhyay, K. (2017). WRKY Transcription factors: Involvement in plant–pathogen interactions. In P. Shukla (Ed.), *Recent advances in applied microbiology* (pp. 229–246). Springer.

Savary, S., Willocquet, L., Pethybridge, S. J., Esker, P., McRoberts, N., & Nelson, A. (2019). The global burden of pathogens and pests on major food crops. *Nature Ecology & Evolution, 3*, 430–439.

Scheublin, T. R., & van der Heijden, M. G. (2006). Arbuscular mycorrhizal fungi colonize nonfixing root nodules of several legume species. *The New Phytologist, 172*, 732–738.

Seidl, R., Thom, D., Kautz, M., Martin-Benito, D., Peltoniemi, M., Vacchiano, G., Wild, J., Ascoli, D., Petr, M., Honkaniemi, J., Lexer, M. J., Trotsiuk, V., Mairota, P., Svoboda, M., Fabrika, M., Nagel, T. A., & Reyer, C. P. O. (2017). Forest disturbances under climate change. *Nature Climate Change, 7*, 395–402.

Shivaprasad, P. V., Chen, H. M., Patel, K., Bond, D. M., Santos, B. A., & Baulcombe, D. C. (2012). A microRNA superfamily regulates nucleotide binding site-leucine-rich repeats and other mRNAs. *Plant Cell, 24*, 859–874.

Singh, A., Singh, S., Panigrahi, K. C., Reski, R., & Sarkar, A. K. (2014). Balanced activity of microRNA166/165 and its target transcripts from the class III homeodomain-leucine zipper family regulates root growth in Arabidopsis thaliana. *Plant Cell Reports, 33*, 945–953.

Song, X., Li, Y., Cao, X., & Qi, Y. (2019). MicroRNAs and Their regulatory roles in plant-environment interactions. *Annual Review of Plant Biology, 70*, 489–525.

Sós-Hegedűs, A., Domonkos, Á., Tóth, T., Gyula, P., Kaló, P., & Szittya, G. (2020). Suppression of NB-LRR genes by miRNAs promotes nitrogen-fixing nodule development in Medicago truncatula. *Plant, Cell & Environment, 43*, 1117–1129.

Soto-Suárez, M., Baldrich, P., Weigel, D., Rubio-Somoza, I., & San Segundo, B. (2017). The Arabidopsis miR396 mediates pathogen-associated molecular pattern-triggered immune responses against fungal pathogens. *Scientific Reports, 7*, 44898.

Staiger, D., Korneli, C., Lummer, M., & Navarro, L. (2013). Emerging role for RNA-based regulation in plant immunity. *The New Phytologist, 197*, 394–404.

Sunkar, R., Li, Y. F., & Jagadeeswaran, G. (2012). Functions of microRNAs in plant stress responses. *Trends in Plant Science, 17*, 196–203.

Tang, D., Wang, G., & Zhou, J. M. (2017). Receptor kinases in plant-pathogen interactions: More than pattern recognition. *Plant Cell, 29*, 618–637.

Tiwari, M., Pandey, V., Singh, B., & Bhatia, S. (2021). Dynamics of miRNA mediated regulation of legume symbiosis. *Plant, Cell & Environment, 44*, 1279–1291.

Tsikou, D., Yan, Z., Holt, D. B., Abel, N. B., Reid, D. E., Madsen, L. H., Bhasin, H., Sexauer, M., Stougaard, J., & Markmann, K. (2018). Systemic control of legume susceptibility to rhizobial infection by a mobile microRNA. *Science (New York, N.Y.), 362*, 233–236.

Tsushima, D., Adkar-Purushothama, C. R., Taneda, A., & Sano, T. (2015). Changes in relative expression levels of viroid-specific small RNAs and microRNAs in tomato plants infected with severe and mild symptom-inducing isolates of Potato spindle tuber viroid. *Journal of General Plant Pathology, 81*, 49–62.

Wang, H., Jiao, X., Kong, X., Hamera, S., Wu, Y., Chen, X., Fang, R., & Yan, Y. (2016). A signaling cascade from miR444 to RDR1 in rice antiviral RNA silencing pathway. *Plant Physiology, 170*, 2365–2377.

Wang, J. W., Wang, L. J., Mao, Y. B., Cai, W. J., Xue, H. W., & Chen, X. Y. (2005). Control of root cap formation by MicroRNA-targeted auxin response factors in Arabidopsis. *Plant Cell, 17*, 2204–2216.

Wang, S., Liu, S., Liu, L., Li, R., Guo, R., Xia, X., & Wei, C. (2020). miR477 targets the phenylalanine ammonia-lyase gene and enhances the susceptibility of the tea plant (Camellia sinensis) to disease during Pseudopestalotiopsis species infection. *Planta, 251*, 59.

Wang, Z., Xia, Y., Lin, S., Wang, Y., Guo, B., Song, X., Ding, S., Zheng, L., Feng, R., Chen, S., Bao, Y., Sheng, C., Zhang, X., Wu, J., Niu, D., Jin, H., & Zhao, H. (2018). Osa-miR164a targets OsNAC60 and negatively regulates rice immunity against the blast fungus Magnaporthe oryzae. *The Plant Journal: For Cell and Molecular Biology, 95*, 584–597.

Wu, J., Yang, R., Yang, Z., Yao, S., Zhao, S., Wang, Y., Li, P., Song, X., Jin, L., Zhou, T., Lan, Y., Xie, L., Zhou, X., Chu, C., Qi, Y., & Cao, X. (2017). ROS accumulation and antiviral defence control by microRNA528 in rice. *Nature Plants, 3*, 16203.

Wu, Y., Wei, B., Liu, H., Li, T., & Rayner, S. (2011). MiRPara: A SVM-based software tool for prediction of most probable microRNA coding regions in genome scale sequences. *BMC Bioinformatics, 12*, 107.

Xu, W., Meng, Y., & Wise, R. P. (2014). Mla- and Rom1-mediated control of microRNA398 and chloroplast copper/zinc superoxide dismutase regulates cell death in response to the barley powdery mildew fungus. *The New Phytologist, 201*, 1396–1412.

Yadav, S., Sarkar Das, S., Kumar, P., Mishra, V., & Sarkar, A. K. (2020). Chapter 3-tweaking microRNA-mediated gene regulation for crop improvement. In N. Tuteja, R. Tuteja, N. Passricha, & S. K. Saifi (Eds.), *Advancement in crop improvement techniques* (pp. 45–66). Woodhead Publishing.

Yang, J.-Y., Iwasaki, M., Machida, C., Machida, Y., Zhou, X., & Chua, N.-H. (2008). BetaC1, the pathogenicity factor of TYLCCNV, interacts with AS1 to alter leaf development and suppress selective jasmonic acid responses. *Genes & Development, 22*, 2564–2577.

Yang, L., Mu, X., Liu, C., Cai, J., Shi, K., Zhu, W., & Yang, Q. (2015). Overexpression of potato miR482e enhanced plant sensitivity to Verticillium dahliae infection. *Journal of Integrative Plant Biology, 57*, 1078–1088.

Yin, H., Hong, G., Li, L., Zhang, X., Kong, Y., Sun, Z., Li, J., Chen, J., & He, Y. (2019). miR156/SPL9 regulates reactive oxygen species accumulation and immune response in Arabidopsis thaliana. *Phytopathology, 109*, 632–642.

Yin, X., Wang, J., Cheng, H., Wang, X., & Yu, D. (2013). Detection and evolutionary analysis of soybean miRNAs responsive to soybean mosaic virus. *Planta, 237*, 1213–1225.

Yu, X., Gong, H., Cao, L., Hou, Y., & Qu, S. (2020). MicroRNA397b negatively regulates resistance of malus hupehensis to botryosphaeria dothidea by modulating MhLAC7 involved in lignin biosynthesis. *Plant Science: An International Journal of Experimental Plant Biology, 292*, 110390.

Zhang, H., Li, Y., & Zhu, J.-K. (2018a). Developing naturally stress-resistant crops for a sustainable agriculture. *Nature Plants, 4*, 989–996.

Zhang, Q., Li, Y., Zhang, Y., Wu, C., Wang, S., Hao, L., Wang, S., & Li, T. (2017). Md-miR156ab and Md-miR395 target WRKY transcription factors to influence apple resistance to leaf spot disease. *Frontiers in Plant Science, 8*.

Zhang, T., Zhao, Y. L., Zhao, J. H., Wang, S., Jin, Y., Chen, Z. Q., Fang, Y. Y., Hua, C. L., Ding, S. W., & Guo, H. S. (2016). Cotton plants export microRNAs to inhibit virulence gene expression in a fungal pathogen. *Nature Plants, 2*, 16153.

Zhang, W., Gao, S., Zhou, X., Chellappan, P., Chen, Z., Zhou, X., Zhang, X., Fromuth, N., Coutino, G., Coffey, M., & Jin, H. (2011). Bacteria-responsive microRNAs regulate plant innate immunity by modulating plant hormone networks. *Plant Molecular Biology, 75*, 93–105.

Zhang, X., Bao, Y., Shan, D., Wang, Z., Song, X., Wang, Z., Wang, J., He, L., Wu, L., Zhang, Z., Niu, D., Jin, H., & Zhao, H. (2018b). Magnaporthe oryzae induces the expression of a microRNA to suppress the immune response in rice. *Plant Physiology, 177*, 352–368.

Zhao, H., Sun, R., Albrecht, U., Padmanabhan, C., Wang, A., Coffey, M. D., Girke, T., Wang, Z., Close, T. J., Roose, M., Yokomi, R. K., Folimonova, S., Vidalakis, G., Rouse, R., Bowman, K. D., & Jin, H. (2013). Small RNA profiling reveals phosphorus deficiency as a contributing factor in symptom expression for citrus huanglongbing disease. *Molecular Plant, 6*, 301–310.

Zhao, J. P., Jiang, X. L., Zhang, B. Y., & Su, X. H. (2012). Involvement of microRNA-mediated gene expression regulation in the pathological development of stem canker disease in Populus trichocarpa. *PloS One, 7*, e44968.

Zhu, Q. H., Fan, L., Liu, Y., Xu, H., Llewellyn, D., & Wilson, I. (2013). miR482 regulation of NBS-LRR defense genes during fungal pathogen infection in cotton. *PloS One, 8*, e84390.

Zhu, Q. H., Shan, W. X., Ayliffe, M. A., & Wang, M. B. (2016). Epigenetic mechanisms: An emerging player in plant-microbe interactions. *Molecular Plant-Microbe Interactions, 29*, 187–196.

9 Plant Response to Bacterial Infections

miRNAomics Approach

Sumira Malik, Shristi Kishore, Nitesh Singh, and Rahul Kumar

9.1 INTRODUCTION

Plants have developed sophisticated mechanisms for adapting to changes in the environment and interacting with several other species. Most of these techniques rely mostly on the repression and activation of large groups of genes, while miRNAs play a key role in this process. MiRNAs can be found in almost every eukaryotic lineage. MicroRNAs regulate the biological system in plants. One of the primary mechanisms of plants' responses to abiotic and biotic stress is the genetic control of miRNAs (Lee et al., 1993; Banerjee & Slack, 2002; Wienholds & Plasterk, 2005; Zhang et al., 2006). They are a small single standard class of non-coding nucleotides (20–24nts). Used to control the 3' untranslated area of targeted transcripts, they act as post-transcriptional regulatory elements (Tili et al., 2008; Winter et al., 2009; Chen et al., 2013). Most plant miRNAs direct endonucleolytic targeted RNA cleavage, resulting in fast degradation of targeted miRNA, by guiding argonaute proteins to recognize target RNAs via perfect sequence matches. The miRNAs were found to limit protein production by inhibiting the target mRNAs translation. The miRNAs control targeted gene expression via DNA methylation as well as histone modifications and protect cells from intrusive "nucleic acids," such as transgenes, transposons, and viruses (Dugas & Bartel, 2008; Lanet et al., 2009; Bin & Wang, 2010; Khraiwesh et al., 2010, 2012).

Although plenty about the mechanisms underpinning miRNAs activity on a cellular level remained unclear, scientists have answered many concerns regarding how miRNAs maintain a regulatory control on post-transcriptional mechanisms, entering a new era in miRNAs study. The current chapter covers the recent information on miRNAs concerning bacterial pathogen response in distinct species of plants at the molecular level. With advancements in technology for detecting miRNAs, the demand for bioinformatics tools to evaluate the data has also been increasing. However, analyzing these data is difficult due to a lack of expertise in computer languages such as Perl, R, Python, and Linux. As a result, this chapter also aims to present a set of web-based technologies at such a level that all scholars can understand it without a prior understanding of programming languages.

9.2 MIRNA DETECTION METHODS

1. Conventional techniques like northern blotting (NB), microarray analysis, and RT-qPCR.
2. Biosensor techniques like electrochemical-based detection and optical-based detection.
3. Other techniques like next generation sequencing (NGS) technology especially Illumina HiSeq 2500 and SOLiD.

9.3 BACTERIAL PATHOGENS AND MIRNAS

Negative and positive correlations occur during the interaction between phytopathogens and plants. Plants have two types of immunological responses to pathogen invasion: effector-triggered

DOI: 10.1201/9781003248453-9

immunity (ETI) and pathogen-associated molecular pattern (PAMP)-pathogen triggered immunity (PTI) (Boller & He, 2009; Jones & Dangl, 2006). When elicitors or flagella are released from bacteria plants activate ETI and PTI. By growing within the host, bacteria cause the inhibition of extremely efficient plant genes that control development and growth. Highly reactive oxygen species (ROS) like superoxide, peroxides, singlet oxygen, hydroxyl radical, etc., nucleotide-binding site (NBS), leucine-rich repeat (LRR) gene expression, hormone signaling, as well as cross-kingdom gene silencing are all regulated by miRNAs, which adds to the continuing arms race among hosts and pathogens (Jin, 2008; Sunkar et al., 2012).

The miRNAs regulate genes in nitrogen fixation like *Bradyrhizobium japonicum* regulates peroxidase via miRNAs (miR4416 and miR2606b). *Xanthomonas axonopodis* and *Pseudomonas syringae* induce miRNAs (such as miR846, miR393, miR160, miR390, and miR167) which reduce auxin signaling in infected plants by negatively modulating F-box protein AFB3, AFB2, and TIR1 auxin receptors (Wong et al., 2014; Jodder et al., 2017). *X. axonopodis, Physcomithrella patens, Botryosphaeria dothidea, P. syringae*, and *Populus euphratica* induce the expression of miR2911, miR1030, miR168a, and miR825, respectively, during bacterial infection interactions (Griffiths-Jones et al., 2007; Yu et al., 2017). In the same way, in the signaling pathway of plant hormones like abscisic acid and jasmonic acid, miRNAs involved are miR159 and miR319, respectively (Li et al., 2010). These miRNAs appear to play a crucial role in modulating defence responses.

The miRNAs have been linked to the regulation of reactive oxygen species (ROS) like hydrogen peroxide, superoxide radicals, and hydroxyl radicals as a defence response to multiple bacterial phytopathogens (Wu et al., 2017; Li et al., 2019; Su et al., 2018). The miRNAs modulate plant immune responses by regulating secretory pathways exocytosis (Zhang et al., 2011; Jones-Rhoades et al., 2006). Plants conserve a large number of miRNAs, and their regulation pattern gets changed to various degrees in the defence mechanism of plant species to bacteria, fungi, and viruses. The exact control of the interactions between immunity, pathogen, and growth reflects the advanced arms race among plants and phytopathogens. New miRNAs will undoubtedly become a target of resistance gene finding in the future. So more research into the biological mechanism of miRNA within plants would be beneficial.

The introduction of next-generation sequencing (NGS) technology has opened up a wide range of possibilities for the biological analysis of sequencing data. With modest bio-informatics skills, here we have outlined the tools which could be used to do an in-depth interpretation of micro-RNA (miRNAs) sequencing information. An overview of various bioinformatics methods for the studies of miRNA is summarized in Figure 9.1.

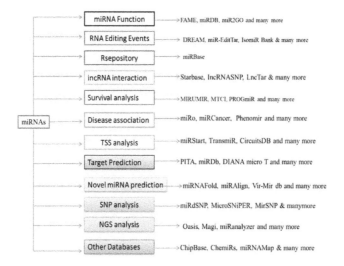

FIGURE 9.1   An overview of various bioinformatics methods for miRNA study.

9.4 DEFENSIVE ROLES OF PLANT MIRNAS IN RESPONSE TO BACTERIAL INFECTIONS

Plants respond to the invasion of pathogens through various mechanisms, viz. chemical defence, physical defence, hypersensitive reactions, and acquired resistance mechanisms. One of these defence mechanisms includes activating self-defence machinery that involves the participation of miRNAs (Xie et al., 2017). Bacterial pathogens are recognized by pathogen-associated molecular patterns (PAMPs) that trigger the primary immune defence system of plants (Thomma et al., 2011). To suppress the plants' PAMP-triggered immunity, pathogens can also counterattack with a variety of virulence factors. To combat this activity of pathogens, plants have developed a secondary defence known as effector-triggered immunity that involves several disease resistance (R) proteins (Hurley et al., 2014). These R proteins can effectively inhibit the growth of invading bacteria by suppressing their important effector proteins (Gijzen et al., 2014). Induction or repression of diverse types of miRNAs leads to the modulation of post-transcriptional gene expression via the regulation of various hormonal signaling pathways, increased callose deposition, etc. (Li et al., 2010; Ludwig-Müller, 2015). The defensive role of plant miRNAs in response to bacterial pathogens is illustrated in Figure 9.2.

The miR393 is the first miRNA that was discovered to be regulated in response to any pathogenic infections in plants (Navarro et al., 2006). To date, this miRNA has been identified in more than 15 plant species (Windels & Vazquez, 2011). In *Arabidopsis*, numerous *AtMIR393* loci encode for miR393. Targets for miR393 are identified as transport inhibitor response 1(TIR1)/auxin-signaling F-box (AFB) auxin hormone co-receptors (Jones-Rhoades & Bartel, 2004). By targeting TIR1, AFB2, and AFB3 mRNAs through the stabilization of Aux/IAA, miR393 suppresses the auxin signaling pathway (Li et al., 2016). It has been reported that the growth of the bacterial pathogen

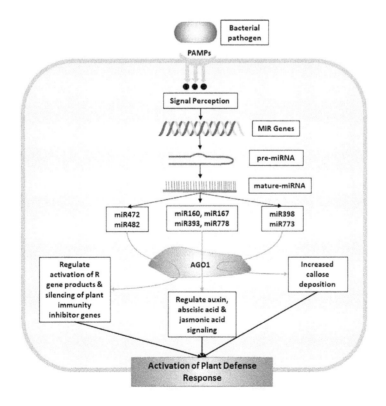

FIGURE 9.2 Diagrammatic illustration of the defensive role of plant miRNAs in response to bacterial pathogens.

Pseudomonas syringae is also inhibited by the overexpression of miR393. The findings of the study also suggest that enhanced resilience to bacterial pathogens is due to the synergistic association between salicylic acid and the miR393 pathway (Mutka et al., 2013). Furthermore, downregulation of the MEMB12 gene encoding vacuole-situated protein has also been reported as a result of over-expression of miR393. Some bacterial effectors such as AvrPtoB downregulate the accumulation of pri-miR393, thus suppressing miRNA biogenesis (Long et al., 2017).

In a study, during infection with the bacterial pathogen *Pseudomonas syringae*, numerous types of miRNAs were reported to be induced in *Arabidopsis thaliana*. The involvement of miR160 and miR167 in targeting genes related to auxin signaling was found, suggesting their roles in plants' defence against bacterial infections (Fahlgren et al., 2007). Some bacterial infections lead to the downregulation of miRNAs like miR825 that do not even participate in the targeting of any genes related to the plant defence system (Barah et al., 2013).

Omics-related techniques such as deep sequential analysis have allowed scientists to discover several other miRNAs that are involved in plants' anti-bacterial defence response. In a study, the miRNAomics approach was applied to identify differentially expressible miRNAs in *Manihot esculenta* infected with *Xanthomonas axonopodis* pv. manihotis (Xam). Deep sequencing of *Manihot esculenta* sRNA libraries was followed with the identification of conserved pre-miRNA and miRNA with the help of miRBase, PMRD, and miRProf. Then, novel pre-miRNAs and miRNAs were identified using various bioinformatics tools including miRcheck, miReap, miPred, and BLAST. In the next step, targets for identified miRNAs were predicted using miRanda, and expression levels of different miRNAs were quantified. The miRNAs that were involved in the regulation of auxin signaling included miR160, miR167, miR390, and miR393. However, some miRNA families such as miR397, miR398, and miR482 were found to regulate copper metabolism. Thus, it also suggests the role of miRNA-mediated copper metabolism in countering bacterial pathogenesis in plants (Pérez-Quintero et al., 2012).

In a study, overexpression of 20 different miRNA families was discovered in *Arabidopsis thaliana* infected with diverse strains of *Pseudomonas* sp. Furthermore, it has also been reported that the targets of nearly all of these miRNA families were various signaling pathways of hormones such as salicylic acid, abscisic acid, and jasmonic acid, which are involved in the plants' antibacterial defence machinery (Zhang et al., 2011). MiRNAs have also been reported to be involved in the regulation of PAMP-induced immune response. PAMP-stimulated callose deposition was found to be enhanced by miR160a in transgenic *Arabidopsis*. However, a reduction in the callose deposition was found to be triggered by miR398b and miR773, thus showing a contrast in the expression of different miRNA families to regulate defence mechanisms (Li et al., 2010).

In some studies, it has been found that the type of bacterial pathogens also affects the miRNAs expressions. For example, nta-miR393 was found to be highly expressed in *Nicotiana tobaccum* infected with *Agrobacterium tumefaciens*, whereas *Bacillus subtilis*-infected tobacco plants showed overexpression of nta-miR167. The expression of flavonoid derivatives is also escalated by the overexpression of nta-miR167 and nta-miR393 (Nazari et al., 2017). MiRNAs involved in anti-bacterial defence pathways and their targets are summarized in **Table 9.1**.

TABLE 9.1

lant miRNAs involved in targeting bacterial pathogens.

Plant	Target bacterial pathogen	Plant miRNA(s) involved	Target genes/proteins	References
Brassica oleracea	*Xanthomonas campestris* pv. *campestris*	miR156, miR167, miR169, miR390	ARF, Aux/IAA	Santos et al., 2019
Arabidopsis thaliana	*Pseudomonas syringae*	miR393a,b	TIR1, AFB2, AFB3	Navarro et al., 2008
Arabidopsis thaliana	*Pseudomonas syringae*	miR167	ARF	Fahlgren et al., 2007

(Continued)

TABLE 9.1 (Continued)
lant miRNAs involved in targeting bacterial pathogens.

Plant	Target bacterial pathogen	Plant miRNA(s) involved	Target genes/proteins	References
Arabidopsis thaliana	*Pseudomonas syringae*	miR160a, miR198b, miR773	ARF, MET2, COX5b.1, CSD1, CSD2.	Li et al., 2010
Nicotiana tobaccum	*Agrobacterium tumefacians*	miR393	TIR1	Pruss et al., 2008
Manihot esculenta	*Xanthomonas axonopodis*	miR160, miR167, miR390, miR393	ARF	Pérez-Quintero et al., 2012
		miR595, miR395, miR482	NB-LRR, LRR	
		miR397, miR398, miR408	TAL	
Arabidopsis thaliana	*Pseudomonas syringae* pv. tomato	miR160, miR167, miR390, miR393	ARF	Zhang et al., 2011
		miR319	TCP	
		miR159	MYB33, MYB101	

TABLE 9.2
Studies involving microRNAomics (transcriptomics) in studies of different microRNA involved in plants' response to bacterial infections

Bioinformatics tool and molecular approach	miRNA family	Bacterial pathogen	Target	Reference
High throughput sequencing, Deep sequencing	miR167	*P. syringae*	NAC	Li et al., 2010 Fahlgren et al., 2007
Deep sequencing	miR169	flg22	HAP	Li et al., 2010
High-throughput sequencing	miR398	*P. syringae*, flg22	CSD	Jagadeeswaran et al., 2009; Fahlgren et al., 2007
High throughput sequencing, Deep sequencing, Comparative transcript profiling	miR393	*P. syringae*, flg22	TIR1	Li et al., 2010 Fahlgren et al., 2007, Navarro et al., 2006

9.4.1 APPLICATION OF MIRNAOMICS IN PLANT'S RESPONSE TO BACTERIAL INFECTIONS

The miRNA-mediated regulatory mechanisms in the defence of plants against different biotic stress specifically against bacterial pathogens are well-established systems. It involves high-throughput sequencing technologies, deep sequencing, and transcript profiling against numerous bacterial pathogens-responsive miRNAs which were identified from different plants. The role of miRNAs in biotic stress in plant-pathogen interactions was tabulated with specific miRNAs, and their target in **Table 9.2** were illustrated in a few well-defined cases.

9.5   CONCLUSIONS AND FUTURE PROSPECTIVE

Molecular, functional, and bioinformatics-based characterization targets attain the deeper knowledge and scope to gain an insight into the effective regulation of causative and involved miRNAs with their respective targets in plants. From a future perspective, it will provide a better understanding and attain a deeper knowledge of the molecular role of miRNAs during plant-bacterial pathogen interactions. Further advances in these studies will elucidate the molecular mechanism underlying bacterial pathogen-based tolerance and related resistance in the deployment of miRNAs-mediated crop improvement strategies against severe bacterial diseases.

REFERENCES

Banerjee, D., & Slack, F. (2002). Control of developmental timing by small temporal RNAs: A paradigm for RNA-mediated regulation of gene expression. Bioessays, *24*(2), 119–129. https://doi.org/10.1002/bies.10046

Barah, P., Winge, P., Kusnierczyk, A., Tran, D. H., & Bones, A. M. (2013). Molecular signatures in Arabidopsis thaliana in response to insect attack and bacterial infection. *PLoS One, 8*, e58987. https://doi.org/10.1371/journal.pone.0058987

Bin, Y., & Wang, H. (2010). Translational inhibition by microRNAs in plants. *Progress in Molecular and Subcellular Biology, 50*, 41–57.

Boller, T., & He, S. Y. (2009). Innate immunity in plants: An arms race between pattern recognition receptors in plants and effectors in microbial pathogens. *Science, 324*, 742–744.

Chen, C. Z., Schaffert, S., Fragoso, R., & Loh, C. (2013). Regulation of immune responses and tolerance: The micro-RNA perspective. *Immunological Reviews, 253*(1), 112–128. https://doi.org/10.1111/imr.12060 PMID:23550642.

Dugas, D. V., & Bartel, B. (2008). Sucrose induction of Arabidopsis miR398 represses two Cu/Zn superoxide dismutases. *Plant Molecular Biology, 67*, 403–417.

Fahlgren, N., Howell, M. D., Kasschau, K. D., Chapman, E. J., Sullivan, C. M., Cumbie, J. S., Givan, S. A., Law, T. F., Grant, S. R., Dangl, J. L., & Carrington, J. C., (2007). High throughput sequencing of Arabidopsis microRNAs: Evidence for frequent birth and death of MIRNA genes. *PLoS One, 2*, e219. https://doi.org/10.1371/journal.pone.0000219

Gijzen, M., Ishmael, C., & Shrestha, S. D. (2014). Epigenetic control of effectors in plant pathogens. *Frontiers in Plant Science, 5*, 638. https://doi.org/10.3389/fpls.2014.00638

Griffiths-Jones, S., Saini, H. K., Van Dongen, S., & Enright, A. J. (2007). miRBase: Tools for microRNA genomics. *Nucleic Acids Research, 36*, (suppl 1), D154–D158.

Hurley, B., Subramaniam, R., Guttman, D. S., & Desveaux, D. (2014). Proteomics of effector-triggered immunity (ETI) in Plants. *Virulence, 5*, 752–760. https://doi.org/10.4161/viru.36329

Jagadeeswaran, G., Saini, A., & Sunkar, R. (2009). Biotic and abiotic stress down-regulate miR398 expression in Arabidopsis. *Planta, 229*, 1009–1014. https://doi.org/10.1007/s00425-009-0889-3.

Jin, H. (2008). Endogenous small RNAs and antibacterial immunity in plants. *FEBS Letters, 582*(18), 2679–2684.

Jodder, J., Basak, S., Das, R., & Kundu, P. (2017). Coherent regulation of miR167a biogenesis and expression of auxin signaling pathway genes during bacterial stress in tomato. *Physiological and Molecular Plant Pathology, 100*, 97–105.

Jones-Rhoades, M. W., & Bartel, D. P. (2004). Computational identification of plant microRNAs and their targets, including a stress-induced miRNA. *Molecular Cell, 14*, 787–799. https://doi.org/10.1016/j.molcel.2004.05.027.

Jones-Rhoades, M. W., Bartel, D. P., & Bartel, B. (2006). MicroRNAs and their regulatory roles in plants. *Annual Review of Plant Biology, 57*, 19–53.

Khraiwesh, B., Arif, M. A., Seumel, G. I., Ossowski, S., Weigel, D., Reski, R., & Frank, W. (2010). Transcriptional control of gene expression by microRNAs. *Cell, 140*, 111–122.

Khraiwesh, B., Zhu, J., & Zhu, J. (2012). Role of miRNAs and siRNAs in biotic and abiotic stress responses of plants. *Biochimica et Biophysica Acta, 1819*, 137–148.

Lanet, E., Delannoy, E., Sormani, R., Floris, M., Brodersen, P., Crété, P., Voinnet, O., & Robaglia, C. (2009). Biochemical evidence for translational repression by Arabidopsis microRNAs. *Plant Cell, 21*, 1762–1768.

Lee, R. C., Feinbaum, R. L., & Ambros, V. (1993). The C. Elegans heterochronic gene lin-4 encodes small RNAs with antisense complementarity to lin-14. *Cell, 75*(5), 843–854.

Li, X., Xia, K., Liang, Z., Chen, K., Gao, C., & Zhang, M. (2016). MicroRNA393 is involved in nitrogen-promoted rice tillering through regulation of auxin signal transduction in axillary buds. *Scientific Reports, 6*, 32158. https://doi.org/10.1038/srep32158

Li, Y., Cao, X. L., Zhu, Y., Yang, X. M., Zhang, K. N., Xiao, Z. Y., Wang, H., Zhao, J. H., Zhang, L. L., Li, G. B., Zheng, Y. P., Fan, J., Wang, J., Chen, X. Q., Wu, X. J., Zhao, J. Q., Dong, O. X., Chen, X. W., Chern, M., & Wang, W. M. (2019). Osa-miR398b boosts H2 O2 production and rice blast disease-resistance via multiple superoxide dismutases. *New Phytologist, 222*, 1507–1522.

Li, Y., Zhang, Q. Q., Zhang, J., Wu, L., Qi, Y., & Zhou, J. M. (2010). Identification of microRNAs involved in pathogen-associated molecular pattern-triggered plant innate immunity. *Plant Physiology, 152*(4), 2222–2231. https://doi.org/10.1104/pp.109.151803

Long, R., Li, M., Li, X., Gao, Y., Zhang, T., Sun, Y., Kang, J., Wang, T., Cong, L., & Yang, Q. (2017). A novel miRNA sponge form efficiently inhibits the activity of miR393 and enhances the salt tolerance and ABA insensitivity in Arabidopsis thaliana. *Plant Molecular Biology Reporter, 35*, 409–415.

Ludwig-Müller, J. (2015). Bacteria and fungi controlling plant growth by manipulating auxin: 899 Balance between development and defense, *Journal of Plant Physiology, 172*, 4–12.

Mutka, A. M., Fawley, S., Tsao, T., & Kunkel, B. N. (2013). Auxin promotes susceptibility to Pseudomonas syringae via a mechanism independent of suppression of salicylic acid-mediated defenses. *The Plant Journal: For Cell and Molecular Biology, 74*, 746–754. https://doi.org/10.1111/tpj.12157

Navarro, L., Dunoyer, P., Jay, F., Arnold, B., Dharmasiri, N., Estelle, M., Voinnet, O., & Jones, J. D. G. (2006). A plant miRNA contributes to antibacterial resistance by repressing auxin signaling. *Science, 312*, 436–439. https://doi.org/10.1126/science.1126088.

Navarro, L., Jay, F., Nomura, K., He, S. Y., & Voinnet, O. (2008). Suppression of the microRNA pathway by bacterial effector proteins. *Science (New York, N.Y.),321*, 964–967. https://doi.org/10.1126/science.1159505

Nazari, F., Safaie, N., Soltani, B. M., Shams-Bakhsh, M., & Sharifi, M. (2017). Bacillus subtilis affects miRNAs and flavanoids production in agrobacterium-tobacco interaction. *Plant Physiology and Biochemistry, 118*, 98–106. https://doi.org/10.1016/j.plaphy.2017.06.010

Park, J. H., & Shin, C. (2005). The role of plant small RNAs in NB-LRR regulation. *Briefings in Functional Genomics, 14*, 268–274.

Pérez-Quintero, Á. L., Quintero, A., Urrego, O., Vanegas, P., & López, C. (2012). Bioinformatic identification of cassava miRNAs differentially expressed in response to infection by xanthomonas axonopodis pv. Manihotis. *BMC Plant Biology, 12*, 29. https://doi.org/10.1186/1471-2229-12-29

Pruss, G. J., Nester, E. W., & Vance, V. (2008). Infiltration with agrobacterium tumefaciens induces host defense and development-dependent responses in the infiltrated zone. *Molecular Plant Microbe Interactions, 21*, 1528–1538. https://doi.org/10.1094/MPMI-21-12–1528.

Santos, L. S., Maximiano, M. R., Megias, E., Pappas, M., Ribeiro, S. G., & Mehta, A. (2019). Quantitative expression of microRNAs in brassica oleracea infected with xanthomonas campestris pv. Campestris. *Molecular Biology Reports, 46*, 3523–3529. https://doi.org/10.1007/s11033-019-04779-7

Su, Y., Li, H., Wang, Y., Li, S., Wang, H., Yu, L., He, F., Yang, Y., Feng, C., Shuai, P., Liu, C., Yin, W., & Xia, X. (2018). Poplar miR472a targeting NBS-LRRs is involved in effective defence against the necrotrophic fungus cytospora chrysosperma. *Journal of Experimental Botany, 69*, 5519–5530.

Sunkar, R., Li, Y. F. & Jagadeeswaran, G. (2012). Functions of microRNAs in plant stress responses. *Trends in Plant Science, 17*(4), 196–203.

Thomma, B. P., Nürnberger, T., & Joosten, M. H. (2011). Of PAMPs and effectors: The blurred PTI-ETI dichotomy. *The Plant Cell, 23*, 4–15. https://doi.org/10.1105/tpc.110.082602

Tili, E., Michaille, J. J., & Calin, G. A. (2008). Expression and function of microRNAs in immune cells during normal or disease state. *International Journal of Medical Sciences, 5*(2), 73–79.

Wienholds, E., & Plasterk, R. H. (2005). MicroRNA function in animal development. *FEBS Letters, 579*(26), 5911–5922. https://doi.org/10.1016/j.febslet.2005.07.070

Windels, D., & Vazquez, F. (2011). MiR393: Integrator of environmental cues in auxin signaling? *Plant Signaling & Behavior, 6*, 1672–1675. https://doi.org/10.4161/psb.6.11.17900

Winter, J., Jung, S., Keller, S., Gregory, R. I., & Diederichs, S. (2009). Many roads to maturity: MicroRNA biogenesis pathways and their regulation. *Nature Cell Biology, 11*, 228–34.

Wong, J., Gao, L., Yang, Y., Zhai, J., Arikit, S., Yu, Y., Duan, S., Chan, V., Xiong, Q., Yan, J., & Li, S. (2014). Roles of small RNA s in soybean defense against Phytophthora sojae infection. *The Plant Journal, 79*(6), 928–40.

Wu, J., Yang, R., Yang, Z., Yao, S., Zhao, S., Wang, Y., Li, P., Song, X., Jin, L., Zhou, T., Lan, Y., Xie, L., Zhou, X., Chu, C., Qi, Y., & Cao, X. (2017). ROS accumulation and antiviral defence control by microRNA528 in rice. *Nature Plants, 3*, 16203

Xie, S., Jiang, H., Xu, Z., Xu, Q., & Cheng, B. (2017). Small RNA profiling reveals important roles for miR-NAs in Arabidopsis response to bacillus velezensis FZB42. *Gene*, *629*, 9–15. https://doi.org/10.1016/j.gene.2017.07.064

Yu, X., Hou, Y., Chen, W., Wang, S., Wang, P., & Qu, S. (2017). Malus hupehensis miR168 targets to argonaute1 and contributes to the resistance against botryosphaeria dothidea infection by altering defense responses. *Plant and Cell Physiology*, *58*(9), 1541–57.

Zhang, B., Pan, X., Cobb, G. P., & Anderson, T. A. (2006). Plant microRNA: A small regulatory molecule with big impact. *Developmental Biology*, *289*, 3–16. https://doi.org/10.1016/j.ydbio.2005.10.036

Zhang, W., Gao, S., Zhou, X., Chellappan, P., Chen, Z., Zhou, X., Zhang, X., Fromuth, N., Coutino, G., Coffey, M., & Jin, H. (2011). Bacteria-responsive microRNAs regulate plant innate immunity by modulating plant hormone networks. *Plant Molecular Biology*, *75*, 93–105. https://doi.org/10.1007/s11103-010-9710-8

Zhang, X., Zhao, H., Gao, S., Wang, W., Katiyar-Agarwal, S., Huang, H., Raikhel, N., & Jin, H. (2011). Arabidopsis argonaute 2 regulates innate immunity via miRNA393∗-mediated silencing of a golgi-localized SNARE gene, MEMB12. *Molecular Cell*, *42*, 356–366.

10 miRNAs

A Novel TARGET for Improving Stress Tolerance in Plants Using Transgenics

C. Deepika, S. R. Venkatachalam,
A. Yuvaraja, and P. Arutchenthil

10.1 INTRODUCTION

Transgenics is one of the novel inventions to be adopted for making the crop to combat extrinsic environmental stresses or to improve intrinsic crop yields. With the use of this innovative technology, various yield-sustaining crop varieties with enhanced resistance had been released. However, the currently available commercialized genetically modified crops are rendered with simple gene insertion, which may lead to the breakdown of resistance with the increasing pest and pathogen biotypes. Shortly, it is anticipated that GM crops will combine desirable traits, such as improved resistance, quality, and grain yields, under intricate polygenic control (Buiatti et al., 2012). So for the proper use of discovery of mankind, alternative new sources of genes are to be isolated, identified, validated, and functionally characterized (Zhou et al., 2013). One such approach is based on miRNA (microRNA), which is becoming an attractive strategy to be integrated with transgenics (Gao et al., 2010; Zhou, 2012). MicroRNAs (miRNAs) are a distinctive class of non-coding, single-stranded riboregulator RNAs with 20–24 nucleotides involved in eukaryotic gene regulation (Shriram et al., 2016). It is a well-known fact that miRNAs are the prime junctures of complex gene regulative networks that is regulating different biological and metabolic processes *via* upregulating or downregulating their target mRNAs, the majority of which are transcription factors (Jones-Rhoades et al., 2006; Khraiwesh et al., 2012; Sunkar et al., 2012). With the availability of high-throughput sequencing technology and various computational tools and databases, the mining of novel miRNAs involved in stress-related functions had become feasible (Zhou et al., 2013; Zhang & Wang, 2015; Shriram et al., 2016). These discovered stress-related miRNAs could be utilized in knockout studies to engineer stress tolerance in plants (Banerjee et al., 2016; Wani et al., 2020). This book chapter herein aims to provide a current understanding of recent information on transgenic plants expressing stress-related miRNAs and a critical analysis of the evolution of miRNAs as novel targets to engineer stress tolerance in plants that has an emphasis on current research and prospects.

10.2 MIRNA – A NOVEL TRANSGENE CANDIDATE

The miRNAs are tiny endogenous RNA molecules with a base pair ranging between 20 and 24 nucleotides. Plant miRNAs are distributed across a wide range of species either as conserved miRNAs or as species-specific ones. But irrespective of whether conserved or species-specific, they are involved in gene regulation during stress (Zhang & Wang, 2015). The gene encoding miRNA is transcribed to primary miRNAs (pri-miRNA) which are double-stranded and characterized by stem-loop structure at one end. This pri-miRNA is converted into pre-miRNA by Dicer-like proteins which

DOI: 10.1201/9781003248453-10

further cleaved to miRNA duplexes (Jones-Rhoades et al., 2006). Short miRNA duplex associates with the RISC complex, thereby forming complementary base pairing with the target miRNAs. In plants, very often silencing is caused by a splicing mechanism rather than translational repression (Brodersen et al., 2008; Voinnet, 2009; Khraiwesh et al., 2012).

Recent investigations showed the existence of numerous stress-responsive miRNAs from bioinformatics studies and by sequencing small RNA libraries generated from materials that are exposed to stress (Meng et al., 2010; Sunkar, 2010; Khraiwesh et al., 2012; Sunkar et al., 2012; Sun, 2012; Zhou, 2012). Hence the isolated miRNAs could be targeted in transgenics for better stress management.

10.3 STRATEGIES FOR DEVELOPING MIRNA BASED TRANSGENIC PLANTS

The illustration of different strategies to be used for developing transgenic plants based on the miRNA approach had been illustrated in Figure 10.1. Research is being undertaken to demonstrate various strategies in various crops.

10.3.1 MIRNA AS POSITIVE REGULATORS ON TARGETS

Constitutive overexpression of target miRNAs had been carried out in various crops for generating increased stress tolerance in transgenic plants (Gao et al., 2010; Zhou et al., 2013; Li et al., 2016; Yu et al., 2019). These miRNAs act as positive regulators for their targets and when there are any undesirable side effects produced by the miRNA overexpression. Another strategy, RNAi, is a newly developed gene-regulatory strategy in functional genomics because it may precisely silence or downregulate the target gene(s) without influencing unrelated genes (Wani et al., 2020), thereby avoiding the possible detrimental pleiotropic effects stimulated by miRNA overexpression. For instance, RNAi constructs for the miR319 targets Osa-PCF5 and Osa-TCP21 in transgenic rice plants showed

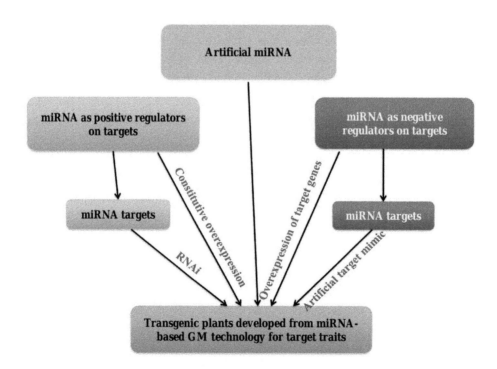

FIGURE 10.1   Strategies for developing miRNA based transgenic plants for stress resistance.

improved cold tolerance in comparison to wild-type controls, similar to miR319-overexpressing lines, but without severe mutant morphological abnormalities (Yang et al., 2013). Also, overexpression of soybean *Gma-miR172c* resulted in decreased leaf water loss, thereby tolerating water deficit conditions, and also increased survival rate under stress conditions were reported in *Arabidopsis* (Li et al., 2016), but the transgenic plants exhibited increased ABA sensitivity by regulating *Glyma01g39520*, which also accelerates flowering under abiotic stresses (Li et al., 2016). Hence RNAi is a potent technology that could be effectively utilized in creating gene-specific knock-out effects.

10.3.2 miRNA as Negative Regulators on Targets

Transgenic plants that overexpress particular miRNA may become vulnerable to various stresses when they function as negative regulators (Li et al., 2008; Zhao et al., 2011). In such a case overexpressing the miRNA-resistant target genes or using a miRNA sponge (Ebert et al., 2007), miRNA target mimicry (TM) (Franco-Zorrilla et al., 2007), and short tandem target mimic (STTM) (Yan et al., 2012; Tang & Tang, 2013) would be an effective strategy for enhancing the plant stress response. Todesco et al. (2010) developed a collection of knockdown mutant transgenic *Arabidopsis* plants using a target mimic. The phenotypes of the transgenic plants overexpressing the target mimic and those overexpressing the miRNA-resistant targets were found to be very similar in morphology, but the transgenics expressing the specific miRNA produced the mutant phenotype. Plants overexpressing the target mimic MIM319 had shrunken leaves (Todesco et al., 2010), which was comparable to transgenics overexpressing the mutant forms of TCP2 and TCP4 (miRNA-resistant variant) (Palatnik et al., 2003) but opposite to *miR319*-overexpressing plants. Recently, several MIR genes have been targeted in investigating the role of miRNAs in crop plants by the STTM approach (Jiang et al., 2018; Ferdous et al., 2016; Hackenberg et al., 2015). For example, in rice STTM strategy has been employed to unravel the function of 35 miRNA families related to important agronomical traits (Zhang et al., 2017a). In other plant species, a different artificial transcript strategy known as miRNA SPONGES with multiple miRNA binding sites was also produced (Todesco et al., 2010; Reichel et al., 2015). Sometimes these miRNAs in the form of SPONGES can be used to restrict the activity of the entire plant family of miRNAs (Zhang et al., 2017a; Jiang et al., 2018).

10.3.3 Artificial miRNAs

A major concern from the traditional transgenic approach is the knockdown effect of nontarget mRNAs by the overexpression of miRNA. This could be solved by creating an alternative strategy called artificial miRNA (Sablok et al., 2011; Zhang et al., 2018b). The mature miRNA might be created using the pre-conserved miRNA's stem-loop structure in the amiRNA method. Duplex miRNA is inserted into the target mRNA with great specificity, forming complementary base pairing without interfering with the activity of non-target genes. Additionally, *Arabidopsis* and tobacco genes for phytopathogens have been studied using knockout techniques using an amiRNA (Niu et al., 2006; Agrawal et al., 2015; Kis et al., 2015; Wagaba et al., 2016). When the transgenic plants harbouring the amiRNA construct were expressed under constitutive promoter or tissue-specific or stress-inducible promoters (Schwab et al., 2006; Warthmann et al., 2008; Khraiwesh et al., 2008), the target mRNA genes are precisely, specifically, and effectively downregulated by miRNA. For example, the effect of silencing the C-3 oxidase-encoded gene *StCPD* on potato drought resistance was studied by Zhou et al. (2018). Recently an amiRNA screen uncovers redundant CBF and ERF34/35 transcription factors that differentially regulate arsenite and cadmium responses in plants (Xie et al., 2021).

10.4 MIRNAS AND STRESS RESPONSE IN TRANSGENIC PLANTS

Numerous studies had been carried out for identifying stress-responsive miRNAs, especially biotic stress through bacteria, viruses, fungi, and nematodes (Katiyar-Agarwal et al., 2006; Navarro et al.,

2006; Bazzini et al., 2007; He et al., 2008; Hewezi et al., 2008; Jagadeeswaran et al., 2009; Katiyar-Agarwal & Jin, 2010; Eckardt, 2012; Romanel et al., 2012), but only a few had focused on transgenic technology for relating miRNA with biotic stress. Researchers reported various target transcription factors of miRNA such as MYB, TCP bHLH, NAC, AP2, NYF, and HD-ZIP having a wide variety of defence genes, including ABC transporters, peroxidase, MAPK genes, and BB-LRRs (Zhang et al., 2017a; Jalmi et al., 2018; Wang et al., 2019). Functional characterization of these biotic stress-related miRNAs through advanced biotechnological tools should be considered soon for the generation of viral/fungal/nematode/bacterial stress-tolerant transgenic varieties. Figure 10.2 represents various stress-related miRNAs and their role in combating abiotic and biotic stress.

10.4.1 BIOTIC RESISTANCE BY miRNA

Insights on miRNA-based biotic stress response in plants have provided a few genetically modified plants based on the miRNA approach (Kasschau et al., 2003; Navarro et al., 2006; Campo et al., 2013; Li et al., 2014a; Jiang et al., 2018; Lee et al., 2015a; Yang et al., 2015; Jeyaraj et al., 2019; Li et al., 2019; Jeyaraj et al., 2020). Plants genetically engineered to improve tolerance to biotic stresses using constitutive overexpression of miRNAs/artificial miRNA are compiled in Table 10.1.

The role of microRNAs in plant anti-viral RNA silencing pathway has been evident from the studies of Kasschau et al., 2003. When RNA silencing suppressor *P1/HC-Pro* was expressed in transgenic plants, it interferes with *miR171* but not with the production of *miR171*. *P1/HC-Pro* increased the concentration of the *miR171* targets (Kasschau et al., 2003). It is really interesting to note that transgenic *Arabidopsis* expressing *P1/HC-Pro* also increased the expression of a few additional miRNAs (*miR156* and *miR164*) (Kasschau et al., 2003). Transgenic plants overexpressing *miR393*

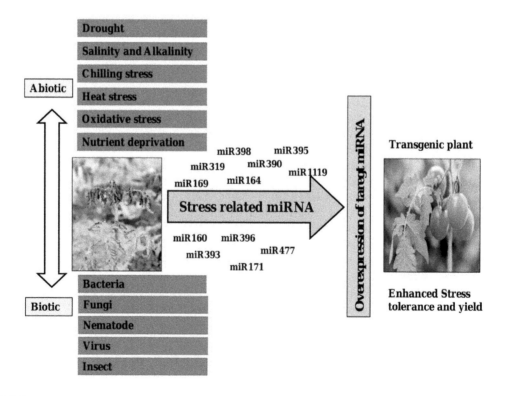

FIGURE 10.2   Various stress-related miRNAs and their role in combating abiotic and biotic stress.

TABLE 10.1

Plants genetically engineered to improve tolerance to biotic stresses using constitutive overexpression of miRNAs/artificial miRNA.

miRNA source	MIR gene or miRNA	Transgenics	Phenotype in transgenic plants	References
Arabidopsis	miR393	*Arabidopsis*	Decrease the transcript level of *TIR1* and show enhanced resistance to bacterium *Pseudomonas syringae*	Navarro et al. (2006)
Tobacco artificial miRNA (*miR2b*)	*miR2b*	Tobacco	Conferred effective resistance to cucumber mosaic virus (CMV) infection	Qu et al. (2007)
Arabidopsis	Ath-miR396a and Ath-miR396b	*Arabidopsis*	Reduced susceptibility to *Heterodera schachtii*	Hewezi et al. (2012)
Rice	osa-miR7696	Rice	Enhanced resistance against blast fungus *Magnaporthe oryzae*	Campo et al. (2013)
Rice	osa-miR160a and osa-miR398b	Rice	Increased resistance to *Magnaporthe oryzae*	Li et al. (2014b)
Tobacco artificial miRNA	Artificial miRNA vector for the gene MpAChE2	Tobacco	Transgenic plants targeting the acetylcholinesterase 2 transcript of the aphid *Myzus persicae* using showed stronger resistance	Guo et al. (2014)
Arabidopsis	Ath-miR844	*Arabidopsis*	Increase the susceptibility to *Pseudomonas syringae* and *Botrytis cinerea*	Lee et al. (2015b)
-	amiR-24	Tobacco	Artificial miRNA targeting chitinase gene from *Helicovepa armigera* improving the tolerance towards caterpillar	Agrawal et al. (2015)
Potato	St-miR482e	Potato	Increases the susceptibility towards *Verticillium dahliae*	Yang et al. (2015)
Hordeum vulgare	Hvu-miR171	*Nicotiana benthamiana* and *Hordeum vulgare*	Increased resistance to Wheat dwarf virus	Kis et al. (2019)
Tea	csn-miRn19, csn-miRn301, csn-miRn392 and csn-miRn421	Tea	*E. oblique*-induced stress-responsive miRNAs are involved in defence-related key metabolite production that resists herbivore attack	Jeyaraj et al. (2017).
Rice	osa –miR171b	Rice	Target mimicry increased the tiller thickness, spikelets, and plants less susceptible to rice stripe virus are attenuated	Tong et al. (2017)
Solanum pimpinellifolium	Sp-miR482b and its target mimicry STTM482	*Solanum lycopersicum*	Overexpressing *Sp-miR482b* resulted in susceptibility, while silencing enhanced the resistance to tomato *Phytophthora infestance*	Jiang et al. (2018)
Tea	csn-miR160, csn-miR169, csn-miR396 and csn-miR477	Tea	*C. gloeosporioides*-induced stress-responsive miRNAs in improving biotic stress tolerance	Jeyaraj et al. (2019)
Rice	osa-miR396f	Rice	Increases the resistance towards *Erwinia chrysanthemi* pv *zeae*	Li et al. (2019)

decrease the transcript level of *TIR1* and show enhanced resistance to bacterial infection (Navarro et al., 2006). Also overexpressing an artificial miRNA (*miR2b*) in transgenic tobacco plants conferred effective resistance to cucumber mosaic virus (CMV) infection by inhibiting the expression of silencing suppressor 2b of CMV.

A comprehensive study was conducted in rice to reveal the role of miRNAs in plant response to the fungus *Magnaporthe oryzae*, which causes a devastating disease blast (Campo et al., 2013). The research led to the identification of a new rice miRNA named *Osa-miR7695*, which is involved in negatively regulating a natural resistance-associated macrophage protein 6 (*OsNramp6*). Transgenic rice plants overexpressing *Osa-miR7696* confer better resistance to rice blast infection (Campo et al., 2013). Hewezi et al. (2012) expressed the MIR genes *Ath-miR396b* and *Ath-miR396a* using CaMV 35S promoter in *Arabidopsis thaliana*. The transgenic plants expressed reduced susceptibility to *Heterodera schachtii*. The transgenic *Nicotiana benthamiana* and *Hordeum vulgare* plants displayed improved resistance to the Wheat dwarf virus when artificial and engineered *Hvu-miR171* targeting viral genes from *Hordeum vulgare* was expressed using CaMV 35S and maize polyubiquitin promoters (Kis et al., 2015). A new technique for miRNA overexpression or knockdown based on a viral vector (Barley stripe mosaic virus) that may be utilized as a probe to look into the functions of miRNAs was published by Jian et al. (2017). Improved resistance against cyst nematodes is another example of a technology involving miRNA overexpression that has been patented (*miR164* and *miR396*; WO2012058266 A1 and WO2012149316 A2) (Basso et al., 2019).

To overcome, tea (*Camellia sinensis* L.) geometrid and anthracnose disease that causes considerable crop loss and tea production, potential target transcripts of *Ectropis oblique*, and *Colletotrichum gloeosporioides* induced stress-responsive miRNAs in *C. sinensis* was reported by Jeyaraj et al. (2017, 2019) respectively. Li et al. (2019) isolated *Osa-miR396f* from rice, and the transgenic rice was found to be with increased resistance against *Erwinia chrysanthemi* pv *zeae*. Undoubtedly from the above examples, miRNA proved to be a potent technology in developing stress-resistant genetically engineered crop varieties.

10.4.2 Abiotic Stress Resistance Transgenics by miRNA

As a consequence of adaptive evolution processes, plants have become acclimatized to challenging environmental conditions. How plants respond *via* altering cellular and molecular pathways is intricate and received great attention to unravel the mechanisms behind them. Multiple stress-responsive genes are used by plants to detect and react to environmental stimuli (Shriram et al., 2016). With the aid of advanced biotechnological tools like microarray, the expression pattern of different protein-coding genes and microRNAs could be analyzed and characterized, which could be used further in modifying plants to withstand various stresses. Deploying miRNA significantly changes the stress tolerance ability of plants; thus, among various miRNAs, some are identified as potential targets for developing transgenics with improved abiotic stress tolerance (Zhang, 2015; Zhang et al., 2018a). Recent information on transgenic plants with modified miRNA expression in primary agricultural plants in response to abiotic stresses is compiled in Table 10.2.

10.4.2.1 Drought Stress

Water deficit is one of the adverse environmental factors that harm agricultural production worldwide (Cattivelli et al., 2008). Therefore, it is crucial to adopt elite varieties with improved water use efficiency that can tolerate severe water stress in crop production to promote sustainable agriculture (Shi et al., 2018). Over the years several miRNAs and their role in managing abiotic stress were evident from various research (Khraiwesh et al., 2012; Zhang, 2015; Zhang et al., 2018a). For instance, under drought stress, transgenic tobacco lines were overexpressed with TaMIR1119. The transgenics behaved with improved phenotypes, showing increased plant biomass, photosynthetic parameters, osmolyte accumulation, and enhanced antioxidant enzyme (AE) activities relative to the wild type. Three main AE genes, *NtFeSOD*, *NtCAT1;3*, and *NtSOD2;1*, encoding superoxide dismutase (SOD),

TABLE 10.2

Plants genetically engineered to improve tolerance to abiotic stresses using constitutive overexpression of miRNAs.

miRNA source	MIR gene or miRNA	Transgenic plants	Target trait	Phenotype in transgenic plants	References
Arabidopsis and Rice	*miR399*	*Arabidopsis*	Pi deprivation	Overexpressing *miR399* accumulated more phosphorus	Fujii et al. (2005)
Arabidopsis	*miR164*	*Arabidopsis*	Drought	*Arabidopsis* overexpressing *miR164*, which targets NAC transcription factors, exhibited reduced lateral roots, whereas overexpression of *miR164*-resistant NAC1 resulted in an increased number of lateral roots	Guo et al. (2005)
Arabidopsis	*miR160*	*Arabidopsis*	Drought	Overexpression of *miR160*, which targets auxin response factors (ARFs) resulted in gravitropic roots and increased the number of lateral roots	Wang et al. (2005)
Arabidopsis and Rice	*miR398*	*Arabidopsis*	Oxidative stress	Overexpressing *miR398*-resistant form of *CSD2*, which led to increased tolerance to high-intensity light, heavy metals, and other oxidative stresses	Sunkar et al. (2006)
Rice	*miR169a*	*Arabidopsis*	Drought	Overexpression of *NF-YA5* enhanced the drought tolerance	Li et al. (2008)
Rice	*osa-miR396*	*Arabidopsis* and Rice	Salinity and alkalinity	Overexpression of *osa-miR396c* decreases salt and alkali stress tolerance	Gao et al. (2010)
Arabidopsis	*miR395*	*Arabidopsis*	Drought and salinity stress	Overexpression of *miR395c* or *miR395e* retarded and accelerated, respectively, the seed germination of *Arabidopsis* under high salt or dehydration stress conditions	Kim et al. (2010a)
Arabidopsis	*miR402*	*Arabidopsis*	Multiple stress	Overexpression of *miR402* accelerated seedling growth under salt stress conditions, while its overexpression deterred the seedling growth dehydration or cold stress conditions	Kim et al. (2010b)
Arabidopsis	*miR395*	*Arabidopsis*	S deprivation	*miR395*-overexpressing transgenic *Arabidopsis* plants over-accumulated S in shoots, they still showed S-deficient symptoms, and the distribution of sulphate from older leaves to younger leaves was impaired	Liang et al. (2010)
Rice	*osa-miR393*	*Arabidopsis* and Rice	Salinity and alkalinity	Overexpression of *osa-miR393* decreased the salt and alkali stress tolerance of transgenic plants	Gao et al. (2011)
Tomato	*miR169c*	Tomato	Drought	Overexpression of *miR169c* enhances drought tolerance	Zhang et al. (2011)
Arabidopsis	*miR828*	Sweet potato	Oxidative stress	Overexpressing *miR828* precursor affected lignin and H_2O_2 contents to participate in defence mechanisms	Lin et al. (2012)

(Continued)

TABLE 10.2 (Continued)

Plants genetically engineered to improve tolerance to abiotic stresses using constitutive overexpression of miRNAs.

miRNA source	MIR gene or miRNA	Transgenic plants	Target trait	Phenotype in transgenic plants	References
Soybean	gma-miR394a	Arabidopsis	Drought	Overexpression of gma-miR394a resulted in plants with lowered leaf water loss and enhanced drought tolerance	Ni et al. (2012)
Wheat	TamiR159	Rice	Heat stress	TamiR159 overexpression rice lines were more sensitive to heat stress	Wang et al. (2012)
Arabidopsis	miR398	Arabidopsis	Heat stress	Loss of function mutants of CSD1 and CCS, knockdown mutant of CSD2 enhanced the thermo-tolerance	Guan et al. (2013)
Soybean	miR169	Arabidopsis	Drought	Overexpression of GmNFYA3 positively modulates drought stress tolerance and increased sensitivity to salinity stress	Ni et al. (2013)
Rice	osa-miR319	Rice	Chilling stress	Overexpression of the Osa-miR319 gene led to increased cold stress tolerance	Yang et al. (2013)
Rapeseed	miR395	Rapeseed	Heavy metal stress	miR395-overexpressing plants showed a lower degree of Cd-induced oxidative stress and were thus involved in the detoxification of Cd	Zhang et. al. (2013a)
Rice	osa-miR319	Creeping bentgrass	Drought and salinity	Overexpressing a rice miR319 gene (Osa-miR319) exhibited enhanced tolerance to drought and salinity	Zhou and Luo (2014)
Arabidopsis	miR173	Arabidopsis	Heat stress	Targets of TAS1viz., HTT1 and HTT2 mediate thermotolerance pathways	Li et al. (2014a)
Arabidopsis	miR156	Arabidopsis	Heat stress	miR156 isoforms showed the strongest upregulation after heat stress and enhanced the tolerance	Stief et al. (2014)
Rice	osa-miR319b	Rice	Chilling stress	Enhanced the tolerance towards cold stress	Wang et al. (2014)
Tomato	miR396a-5p	Tobacco	Multiple stress	Overexpression of miR396a-5p increased plant tolerance towards salinity, drought, and cold stresses primarily by targeting growth-regulating factors (GRFs)	Chen et al. (2015)
Rice	OsmiR156k	Rice	Chilling stress	Overexpression of OsmiR156k reduced the cold tolerance	Cui et al. (2015)
Arabidopsis	miR399	Tomato	Chilling stress and Pi deprivation	Overexpression of miR399 causes P toxicity but transgenic tomato with ath-miR399d increased tolerance to cold and phosphorus deficiency	Gao et al. (2015)
Arabidopsis	miR408	Chickpea	Drought	Overexpression of Athpre-miR408 enhanced the drought tolerance	Hajyzadeh et al. (2015)
Arabidopsis	Ath-miR408	Arabidopsis	Multiple stress	Overexpressing enhances the tolerance to cold, oxidative, and salinity stress but reduces the tolerance against drought	Ma et al. (2015)
Arabidopsis	Ath-miR399f	Arabidopsis	Drought and salinity	Increases the tolerance towards salinity but causes a hypersensitive reaction to drought stress	Baek et al. (2016)

(Continued)

TABLE 10.2 (Continued)

Plants genetically engineered to improve tolerance to abiotic stresses using constitutive overexpression of miRNAs.

miRNA source	MIR gene or miRNA	Transgenic plants	Target trait	Phenotype in transgenic plants	References
Rice, Maize	miR390	Rice	Heavy metal stress	Overexpressing miR390 displayed reduced Cd tolerance and increased susceptibility to Cd	Ding et al. (2016)
Arabidopsis and barley	miR827	Barley	Drought	The recovery of Hv-miR827 overexpressing plants also improved following severe drought stress	Ferdous et al. (2016)
Cotton	miRNVL5	Arabidopsis	Salinity	Transgenic plants constitutively expressing miRNVL5 showed hypersensitivity to salt stress whereas the expression of GhCHR in Arabidopsis enhanced the salt stress tolerance	Gao et al. (2016)
Soybean	gma-miR172	Arabidopsis	Drought and salinity	Overexpression of Gma-miR172 in Arabidopsis Revealed the enhanced water deficit and salt tolerance	Li et al. (2016)
Populus suaveolens	Psu-miR475b	Populus suaveolens	Chilling stress	Improved tolerance towards cold stress	Niu et al. (2016)
Arabidopsis	miR394a	Arabidopsis	Chilling stress	Overexpression of miR394a/LCR loss of function mutant enhanced the cold tolerance	Song et al. (2016)
Soybean	gma-miR172a	Soybean	Salinity	Enhance the tolerance to salinity	Pan et al. (2017)
Wheat	TaemiR408	Tobacco	Pi deprivation and salinity	Tobacco lines with TaemiR408 overexpression exhibited enhanced stress tolerance under both Pi starvation and salt treatments	Bai et al. (2018)
Wheat	TaMIR1119	Tobacco	Drought	TaMIR1119 overexpression regulates plant drought tolerance through transcriptionally regulating the target genes that modulate osmolyte accumulation, and photosynthetic function, and improve cellular ROS homeostasis of plants	Shi et al. (2018)
Rice	osa-miR166 knock down STTM 166	Rice	Drought	Knockdown of miR166 using STTM increases the tolerance to drought with the altered phenotype of the plant	Zhang et al. (2018a)
Banana	miR397	Banana	Salinity	Overexpression in bananas significantly increased plant growth by 2–3 fold but did not compromise tolerance towards Cu deficiency and NaCl stress	Patel et al. (2019)
Soybean	gma -miR169c	Arabidopsis	Drought	Overexpressing Gma-miR169c increases sensitivity to drought stress	Yu et al. (2019)

(Continued)

TABLE 10.2 (Continued)
Plants genetically engineered to improve tolerance to abiotic stresses using constitutive overexpression of miRNAs.

miRNA source	MIR gene or miRNA	Transgenic plants	Target trait	Phenotype in transgenic plants	References
Rice	osa- miR393a	Agrostis stolonifera	Multiple stress	Altered the phenotype of the plant and improves the tolerance to drought, heat, and salinity, and increases the uptake of potassium	Zhao et al. (2019)
Tomato	SlmiR168	Tomato	Potassium deficiency stress	Transgenic tomato plants constitutively expressing pri-SlmiR168a showed higher K+ contents in roots	Liu et al. (2020)
Hickory (Carya cathayensis)	cca-miR398	Arabidopsis	Copper sulphate stress	An overexpression of cca-miR398 in Arabidopsis caused a reduction in root length and increases the sensitivity	Sun et al. (2020)
Medicago truncatula	miR156	Poplar	Anthocyanin involved in stress response	The levels of anthocyanins, flavones, and flavonols were substantially elevated in transgenic poplar plants overexpressing miR156 whereas the total lignin content was reduced	Wang et al. (2020)
Potato	miR160a-5p	Potato	Drought and Heat Stress	Overexpression of miR160a-5p could alter mechanisms for auxin response factor pathways and enhances stress tolerance	Shen et al. (2022)

catalase (CAT), and peroxidase (POD) proteins, respectively, were upregulated in transgenic lines, suggesting that they are involved in the regulation of AE activities and contribution to the improved cellular reactive oxygen species (ROS) homeostasis in drought-challenged transgenic lines (Shi et al., 2018). Li et al. (2008) worked on *miR169* having *NF-YA5* as a target, and overexpression of the miRNA target led to transgenic plants with higher drought tolerance. When *miR169* from soybean was overexpressed in *Arabidopsis*, the miRNA target *GmNFYA3* positively modulates drought stress tolerance and increased sensitivity to salinity stress and exogenous ABA (Ni et al., 2013). Transgenic tomato overexpressing *miR169c* enhances tolerance towards heavy water deficit (Zhang et al., 2011). Yet another *Osa-miR319* isolated from the rice was made to overexpress in creeping bentgrass; positive results with improved drought and salinity tolerance were reported by Zhou et al. (2013). Similarly, studies on transgenics to improve the water use efficiency of various crops were reported by various scientists (Wang et al., 2005; Guo et al., 2005; Ni et al., 2012; Hajyzadeh et al., 2015; Ferdous et al., 2016; Li et al., 2016; Zhang et al., 2018a). When soybean *Gma-miR169c* was overexpressed in *Arabidopsis*, it increases the sensitivity against drought stress (Yu et al., 2019). Şanlı and Gökçe (2021) constructed transgenic potatoes to study the expression of *miR160a-5p* and the results revealed that overexpression of *miR160a-5p* could alter mechanisms for auxin response factor pathways which enabled transgenic plants to have a higher tolerance to abiotic stress conditions especially drought and heat stress. The above-discussed information from genetic engineering studies reveals conclusively that miRNAs have emerged as a novel and effective target for creating drought stress-tolerant cultivars.

10.4.2.2 Salinity and Alkalinity

Crops that are grown in saline soils frequently encounter the problem of ion toxicity and cellular desiccation that ultimately reduces production and productivity (Turkan & Demiral, 2009). Plants have developed several adaptive mechanisms such as ion exclusion and compartmentalization to combat salt stress. Several genes and regulatory mechanisms are involved in the process. Recent years have focused on microRNAs and their deployment in transgenics to impart salinity stress tolerance to various crops (Kim et al., 2010a; Gao et al., 2010; Zhou et al., 2013; Gao et al., 2011; Li et al., 2016; Patel et al., 2019). Cotton was subjected to salt stress at the concentration of 50–400 mM NaCl, which decreased the expression of *miRNVL5* while increasing the expression of *GhCHR*. By lowering Na+ accumulation in plants and enhancing primary root growth and biomass, ectopic expression of *GhCHR* in *Arabidopsis* conferred salt stress tolerance. A striking discovery was that *Arabidopsis* constitutively expressing *miRNVL5* displayed hypersensitivity to salt stress (Gao et al., 2016). Patel et al. (2019) isolated *miR397* from bananas, and it was overexpressed in bananas. The results showed that plant growth was increased by two to three fold without compromising salinity tolerance. *Osa-miR396* from the rice was overexpressed in *Arabidopsis* and rice, but it increases the sensitivity towards salinity and alkalinity (Gao et al., 2010). Interestingly, the target prediction study found that the target genes were involved in signaling, ion-homeostasis regulation, and controlling the reduced plant growth caused by salt stress (Sun et al., 2015). Long non-coding RNAs (lncR-NAs) play important role in response to biotic and abiotic stress by acting as competing endogenous RNAs (ceRNAs) to decoy mature miRNAs. The results of endogenous target mimic (eTM) analysis showed that *lncRNA354* had a potential binding site for *miR160b*. Silencing *lncRNA354* resulted in taller cotton plants and enhanced their resistance to salt stress. The results also indicated that *lncRNA354-miR160b* had an impact on its target *GhARF17/18* expression and may modulate auxin signaling, thus opening new avenues on a mechanism of lncRNA-associated responses to salt stress (Zhang et al., 2021).

10.4.2.3 Extreme Temperature (Heat and Cold) Stress

The other major abiotic stress that adversely affects the growth and productivity of plants is geographical and seasonal variations in extreme temperature stress. Plants differentially express their genes and mRNA to adapt to those inconstant environments (Shriram et al., 2016). Also, there are enormous heat or cold-responsive miRNAs were observed in certain plant species (Cao et al., 2014). A new *Osa-miR319* was allowed to overexpress in rice which ultimately increases the chilling stress tolerance (Yang et al., 2013). *OsmiR156k* overexpression, however, decreased the ability of rice plants to tolerate cold (Cui et al., 2015). To improve cold tolerance, Song et al. (2016) created transgenic *Arabidopsis* with higher expression of the *miR394a/LCR* loss of function mutant. The heat stress tolerance gene *miR398* was overexpressed in *Arabidopsis*. Thermostability in the transgenic plant was improved by *CSD2* knockdown mutants, and loss-of-function mutants of *CSD1* and *CCS* (Guan et al., 2013). In another study, *miR156* isoforms are highly induced after heat stress (HT). SPL transcription factor genes, which operate as primary regulators of developmental changes, are the targets of *miR156*. After HS, *miR156* post-transcriptionally downregulates SPL genes (Stief et al., 2014). Similarly, miRNAs and their response to heat stress were investigated by Li et al. (2014a) and Wang et al. (2014) using transgenic technology. Along with miRNAs, scientists hypothesized the miRNA targets, some of which were verified using qRT-PCR. Additionally, functional investigation of the target genes demonstrated that the majority of them positively influence the chilling response by controlling the expression of antioxidants and anti-stress proteins (Cao et al., 2014).

10.4.2.4   Other Minor Stress

Recent studies suggested several miRNAs as hypoxia-responsive (Sunkar et al., 2006; Lin et al., 2012), nutrient deficiency tolerant (Fujii et al., 2005), heavy metal stress-responsive (Ding et al., 2016), and they are essential post-transcriptional modulators. Waterlogging in the root rhizosphere causes hypoxia or low oxygen conditions, which deprive plants of oxygen (Agarwal & Grover, 2006). By overexpressing *the miR398*-resistant form of *CSD2* the transgenic *Arabidopsis* exhibited enhanced resistance to heavy metals, high-intensity light, and other oxidative stressors (Sunkar et al., 2006). Similarly, when *the miR828* precursor from *Arabidopsis* was overexpressed in sweet potato, it affected lignin and H_2O_2 contents to participate in defence mechanisms (Lin et al., 2012).

Mineral nutrient homeostasis is crucial for the growth of the plant, development, and crop yield. Recent advances have focused on miRNAs and the critical role played by them in maintaining nutrient homeostasis (Kehr, 2013; Paul et al., 2015). In one of the studies, overexpression of *miR399* causes P toxicity and results in retarded growth. Thus, when a transgenic tomato (*Solanum lycopersicum* L., C1), which contains the foreign gene *Ath-miR399d* under the control of *rd29A* promoter was constructed, the *miR399d* was moderately upregulated in response to salinity and cold stresses (Gao et al., 2015). Bai et al. (2018) overexpressed tobacco lines with *TaemiR408* from wheat, which eventually increased stress resistance in high salinity and Pi deficiency situations. Fujii et al. (2005) and Liang et al. (2010) did transgenic experiments in *Arabidopsis* with *miR399* and *miR395* respectively to decipher their role in generating nutrient-deficient tolerant transgenic lines.

Heavy metal toxicity is yet another abiotic stress that causes hazardous effects in plants that ultimately reduce yield. Numerous miRNAs and their response to different heavy metal toxicity were reported by various scientists (Valdes-Lopez et al., 2010; Ding et al., 2011; Gupta et al., 2014). *OsSRK*, the *miR390* target gene, was markedly increased by Cd treatment. *MiR390*-overexpressing transgenic rice plants showed decreased Cd tolerance and increased Cd accumulation. Simultaneously, the expression of *OsSRK* was less pronounced, and these findings showed the ability of rice to tolerate Cd stress, which was negatively regulated by *miR390* (Ding et al., 2016). Several other miRNAs are involved in developing multiple stress resistance in various crop plants (Kim et al., 2010b; Chen et al., 2015; Zhao et al., 2019). The same miRNA may respond differently to different stress. For example, overexpression of *miR402* accelerated both seed germination and seedling growth of *Arabidopsis* under salt stress conditions, while the seedling growth was arrested under dehydration or cold stress conditions (Kim et al., 2010b). The levels of anthocyanins, flavones, and flavonols were substantially elevated in transgenic poplar plants overexpressing *miR156* whereas the total lignin content was reduced suggesting the role played in creating stress-resistant plants.

10.5   DATABASES FOR IDENTIFICATION, FUNCTIONAL
CHARACTERIZATION, AND TARGET PREDICTION

miRNAs and their potential role in defending against various stresses had been predicted by different computational tools and databases. Detecting a huge number of miRNAs is not so easy, because of their manifold existences, smaller size, and methylation condition. But in the last few years, improvements in cloning techniques, sequencing technology, and computational algorithms led to the easy recognition of miRNAs in plants (Tripathi et al., 2015). A detailed set of computational tools and databases to identify and functionally characterize the miRNAs is enlisted in Table 10.3. Different miRNA databases and the roles played by them are represented in Figure 10.3.

In addition to finding the miRNAs, Lorenzetti et al. (2016) suggested creating a database of plant TE-related miRNAs (PlanTE-MIR DB) which enabled the discovery of more than 150 miRNAs that duplicated TEs in the genomes of ten plant species. The miRDB is one of the recent databases developed for miRNA target prediction and functional annotations which includes 3.5 million expected targets in five species that are regulated by 7,000 miRNAs (Chen & Wang 2020). The issues faced by the databases developed so far include lack of entries for many important species, uneven annotation standards across different species, abundant questionable entries, and limited annotation. To address these issues, a knowledge-based database called Plant miRNA Encyclopedia that contains 16,422 high-confidence novel miRNA loci in 88 plant species and 3,966 retrieved from miRBase was developed by Guo et al. (2020). TarDB provides a user-friendly interface that enables users to easily search, browse, and retrieve miRNA targets and miRNA-initiated phasiRNAs in a broad variety of plants. Even in the well-researched model species, TarDB offers an extensive library of reliable plant miRNA targets that includes previously unreported miRNA targets and miRNA-triggered phasiRNAs. The majority of these new miRNA targets are important for species or lineage-specific miRNAs (Liu et al., 2021).

FIGURE 10.3   Different roles of identified miRNA databases.

TABLE 10.3

Computational tools/databases used in bioinformatics for plant miRNAs, their target identification/prediction with description.

Database/Tool	Description	References
miRCheck	The algorithm is scripted in PERL to identify the 20 mers which can be potential miRNAs.	Jones-Rhoades and Bartel (2004)
miRdeep -P	The algorithm is scripted in PERL script, contains in-built plant miRNA-specific criteria, and predicts novel miRNAs from deep sequencing data.	Friedlander et al. (2008)
TAPIR	Tool for prediction of miRNA targets with help of FASTA search engine or RNA-hybrid search engine.	Bonnet et al. (2010)
miRTour	Freely available computational tool for homology-based discovery of plant miRNAs as well as their targets by retrieving data from sequence datasets. Automates most of the programs for similarity search, precursor selection, target prediction, and annotation for miRNAs, using a single input.	Milev et al. (2011)
C-mii	Homology-based blast search of EST sequences of query plant species against the known miRNAs from different plant species.	Numnark et al. (2012)
PMTED	Specially designed database for retrieval and analysis of expression profiling of miRNA targets. Provides basic query function regarding miRNAs and respective targets, gene ontology, and variable expression profile.	Sun et al. (2013)
PASmiR	Provides a strong platform for collection, standardization, and querying miRNA-abiotic stress regulation data in plants.	Zhang et al. (2013b)
PolymiRTS	Database of naturally arising DNA variations in miRNA seed region as well as miRNA target sites.	Bhattacharya et al. (2014)
miRPlant	Tool for plant miRNA identification from RNA-sequencing data, strategically developed for identification of hairpin excision regions, and hairpin structure filtering for plants.	An et al. (2014)
miRBase	Online repository of prominent miRNA from various sources along with their associated annotations. Freely available for searching and browsing purpose along with easy entry retrieval by using the name, keywords, references, and annotations.	Kozomara and Griffiths-Jones (2014)
miRNEST	Integrative miRNA resources vitally include information about sequences, polymorphism, expression, and promoters. Also provides useful degradome support for plant miRNA targets which offers detailed information regarding plant miRNAs along with the source, sequence, alignment, transcript, cleavage position, abundance, etc.	Szczesniak and Makalowska (2014)
PNRD	An updated version of PMRD combining the information regarding non-coding RNAs, from 166 species.	Yi et al. (2015)
miRge	A quick multifaceted way for sRNA data processing for miRNA entropy determination and identification differential miRNA-isomiRs.	Baras et al. (2015)
PmiRExAt	Online web resource focusing on plant miRNA expressions levels in variable tissues, at developmental stages of plants. Supplies pre-computed expression matrices of miRNAs from various plants, which act as proxy for relative expression-levels of miRNA sequences.	Gurjar et al. (2016)
P-SAMS	Web tool for study of artificial miRNAs, synthetic trans-acting small interfering RNAs.	Fahlgren et al. (2016)
PlanTE-MIR DB	A database for transposable element-related microRNAs in plant genomes.	Lorenzetti et al. (2016)
PlantCircNet	Database to visualize circRNA-miRNA-mRNA regulatory networks from plants containing acknowledged circRNAs from eight model plants.	Zhang et al. (2017b)

(*Continued*)

TABLE 10.3 (Continued)

Computational tools/databases used in bioinformatics for plant miRNAs, their target identification/prediction with description.

Database/Tool	Description	References
miRTarBase	Database with numerous experimentally validated MTIs.	Chou et al. (2018)
miRToolsGallery	Comprehensive database for miRNA tools, for searching, sieving, and positing all the available miRNA tools.	Chen et al. (2018)
miRDB	An online database for prediction of functional microRNA targets.	Chen & Wang (2020)
PmiREN	A comprehensive encyclopaedia of plant miRNAs.	Guo et al. (2020)
TarDB	An online database for plant miRNA targets and miRNA-triggered phased siRNAs.	Liu et al. (2021)

10.6 CONCLUSION AND FUTURE PROSPECTS

Once miRNAs were thought of as gene silencers, but recent advances have thrown light on plant miRNAs and their diverse roles in responding to various stresses in almost all biological networks. They are thus a good choice for increasing crop varieties' capacity to withstand stress. Agricultural researchers and scientists would be able to implement particular specific features in crops if they had a comprehensive knowledge of the genetic/molecular mechanism governed by miRNA. Deciphering pri-miRNA-mediated regulatory networks requires effective tools/databases, nevertheless, to predict the functional implications of synonymous variations. Therefore, miRNA-based techniques have given scientists great hope for improving crops by using transgenic technology to enhance crop plants' resistance to biotic and abiotic stress. Apart from the innate miRNAs construction of artificial miRNA (amiRNA) could be successful as suggested by some scientists (Agrawal et al., 2015; Zhou et al., 2018; Xie et al., 2021). Nevertheless, it is also pertinent to exploit more before being successful and to get rid of undesirable side effects/inappropriate results obtained in the near term by employing miRNA-based genetic editing techniques.

KEYWORDS: gene regulation, microRNA, overexpression, stress, transgenics

REFERENCES

Agarwal, S., & Grover, A. (2006). Molecular biology, biotechnology and genomics of flooding-associated low O_2 stress response in plants. *Critical Reviews in Plant Sciences*, 25(1), 1–21.

Agrawal, A., Rajamani, V., Reddy, V. S., Mukherjee, S. K., & Bhatnagar, R. K. (2015). Transgenic plants over-expressing insect-specific microRNA acquire insecticidal activity against *Helicoverpa armigera*: An alternative to Bt-toxin technology. *Transgenic Research*, 24(5), 791–801.

An, J., Lai, J., Sajjanhar, A., Lehman, M. L., & Nelson, C. C. (2014). miRPlant: An integrated tool for identification of plant miRNA from RNA sequencing data. *BMC Bioinformatics*, 15(1), 1–4.

Baek, D., Chun, H. J., Kang, S., Shin, G., Park, S. J., Hong, H., Kim, C., Kim, D. H., Lee, S. Y., Kim, M. C., & Yun, D. J. (2016). A role for Arabidopsis *miR399f* in salt, drought, and ABA signaling. *Molecules and Cells*, 39(2), 111.

Bai, Q., Wang, X., Chen, X., Shi, G., Liu, Z., Guo, C., & Xiao, K. (2018). Wheat miRNA *TaemiR408* acts as an essential mediator in plant tolerance to Pi deprivation and salt stress via modulating stress-associated physiological processes. *Frontiers in Plant Science*, 9, 499.

Banerjee, A., Roychoudhury, A., & Krishnamoorthi, S. (2016). Emerging techniques to decipher microRNAs (miRNAs) and their regulatory role in conferring abiotic stress tolerance of plants. *Plant Biotechnology Reports*, 10(4), 185–205.

Baras, A. S., Mitchell, C. J., Myers, J. R., Gupta, S., Weng, L. C., Ashton, J. M., Cornish, T. C., Pandey, A., & Halushka, M. K. (2015). miRge-A multiplexed method of processing small RNA-seq data to determine microRNA entropy. *PLoS One*, 10(11), e0143066.

Basso, M. F., Ferreira, P. C. G., Kobayashi, A. K., Harmon, F. G., Nepomuceno, A. L., Molinari, H. B. C., & Grossi-de-Sa, M. F. (2019)Micro RNA s and new biotechnological tools for its modulation and improving stress tolerance in plants. *Plant Biotechnology Journal*, *17*(8), 1482–1500.

Bazzini, A. A., Hopp, H. E., Beachy, R. N., & Asurmendi, S. (2007). Infection and coaccumulation of tobacco mosaic virus proteins alter microRNA levels, correlating with symptom and plant development. *Proceedings of the National Academy of Sciences*, *104*(29), 12157–12162.

Bhattacharya, A., Ziebarth, J. D., & Cui, Y. (2014). PolymiRTS database 3.0: Linking polymorphisms in microRNAs and their target sites with human diseases and biological pathways. *Nucleic Acids Research*, *42*,86–91.

Bonnet, E., He, Y., Billiau, K., & Van de Peer, Y. (2010). TAPIR, a web server for the prediction of plant microRNA targets, including target mimics. *Bioinformatics*, *26*(12), 1566–1568

Brodersen, P., Sakvarelidze-Achard, L., Bruun-Rasmussen, M., Dunoyer, P., Yamamoto, Y. Y., Sieburth, L., & Voinnet, O. (2008). Widespread translational inhibition by plant miRNAs and siRNAs. *Science, 320*(5880), 1185–1190.

Buiatti, M., Christou, P., & Pastore, G. (2012). The application of GMOs in agriculture and in food production for better nutrition: Two different scientific points of view. *Genes & Nutrition*, *8*, 255–270.

Campo, S., Peris-Peris, C., Siré, C., Moreno, A. B., Donaire, L., Zytnicki, M., Notredame, C., Llave, C., & San Segundo, B. (2013). Identification of a novel micro RNA (mi RNA) from rice that targets an alternatively spliced transcript of the *Nramp6* (Natural resistance-associated macrophage protein 6) gene involved in pathogen resistance. *New Phytologist*, *199*(1), 212–227.

Cao, X., Wu, Z., Jiang, F., Zhou, R., & Yang, Z. (2014). Identification of chilling stress-responsive tomato microRNAs and their target genes by high-throughput sequencing and degradome analysis. *BMC Genomics*, *15*(1), 1–16.

Cattivelli, L., Rizza, F., Badeck, F. W., Mazzucotelli, E., Mastrangelo, A. M., Francia, E., Marè, C., Tondelli, A., & Stanca, A. M. (2008). Drought tolerance improvement in crop plants: An integrated view from breeding to genomics. *Field Crops Research*, *105*(1–2). 1–14.

Chen, L., Heikkinen, L., Wang, C., Yang, Y., Knott, K. E., & Wong, G. (2018). miRToolsGallery: A tag-based and rankable microRNA bioinformatics resources database portal. *Database: The Journal of Biological Databases and Curation*, 2018, bay004.

Chen, L., Luan, Y., & Zhai, J. (2015). *Sp-miR396a-5p* acts as a stress-responsive genes regulator by conferring tolerance to abiotic stresses and susceptibility to *Phytophthora nicotianae* infection in transgenic tobacco. Plant Cell Reports, *34*(12), 2013–2025.

Chen, Y., & Wang, X. (2020). miRDB: An online database for prediction of functional microRNA targets. *Nucleic Acids Research*, *48*, 127–131.

Chou, C. H., Shrestha, S., Yang, C. D., Chang, N. W., Lin, Y. L., Liao, K. W., Huang, W. C., Sun, T. H., Tu, S. J., Lee, W. H., & Chiew, M. Y. (2018). miRTarBase update 2018: A resource for experimentally validated microRNA-target interactions. *Nucleic Acids Research*, *46*, 296–302.

Cui, N., Sun, X., Sun, M., Jia, B., Duanmu, H., Lv, D., Duan, X., & Zhu, Y. (2015). Overexpression of *OsmiR156k* leads to reduced tolerance to cold stress in rice (*Oryza Sativa*). *Molecular Plant Breeding, 35*(11), 1–11.

Ding, Y., Chen, Z., & Zhu, C. (2011). Microarray-based analysis of cadmium-responsive microRNAs in rice (*Oryza sativa*). *Journal of Experimental Botany, 62*(10), 3563–3573.

Ding, Y., Ye, Y., Jiang, Z., Wang, Y., & Zhu, C. (2016). MicroRNA390 is involved in cadmium tolerance and accumulation in rice. *Frontiers in Plant Science*, *7*, 235.

Ebert, M. S., Neilson, J. R., & Sharp, P. A. (2007). MicroRNA sponges: Competitive inhibitors of small RNAs in mammalian cells. *Nature Methods*, *4*(9), 721–726.

Eckardt, A. A. (2012). A microRNA cascade in plant defense. *Plant Cell*, *24*, 840–840.

Fahlgren, N., Hill, S. T., Carrington, J. C., & Carbonell, A. (2016). P-SAMS: A web site for plant artificial microRNA and synthetic trans-acting small interfering RNA design. *Bioinformatics*, *32*(1), 157–158.

Ferdous, J., Sanchez-Ferrero, J. C., Langridge, P., Milne, L., Chowdhury, J., Brien, C., & Tricker, P. J. (2016). Differential expression of microRNAs and potential targets under drought stress in barley. *Plant, Cell & Environment, 40*, 11–24.

Franco-Zorrilla, J. M., Valli, A., Todesco, M., Mateos, I., Puga, M. I., Rubio-Somoza, I., Leyva, A., Weigel, D., García, J. A., Paz-Ares, J. (2007). Target mimicry provides a new mechanism for regulation of microRNA activity. *Nature Genetics, 39*(8), 1033–1037.

Friedlander, M. R., Chen, W., Adamidi, C., Maaskola, J., Einspanier, R., Knespel, S., & Rajewsky, N. (2008). Discovering microRNAs from deep sequencing data using miRDeep. *Nature Biotechnology, 26*(4), 407–415

Fujii, H., Chiou, T. J., Lin, S. I., Aung, K., & Zhu, J. K. (2005). A miRNA involved in phosphate-starvation response in Arabidopsis. *Current Biology*, *15*(22), 2038–2043.

Gao, N., Qiang, X. M., Zhai, B. N., Min, J., & Shi, W. M. (2015). Transgenic tomato overexpressing *ath-miR399d* improves growth under abiotic stress conditions. *Russian Journal of Plant Physiology*, *62*(3), 360–366.

Gao, P., Bai, X., Yang, L., Lv, D., Li, Y., Cai, H., Ji, W., Guo, D., & Zhu, Y. (2010). Over-expression of *osa-MIR396c* decreases salt and alkali stress tolerance. *Planta*, *231*, 991–1001.

Gao, P., Bai, X., Yang, L., Lv, D., Pan, X., Li, Y., Cai, H., Ji, W., Chen, Q., & Zhu, Y. (2011). *Osa-MIR393*: A salinity-and alkaline stress-related microRNA gene. Molecular Biology Reports, *38*(1), 237–242.

Gao, S., Yang, L., Zeng, H. Q., Zhou, Z. S., Yang, Z. M., Li, H., Sun, D., Xie, F., & Zhang, B. (2016). A cotton miRNA is involved in regulation of plant response to salt stress. *Scientific Reports*, *6*(1), 1–14

Guan, Q., Lu, X., Zeng, H., Zhang, Y., & Zhu, J. (2013). Heat stress induction of *miR398* triggers a regulatory loop that is critical for thermotolerance in Arabidopsis. *Plant Journal*, *74*(5), 840–851.

Guo, H., Song, X., Wang, G., Yang, K., Wang, Y., Niu, L., Chen, X., & Fang, R. (2014). Plant-generated artificial small RNAs mediated aphid resistance. *PloS One*, *9*(5), e97410.

Guo, H. S., Xie, Q., Fei, J. F., & Chua, N. H. (2005). MicroRNA directs mRNA cleavage of the transcription factor NAC1 to downregulate auxin signals for Arabidopsis lateral root development. *Plant Cell*, *17*(5), 1376–1386.

Guo, Z., Kuang, Z., Wang, Y., Zhao, Y., Tao, Y., Cheng, C., Yang, J., Lu, X., Hao, C., Wang, T., & Cao, X. (2020). PmiREN: A comprehensive encyclopedia of plant miRNAs. *Nucleic Acids Research*, *48*, 1114–1121.

Gupta, O. P., Sharma, P., Gupta, R. K., & Sharma, I. (2014). MicroRNA mediated regulation of metal toxicity in plants: Present status and future perspectives. *Plant Molecular Biology, 84*(1–2), 1–18

Gurjar, A. K. S., Panwar, A. S., Gupta, R., & Mantri, S. S. (2016). PmiRExAt: Plant miRNA expression atlas database and web applications. *Database*.

Hackenberg, M., Gustafson, P., Langridge, P., & Shi, B. (2015). Differential expression of micro RNA s and other small RNA s in barley between water and drought conditions. *Plant Biotechnology Journal*, *13*, 2–13.

Hajyzadeh, M., Turktas, M., Khawar, K. M., & Unver, T. (2015). *miR408* overexpression causes increased drought tolerance in chickpea. *Gene*, *555*(2), 186–193.

He, X. F., Fang, Y. Y., Feng, L., & Guo, H. S. (2008). Characterization of conserved and novel microRNAs and their targets, including a TuMV-induced TIR–NBS–LRR class R gene-derived novel miRNA in brassica. *FEBS Letters*, *582*(16), 2445–2452.

Hewezi, T., Howe, P., Maier, T. R., & Baum, T. J. (2008). Arabidopsis small RNAs and their targets during cyst nematode parasitism. *Molecular Plant Microbe Interactions*, *21*(12), 1622–1634.

Hewezi, T., Maier, T. R., Nettleton, D., & Baum, T. J. (2012). The Arabidopsis *microRNA396-GRF1/GRF3* regulatory module acts as a developmental regulator in the reprogramming of root cells during cyst nematode infection. *Plant Physiology, 159*(1), 321–335.

Jagadeeswaran, G., Zheng, Y., Li, Y. F., Shukla, L. I., Matts, J., Hoyt, P., Macmil, S. L., Wiley, G. B., Roe, B. A., Zhang, W., & Sunkar, R. (2009). Cloning and characterization of small RNAs from *Medicago truncatula* reveals four novel legume-specific microRNA families. *New Phytologist*, *184*(1), 85–98.

Jalmi, S. K., Bhagat, P. K., Verma, D., Noryang, S., Tayyeba, S., Singh, K., Sharma, D., & Sinha, A. K. (2018). Traversing the links between heavy metal stress and plant signaling. *Frontiers of Plant Science*, *9*, 12.

Jeyaraj, A., Elango, T., Li, X., & Guo, G. (2020). Utilization of microRNAs and their regulatory functions for improving biotic stress tolerance in tea plant [*Camellia sinensis* (L.) O. Kuntze]. *RNA Biology*, *17*(10), 1365–1382.

Jeyaraj, A., Liu, S., Zhang, X., Zhang, R., Shangguan, M., & Wei, C. (2017). Genome-wide identification of microRNAs responsive to *Ectropis* oblique feeding in tea plant (*Camellia sinensis* L.). *Scientific Reports*, *7*(1), 1–16

Jeyaraj, A., Wang, X., Wang, S., Liu, S., Zhang, R., Wu, A., & Wei, C. (2019). Identification of regulatory networks of microRNAs and their targets in response to *Colletotrichum gloeosporioides* in tea plant (*Camellia sinensis* L.). *Frontiers of Plant Science*, 10, 1096.

Jian, C., Han, R., Chi, Q., Wang, S., Ma, M., Liu, X., & Zhao, H. (2017). Virus-based microRNA silencing and overexpressing in common wheat (*Triticum aestivum* L.). *Frontiers in Plant Science*, *8*, 500.

Jiang, N., Meng, J., Cui, J., Sun, G., & Luan, Y. (2018). Function identification of miR482b, a negative regulator during tomato resistance to *Phytophthora infestans*. *Horticulture Research*, *5*, 1–11.

Jones-Rhoades, M. W., & Bartel, D. P. (2004). Computational identification of plant microRNAs and their targets, including a stress-induced miRNA. *Molecules and Cells*, *14*(6), 787–799.

Jones-Rhoades, M. W., Bartel, D. P., & Bartel, B. (2006). MicroRNAs and their regulatory roles in plants. *Annual Review of Plant Biology, 57*, 19–53.

Kasschau, K. D., Xie, Z., Allen, E., Llave, C., Chapman, E. J., Krizan, K. A., & Carrington, J. C. (2003). P1/HC-Pro, a viral suppressor of RNA silencing, interferes with Arabidopsis development and miRNA function. *Developmental Cell, 4,* 205–217

Katiyar-Agarwal, S., & Jin, H. (2010). Role of small RNAs in host-microbe interactions. *Annual Review of Phytopathology, 48,* 225–246.

Katiyar-Agarwal, S., Morgan, R., Dahlbeck, D., Borsani, O., Villegas, A., Zhu, J. K., Staskawicz, B. J., & Jin, H. (2006). A pathogen-inducible endogenous siRNA in plant immunity. *Proceedings of the National Academy of Sciences, 103*(47), 18002–18007.

Kehr, J. (2013). Systemic regulation of mineral homeostasis by micro RNAs. *Frontiers of Plant Science,* 4, 145.

Khraiwesh, B., Ossowski, S., Weigel, D., Reski, R., & Frank, W. (2008). Specific gene silencing by artificial MicroRNAs in Physcomitrella patens: An alternative to targeted gene knockouts. *Plant Physiology, 148*(2), 684–693.

Khraiwesh, B., Zhu, J. K., & Zhu, J. (2012). Role of miRNAs and siRNAs in biotic and abiotic stress responses of plants. *Biochimica et Biophysica Acta, 1819,* 137–148

Kim, J. Y., Kwak, K. J., Jung, H. J., Lee, H. J., & Kang, H. (2010b). MicroRNA402 affects seed germination of *Arabidopsis thaliana* under stress conditions via targeting demeter-like protein3 mRNA. *Plant and Cell Physiology, 51*(6), 1079–1083.

Kim, J. Y., Lee, H. J., Jung, H. J., Maruyama, K., Suzuki, N., & Kang, H. (2010a). Overexpression of *microRNA395c* or *395e* affects differently the seed germination of *Arabidopsis thaliana* under stress conditions. *Planta, 232*(6), 1447–1454.

Kis, A., Hamar, É., Tholt, G., Bán, R., & Havelda, Z. (2019). Creating highly efficient resistance against wheat dwarf virus in barley by employing CRISPR/Cas9 system. *Plant Biotechnology Journal, 17*(6), 1004–1006.

Kis, A., Tholt, G., Ivanics, M., Várallyay, É., Jenes, B., & Havelda, Z. (2015). Polycistronic artificial miRNA-mediated resistance to wheat dwarf virus in barley is highly efficient at low temperature. *Molecular Plant Pathology, 17,* 427–437.

Kozomara, A., & Griffiths-Jones, S. (2014). miRBase: Annotating high confidence microRNAs using deep sequencing data. *Nucleic Acids Research, 42,* 68–73

Lee, H. J., Park, Y. J., Kwak, K. J., Kim, D., Park, J. H., Lim, J. Y., Shin, C., Yang, K. Y., & Kang, H. (2015b). MicroRNA844-guided downregulation of cytidinephosphate diacylglycerol synthase3 (CDS3) mRNA affects the response of *Arabidopsis thaliana* to bacteria and fungi. *Molecular Plant Microbe Interactions, 28*(8), 892–900.

Lee, J. H., Kim, J. A., Kwon, M. H., Kang, J. Y., & Rhee, W. J. (2015a). In situ single step detection of exosome microRNA using molecular beacon. *Biomaterials, 54,* 116–125.

Li, S., Liu, J., Liu, Z., Li, X., Wu, F., & He, Y. (2014a). Heat-induced TAS1 TARGET1 mediates thermotolerance via heat stress transcription factor A1a–directed pathways in Arabidopsis. *Plant Cell, 26*(4), 1764–1780.

Li, W., Jia, Y., Liu, F., Wang, F., Fan, F., Wang, J., Zhu, J., Xu, Y., Zhong, W., & Yang, J. (2019). Integration analysis of small RNA and degradome sequencing reveals microRNAs responsive to *Dickeya zeae* in resistant rice. *International Journal of Molecular Sciences, 20,* 222.

Li, W., Wang, T., Zhang, Y., & Li, Y. (2016). Overexpression of soybean *miR172c* confers tolerance to water deficit and salt stress, but increases ABA sensitivity in transgenic *Arabidopsis thaliana*. *Journal of Experimental Botany, 67*(1), 175–194.

Li, W. X., Oono, Y., Zhu, J., He, X. J., Wu, J. M., Iida, K., Lu, X. Y., Cui, X., Jin, H., & Zhu, J. K. (2008). The Arabidopsis *NFYA5* transcription factor is regulated transcriptionally and post transcriptionally to promote drought resistance. *Plant Cell, 20*(8), 2238–2251.

Li, Y., Lu, Y. G., Shi, Y., Wu, L., Xu, Y. J., Huang, F., Guo, X. Y., Zhang, Y., Fan, J., Zhao, J. Q., & Zhang, H. Y. (2014b). Multiple rice microRNAs are involved in immunity against the blast fungus *Magnaporthe oryzae*. *Plant Physiology, 164*(2), 1077–1092.

Liang, G., Yang, F., & Yu, D. (2010). MicroRNA395 mediates regulation of sulfate accumulation and allocation in *Arabidopsis thaliana*. *Plant Journal, 62*(6), 1046–1057.

Lin, J. S., Lin, C. C., Lin, H. H., Chen, Y. C., & Jeng, S. T. (2012). Micro R 828 regulates lignin and H_2O_2 accumulation in sweet potato on wounding. *New Phytologist, 196*(2), 427–440.

Liu, J., Liu, X., Zhang, S., Liang, S., Luan, W., & Ma, X. (2021). TarDB: An online database for plant miRNA targets and miRNA-triggered phased siRNAs. *BMC Genomics, 22*(1), 1–12.

Liu, X., Tan, C., Cheng, X., Zhao, X., Li, T., & Jiang, J. (2020). miR168 targets Argonaute1A mediated miRNAs regulation pathways in response to potassium deficiency stress in tomato. *BMC Plant Biology, 20*(1), 477.

Lorenzetti, A. P. R., Ade Antonio, G. Y., Paschoal, A. R., & Domingues, D. S. (2016). PlanTE-MIRDB: A database for transposable element-related microRNAs in plant genomes. *Functional & Integrative Genomics, 16,* 235–242

Ma, C., Burd, S., & Lers, A. (2015). *miR408* is involved in abiotic stress responses in Arabidopsis. The *Plant Journal, 84*, 169–187.

Meng, Y., Gou, L., Chen, D., Mao, C., Jin, Y., Wu, P., & Chen, M. (2010). PmiRKB: A plant microRNA knowledge base. *Nucleic Acids Research, 39*, 181-D187.

Milev, I., Yahubyan, G., Minkov, I., & Baev, V. (2011). miRTour: Plant miRNA and target prediction tool. *Bioinformation, 6*(6), 248.

Navarro, L., Dunoyer, P., Jay, F., Arnold, B., Dharmasiri, N., Estelle, M., Voinnet, O., & Jones, J. D. (2006). A plant miRNA contributes to antibacterial resistance by repressing auxin signaling. *Science, 312*(5772), 436–439.

Ni, Z., Hu, Z., Jiang, Q., & Zhang, H. (2012). Overexpression of *gma-MIR394a* confers tolerance to drought in transgenic *Arabidopsis thaliana. Biochemical and Biophysical Research Communications, 427*(2), 330–335.

Ni, Z., Hu, Z., Jiang, Q., & Zhang, H. (2013). *GmNFYA3*, a target gene of *miR169*, is a positive regulator of plant tolerance to drought stress. *Plant Molecular Biology, 82*(1–2), 113–129.

Niu, J., Wang, J., Hu, H., Chen, Y., An, J., Cai, J., Sun, R., Sheng, Z., Liu, X., & Lin, S. (2016). Cross-talk between freezing response and signaling for regulatory transcriptions of *MIR475b* and its targets by *miR475b* promoter in *Populus suaveolens. Scientific Reports, 6*(1), 1–11.

Niu, Q. W., Lin, S. S., Reyes, J. L., Chen, K. C., Wu, H. W., Yeh, S. D., & Chua, N. H. (2006). Expression of artificial microRNAs in transgenic Arabidopsis thaliana confers virus resistance. *Nature Biotechnology, 24*, 1420–1428.

Numnark, S., Mhuantong, W., Ingsriswang, S., & Wichadakul, D. (2012). C-mii: A tool for plant miRNA and target identification. *BMC Genomics, 13*(7), 1–10.

Palatnik, J. F., Allen, E., Wu, X., Schommer, C., Schwab, R., Carrington, J. C., & Weigel, D. (2003). Control of leaf morphogenesis by microRNAs. *Nature, 425*(6955), 257–263.

Pan, L., Zhao, H., Yu, Q., Bai, L., & Dong, L. (2017). *miR397/Laccase* gene mediated network improves tolerance to fenoxaprop-P-ethyl in *Beckmannia syzigachne* and *Oryza sativa. Frontiers in Plant Science, 8*, 1–14.

Patel, P., Yadav, K., Srivastava, A. K., Suprasanna, P., & Ganapathi, T. R. (2019). Overexpression of native musa-miR397 enhances plant biomass without compromising abiotic stress tolerance in banana. *Scientific Reports, 9*(1), 1–15.

Paul, S., Datta, S. K., & Datta, K. (2015). miRNA regulation of nutrient homeostasis in plants. *Frontiers in Plant Science, 6*, 232.

Qu, J., Ye, J., & Fang, R. (2007). Artificial microRNA-mediated virus resistance in plants. *Journal of Virology, 81*, 6690–6699.

Reichel, M., Li, Y., Li, J., & Millar, A. A. (2015). Inhibiting plant microRNA activity: Molecular SPONGEs, target MIMICs and STTMs all display variable efficacies against target microRNAs. *Plant Biotechnology Journal,* 13, 915–926.

Romanel, E., Silva, T. F., Corrêa, R. L., Farinelli, L., Hawkins, J. S., Schrago, C. E., & Vaslin, M. F. (2012). Global alteration of microRNAs and transposon-derived small RNAs in cotton (*Gossypium hirsutum*) during Cotton leafroll dwarf polerovirus (CLRDV) infection. *Plant Molecular Biology, 80*(4), 443–460.

Sablok, G., Pérez-Quintero, Á. L., Hassan, M., Tatarinova, T. V., & López, C. (2011). Artificial microRNAs (amiRNAs) engineering—On how microRNA-based silencing methods have affected current plant silencing research. *Biochemical and Biophysical Research Communications, 406*, 315–319.

Şanlı, B. A., & Gökçe, Z. N. O. (2021). Investigating effect of *miR160* through overexpression in potato cultivars under single or combination of heat and drought stresses. *Plant Biotechnology Reports*, 1–14.

Schwab, R., Ossowski, S., Riester, M., Warthmann, N., & Weigel, D. (2006). Highly specific gene silencing by artificial microRNAs in Arabidopsis. *Plant Cell, 18*(5), 1121–1133.

Shen, X., He, J., Ping, Y., Guo, J., Hou, N., Cao, F., Li, X., Geng, D., Wang, S., Chen, P., Qin, G., Ma, F., & Guan, Q. (2022). The positive feedback regulatory loop of miR160-auxin response factor 17-HYPONASTIC LEAVES 1 mediates drought tolerance in apple trees. *Plant Physiology, 188*(3), 1686–1708.

Shi, G. Q., Fu, J. Y., Rong, L. J., Zhang, P. Y., Guo, C. J., & Xiao, K. (2018). TaMIR1119, a miRNA family member of wheat (*Triticum aestivum*), is essential in the regulation of plant drought tolerance. *Journal of Integrative Agriculture, 17*(11), 2369–2378.

Shriram, V., Kumar, V., Devarumath, R. M., Khare, T. S., & Wani, S. H. (2016). MicroRNAs as potential targets for abiotic stress tolerance in plants. *Frontiers in Plant Science, 7*, 817.

Song, J. B., Gao, S., Wang, Y., Li, B. W., Zhang, Y. L., & Yang, Z. M. (2016). *miR394* and its target gene LCR are involved in cold stress response in Arabidopsis. *Plant Gene, 5*, 56–64.

Stief, A., Altmann, S., Hoffmann, K., Pant, B. D., Scheible, W. R., & Baurle, I. (2014). Arabidopsis *miR156* regulates tolerance to recurring environmental stress through SPL transcription factors. *Plant Cell, 26*(4), 1792–1807.

Sun, G. (2012). MicroRNAs and their diverse functions in plants. *Plant Molecular Biology*, *80*,17–36.

Sun, X., Dong, B., Yin, L., Zhang, R., Du, W., Liu, D., Shi, N., Li, A., Liang, Y., & Mao, L. (2013). PMTED: A plant microRNA target expression database. *BMC Bioinformatics*, *14*(1), 1–7.

Sun, X., Xu, L., Wang, Y., Yu, R., Zhu, X., Luo, X., Gong, Y., Wang, R., Limera, C., Zhang, K., & Liu, L. (2015). Identification of novel and salt-responsive miRNAs to explore miRNA-mediated regulatory network of salt stress response in radish (*Raphanus sativus* L.). *BMC Genomics*, *16*(1), 1–16.

Sun, Z., Shu, L., Zhang, W., & Wang, Z. (2020). Cca-miR398 increases copper sulfate stress sensitivity via the regulation of CSD mRNA transcription levels in transgenic Arabidopsis thaliana. *PeerJ, 8*, e9105.

Sunkar, R. (2010). MicroRNAs with macro-effects on plant stress responses. *Seminars in Cell & Developmental Biology*, *21*(8), 805–811.

Sunkar, R., Kapoor, A., & Zhu, J. K. (2006). Posttranscriptional induction of two Cu/Zn superoxide dismutase genes in Arabidopsis is mediated by downregulation of *miR398* and important for oxidative stress tolerance. *Plant Cell*, *18*(8), 2051–2065.

Sunkar, R., Li, Y. F., & Jagadeeswaran, G. (2012). Functions of microRNAs in plant stress responses. *Trends in Plant Science, 17*, 196–203.

Szczesniak, M. W., & Makalowska, I. (2014). miRNEST 2.0: A database of plant and animal microRNAs. *Nucleic Acids Research, 42*, 74–77.

Tang, G., & Tang, X. (2013). Short tandem target mimic: A long journey to the engineered molecular landmine for selective destruction/blockage of microRNAs in plants and animals. *Journal of Genetics and Genomics*, *40*(6), 291–296.

Todesco, M., Rubio-Somoza, I., Paz-Ares, J., & Weigel, D. (2010). A collection of target mimics for comprehensive analysis of microRNA function in Arabidopsis thaliana. *PLoS Genetics*, *6*, e1001031.

Tong, A., Yuan, Q., Wang, S., Peng, J., Lu, Y., Zheng, H., Lin, L., Chen, H., Gong, Y., Chen, J., & Yan, F. (2017). Altered accumulation of *osa-miR171b* contributes to rice stripe virus infection by regulating disease symptoms. *Journal of Experimental Botany, 68*(15), 4357–4367.

Tripathi, A., Goswami, K., & Sanan-Mishra, N. (2015). Role of bioinformatics in establishing microRNAs as modulators of abiotic stress responses: The new revolution. *Frontiers in Physiology, 6*, 286.

Turkan, I., & Demiral, T. (2009). Recent developments in understanding salinity tolerance. *Environmental and Experimental Botany*, *67*, 2–9.

Valdes-Lopez, O., Yang, S. S., Aparicio-Fabre, R., Graham, P. H., Reyes, J. L., Vance, C. P., & Hernandez, G. (2010). MicroRNA expression profile in common bean (*Phaseolus vulgaris*) under nutrient deficiency stresses and manganese toxicity. *New Phytologist*, *187*(3), 805–818.

Voinnet, O. (2009). Origin, biogenesis and activity of plant microRNAs. *Cell*, *136*, 669–687.

Wagaba, H., Patil, B. L., Mukasa, S., Alicai, T., Fauquet, C. M., & Taylor, N. J. (2016). Artificial microRNA-derived resistance to cassava brown streak disease. *Journal of Virological Methods, 231*, 38–43.

Wang, F., Ren, X., Zhang, F., Qi, M., Zhao, H., Chen, X., Ye, Y., Yang, J., Li, S., & Zhang, Y. (2019). A R2R3-type MYB transcription factor gene from soybean, *GmMYB12*, is involved in flavonoids accumulation and abiotic stress tolerance in transgenic Arabidopsis. *Plant Biotechnology Reports, 13*, 1–15.

Wang, J. W., Wang, L. J., Mao, Y. B., Cai, W. J., Xue, H. W., & Chen, X. Y. (2005). Control of root cap formation by microRNA-targeted auxin response factors in Arabidopsis. *Plant Cell*, *17*(8), 2204–2216.

Wang, S. T., Sun, X. L., Hoshino, Y., Yu, Y., Jia, B., Sun, Z. W., Sun, M. Z., Duan, X. B., & Zhu, Y. M. (2014). MicroRNA319 positively regulates cold tolerance by targeting OsPCF6 and OsTCP21 in rice (*Oryza sativa* L.). *PLoS One*, *9*(3), 91357.

Wang, Y., Liu, W., Wang, X., Yang, R., Wu, Z., Wang, H., Wang, L., Hu, Z., Guo, S., Zhang, H., Lin, J., & Fu, C. (2020). MiR156 regulates anthocyanin biosynthesis through SPL targets and other microRNAs in poplar. *Horticulture Research, 7*, 118.

Wang, Y., Sun, F., Cao, H., Peng, H., Ni, Z., Sun, Q., & Yao, Y. (2012). TamiR159 directed wheat TaGAMYB cleavage and its involvement in anther development and heat response. *PloS One*, *7*(11), e48445.

Wani, S. H., Kumar, V., Khare, T., Tripathi, P., Shah, T., Ramakrishna, C., Aglawe, S., & Mangrauthia, S. K. (2020). miRNA applications for engineering abiotic stress tolerance in plants. *Biologia*, *75*(7), 1063–1081.

Warthmann, N., Chen, H., Ossowski, S., Weigel, D., & Herve, P. (2008). Highly specific gene silencing by artificial miRNAs in rice. *PLoS One*, *3*(3), e1829.

Xie, Q., Yu, Q., Jobe, T. O., Pham, A., Ge, C., Guo, Q., Liu, J., Liu, H., Zhang, H., Zhao, Y., & Xue, S. (2021). An amiRNA screen uncovers redundant CBF and ERF34/35 transcription factors that differentially regulate arsenite and cadmium responses. *Plant, Cell & Environment*, *44*(5), 1692–1706.

Yan, J., Gu, Y., Jia, X., Kang, W., Pan, S., Tang, X., Chen, X., & Tang, G. (2012). Effective small RNA destruction by the expression of a short tandem target mimic in Arabidopsis. *Plant Cell*, *24*(2), 415–427.

Yang, C., Li, D., Mao, D., Liu, X. U. E., Ji, C., Li, X., Zhao, X., Cheng, Z., Chen, C., & Zhu, L. (2013). Overexpression of micro RNA 319 impacts leaf morphogenesis and leads to enhanced cold tolerance in rice (*Oryza sativa* L.). *Plant, Cell & Environment, 36*(12), 2207–2218.

Yang, L., Mu, X., Liu, C., Cai, J., Shi, K., Zhu, W., & Yang, Q. (2015). Overexpression of potato *miR482e* enhanced plant sensitivity to *Verticillium dahliae* infection. *Journal of Integrative Plant Biology, 57*, 1078–1088.

Yi, X., Zhang, Z., Ling, Y., Xu, W., & Su, Z. (2015). PNRD: A plant non-coding RNA database. Nucleic Acids Research, *43*, 982–989.

Yu, Y., Ni, Z., Wang, Y., Wan, H., Hu, Z., Jiang, Q., Sun, X., & Zhang, H. (2019). Overexpression of soybean *miR169c* confers increased drought stress sensitivity in transgenic *Arabidopsis thaliana*. *Plant Science, 285*, 68–78.

Zhang, B. (2015). MicroRNA: A new target for improving plant tolerance to abiotic stress. *Journal of Experimental Botany, 66*, 1749–1761.

Zhang, H., Zhang, J., Yan, J., Gou, F., Mao, Y., Tang, G., Botella, J. R., & Zhu, J. K. (2017a). Short tandem target mimic rice lines uncover functions of miRNAs in regulating important agronomic traits. *Proceedings of the National Academy of Sciences of the United States of America, 114*, 5277–5282.

Zhang, J., Zhang, H., Srivastava, A. K., Pan, Y., Bai, J., Fang, J., & Shi, H. (2018a). Knockdown of rice *MicroRNA166* confers drought resistance by causing leaf rolling and altering stem xylem development. *Plant Physiology, 176*, 2082–2094.

Zhang, L. W., Song, J. B., Shu, X. X., Zhang, Y., & Yang, Z. M. (2013a). *miR395* is involved in detoxification of cadmium in *Brassica napus*. *Journal of Hazardous Materials, 250*, 204–211.

Zhang, N., Zhang, D., Chen, S. L., Gong, B. Q., Guo, Y., Xu, L., Zhang, X. N., & Li, J. F. (2018b). Engineering Artificial MicroRNAs for Multiplex Gene Silencing and Simplified Transgenic Screen. *Plant Physiology, 178*, 989–1001

Zhang, P., Meng, X., Chen, H., Liu, Y., Xue, J., Zhou, Y., & Chen, M. (2017b). PlantCircNet: A database for plant circRNA–miRNA–mRNA regulatory networks. *Database*.

Zhang, S., Yue, Y., Sheng, L., Wu, Y., Fan, G., Li, A., Hu, X., ShangGuan, M., & Wei, C. (2013b). PASmiR: A literature-curated database for miRNA molecular regulation in plant response to abiotic stress. *BMC Plant Biology, 13*(1),1–8.

Zhang, X., Shen, J., Xu, Q., Dong, J., Song, L., Wang, W., & Shen, F. (2021). Long noncoding RNA *lncRNA354* functions as a competing endogenous RNA of *miR160b* to regulate ARF genes in response to salt stress in upland cotton. *Plant, Cell & Environment, 44*(10), 3302–3321.

Zhang, X., Zou, Z., Gong, P., Zhang, J., Ziaf, K., Li, H., Xiao, F., & Ye, Z. (2011). Over-expression of *microRNA169* confers enhanced drought tolerance to tomato. *Biotechnology Letters, 33*(2), 403–409.

Zhang, B., & Wang, Q. (2015). MicroRNA-based biotechnology for plant improvement. *Journal of Cellular Physiology, 230*, 1–15.

Zhao, J., Yuan, S., Zhou, M., Yuan, N., Li, Z., Hu, Q., Bethea Jr., F. G., Liu, H., Li, S., & Luo, H. (2019). Transgenic creeping bentgrass overexpressing *Osa-miR393a* exhibits altered plant development and improved multiple stress tolerance. *Plant Biotechnology Journal, 17*(1), 233–251.

Zhao, M., Ding, H., Zhu, J. K., Zhang, F., & Li, W. X. (2011). Involvement of *miR169* in the nitrogen-starvation responses in Arabidopsis. *New Phytologist, 190*(4), 906–915.

Zhou, M. (2012). *Genetic engineering of turfgrass for enhanced performance under environmental stress* [Dissertation, Clemson University].

Zhou, M., & Luo, H. (2013). MicroRNA-mediated gene regulation: Potential applications for plant genetic engineering. *Plant Molecular Biology, 83*(1–2), 59–75.

Zhou, M., & Luo, H. (2014). Role of microRNA319 in creeping bentgrass salinity and drought stress response. *Plant Signaling & Behavior, 9*(4), e28700.

Zhou, X., Yang, J., Tang, X., Wen, Y., Zhang, N., & Si, H. (2018). Effect of silencing C-3 oxidase encoded gene *StCPD* on potato drought resistance by amiRNA technology. *Acta Agronomica Sinica, 44*(4), 512–521.

11 Engineering Stress Tolerance in Plants Using miRNAomics Approach
Challenges and Future Perspectives

Mubeen Fatima, Safdar Hussain, Sidqua Zafar, and Noureen Zahra

11.1 INTRODUCTION

In the current century, climate change has emerged as a fundamental challenge with a significant impact on weather patterns and temperatures fluctuating the ecosystem for many organisms (Patil et al., 2021). As plants are sessile, hence they are more susceptible to climate change. They face various environmental and agricultural stress daily. These stresses are known as abiotic stress. They adversely affect plant productivity and growth, leading to changes in its morphology, molecular mechanisms, chemical compositions, and biological processes. The abiotic stress at the physiological level leads to a reduction in water uptake, reduced photosynthesis, changes in transpiration and respiration rates, decrease in nitrogen assimilation and metabolic toxicity. At the molecular and biochemical levels, it affects the expression of genes and decreases essential enzyme activities and protein synthesis. It also disrupts membrane systems and macromolecules (Chahal et al., 2022). Water shortage, soil salinity, high temperatures, water logging, and metal toxicity are some examples of abiotic stress (Bhatnagar-Mathur et al., 2008, Seleiman et al., 2021, Shriram et al., 2016, Shabbir et al., 2022).

The devastating effect of abiotic stress subjects a major challenge to the production of crops, livestock, and other food commodities. It is the main cause of reduction (more than 50%) in average crop yield. The major share in the above percentage is by high temperature. It causes a 20% reduction in average crop yield. The share of other kinds of stress like low temperature, and drought is 7% and 9%, respectively. It is reported that 91% of agricultural land is under stress throughout the world, hence producing low-quality food in reduced quantities. Hundreds of millions of dollars have been lost every year due to failures of crops and production reduction (Minhas et al., 2017).

On the other side, the growing world population is expected to rise to 9.6 billion by 2050. By then 70% more food will be in demand to fulfil the feed requirements for that huge population. Amid all climatic changes and population explosion, the pressure is on the agricultural industry to somehow increase food production to meet the nutritional and food needs of the world (Farooq et al., 2018; Sarkar et al., 2021). The development of stress-tolerant crops with enhanced yield and climate resilience is the only way forward to cope with changing climate and the growing population globally (Patil et al., 2021). There are various traditional breeding techniques such as hybridization, selection from landraces, and mutation breeding to introduce favourable and stress-resistant genes in desired crops. But traditional breeding methods have not been proven very successful in achieving advantageous results because they are limited by various factors like challenging environments (Rai et al., 2011; Patil et al., 2021). Moreover, conventional breeding methods are also tedious and time-taking

DOI: 10.1201/9781003248453-11

(Haroon et al., 2022). Therefore, we need modern biotechnological tools to address problems related to crop development for sustainable agriculture (Rai et al., 2011).

In recent times, several biotechnological techniques are available to get to know and understand different stress-coping pathways and mechanisms in plants. "Omics" approaches like phenomics (related to plant phenotype), miRNAomics (non-coding RNA), genomics (whole genes or complete DNA), proteomics (proteins), transcriptomics (coding RNA and its types), metabolomics (metabolites' profiling), and ionomics (macro and micro ion profiling) are being frequently used for this aim (Raza et al., 2022; Patil et al., 2021).

In stress conditions, plants modify the expression of a group of genes through post-transcriptional or translational regulations. Some genes, mainly the protective ones, are overexpressed, and negative regulators are suppressed. During this up and downregulation of important genes at post-transcriptional or translational levels, an intricate mechanism plays a vital role in adaption to stress conditions. This mechanism consists of regulatory microRNAs (Yousuf et al., 2021). They modify the transcribed mRNA by cleavage or neutralization according to the need of plants in particular stress (Raza et al., 2022). MicroRNAs (miRNAs) are non-coding RNAs comprising a length of 20 to 24 nucleotides. They are highly conserved and widely distributed among the plant kingdom from non-vascular flowerless plants to higher flowering plants (Jamla et al., 2022, Zhang, 2015). Along with the conserved miRNA, species-specific miRNAs have also been found in large numbers. They are also suspected to control species-specific traits like morphology and different growth stages in plants. According to research through advanced tools, it is also found that they may regulate different unique processes like the initiation of fibre and its development in cotton crops (Zhang, 2015). After discovering the major role of miRNAs in plants under different stress conditions, it is recommended that miRNA can be a potential target that can be manipulated for stress-tolerant plants (Zhang et al., 2015).

11.2 BIOSYNTHESIS OF MIRNA IN THE PLANTS

MicroRNA biogenesis begins within the nuclear cytoplasm. The MIR genes that code for plant miRNAs are often located in intergenic regions. RNA polymerase II transcribes the MIR genes to produce a lengthy RNA transcript termed pri-miRNA. The pri-miRNAs, like other genes, have a 5' cap and a 3' polyadenylation. The pre-miRNA is formed when a segment of long, pri-miRNA folds (single-stranded) into a perfect stem-loop-like structure that is maintained by the RNA-binding protein, known as DAWDLE (DDL) (Jones-Rhoades et al., 2006; Khraiwesh et al., 2012). Dicer-like (DCL1) (an RNAIII-type enzyme) works in concert with other proteins to detect the mature miRNA region in the stem-loop topology of pre-miRNA (Bologna & Voinnet, 2014). The nucleus contains miRNA:miRNA* duplex structures, which are produced when pre-stem-loop miRNA's structure is processed by the DCL, HYL1, and SE (Bartel, 2004). HUA ENHANCER 1 (HEN1) (a small RNA methyltransferase protein) methylates the 3'-terminus of this freshly generated miRNA:miRNA* duplex to stabilize and protect it from destruction (Ramachandran & Chen, 2008). HASTY (HST1). In the cytoplasm, the mature miRNA unwinds and is then loaded onto the RNA-induced silencing complex (RISC), where it then functions as part of the miR-RISC complex to control gene expression. The ARGONAUTE 1 (AGO1) protein acts as a stabilizer for the miR-RISC complex. One strand of the miRNA:miRNA* duplex is then predominantly unwound with aid of the AGO1 protein and sent toward the exosomes for destruction, while the second strand of the mature miRNA stays linked to the RISC with AGO1 protein (Iki, 2017). The mature miRNA then directs the AGO1-containing RISC complex to either cleave highly homologous complementary mRNA at a particular place or to prevent translation of the targeted mRNA by incorrect base pairing. Previously, it was thought that only mature miRNA could limit mRNA and translation when it comes to miRNA-mediated gene regulation. However, new research shows that additional miRNA* strands each control the expression of their own set of genes by targeting their own unique mRNA (B. Zhang & Unver, 2018).

Recently, there has been a rise in the number of studies looking at how miRNAs control gene expression, especially at the molecular level. By perfectly complementing mRNAs, plant miRNAs control gene expression both during transcription and after it has occurred. MiRNAs may function in one of two ways, the first of which destabilizes the target mRNA, by removing its poly-(A) tail (Sun, 2012). Microscopic RNAs, or miRNAs, also have the capacity in regulating a wide range of cellular functions by dampening transcription activity and lowering the variance of transcript abundance (Ebert & Sharp, 2012). The miRNAs have also regulated the expression of a wide variety of class regulatory genes by either downregulating their expression or upregulating them via overexpression or loss-of-function experiments. The main class of R (resistance) genes and transcription factors (TFs), which account for approximately 66 and 24.2% of miRNA targets, respectively, show the importance of miRNAs in diverse gene regulatory networks and the plant immune system (Tang & Chu, 2017).

11.3   MIRNA APPLICATION FOR ENVIRONMENTAL STRESS TOLERANCE IN PLANTS

Scientists have recently been interested in deciphering the function of the miRNA in plant responses to a variety of stressors. Small RNA sequencing is a relatively new subject that has benefited greatly from the deployment of deep sequencing technology for the investigation of miRNA expression profiles, identification of stress-responsive, and unique miRNAs. Hundreds of miRNAs essential for plant growth and stress responses have been discovered so far (B. Zhang & Wang, 2015). A large number of miRNAs may be analyzed at once using the miRNA microarrays. Stress-responsive miRNAs are being sequenced and identified using next generation sequencing (NGS) methods, which provide high-throughput deep-sequencing data. Plants are known to make modifications at the transcriptional and post-transcriptional levels, mostly via RNAi (RNA interference), to survive and adapt to their harsh surroundings (RNAi). However, different genotypes may show a different pattern of expressive genes even within a species of plant. The microRNAs show a critical duty in regulating plant responses against stressors and adverse environmental conditions. MiRNAs, like protein-coding genes, may exhibit genotype- and species-specific expression patterns in response to the changes within the surrounding environment. Many plants' genotypes and miRNA expression levels (stress-specific) have been studied using methods including deep sequencing, RNA microarrays, real-time PCR, NGS, and other transgenic methodologies. *Arabidopsis* seedlings were treated with cold, NaCl salt drought, dehydration, and abscisic acid, and small RNA libraries were generated to compare the treatments to the controls (Fahad et al., 2015). A response to all four classes of abiotic factors revealed that miR393 is inducible, as predicted. Surprisingly, other miRNAs responded differently to the various stressors tested. miR319, which was induced only by the cold conditions, but not by the salinity, drought, and/or ABA treatments (Gao et al., 2016). Numerous studies have revealed that the same miRNA entity can give different responses to the same abiotic stress factors in different plants or multiple types of abiotic stresses in the same plant. It has been shown that miR396 and miR168 exhibit abnormal expression in *Nicotiana* and *Arabidopsis* (Liu et al., 2008) under drought, but not specifically in *Oryza sativa*. The microRNA miR408 was shown to be upregulated in *Medicago* when the plant was subjected to drought conditions.

Upregulation of the same genes has been shown in rice, *Arabidopsis*, and barley, but downregulation has been observed in peach. While *Arabidopsis* saw an increase in miR156 levels, maize had the reverse trend. *Arabidopsis* but not *Oryza sativa* showed increased miR396 expression in response to a very salty environment (Wani et al., 2020). As has been observed, miRNA specificity varies with plant species and abiotic stress, shedding light on the fact that miRNA regulation mechanisms might vary between plant species. MiRNAs have a major role in plant responses to certain stresses. *Arabidopsis* has been shown to have variable miRNA expression in response to drought (miRNAs miR156, miR319, miR393, and miR408), whereas miRNAs miR156, miR159, miR169, miR319, and miR397 all showed significant upregulation in response to salt stress (Li et al., 2016).

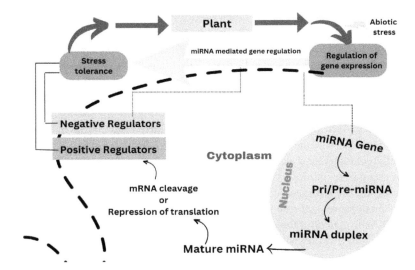

FIGURE 11.1 A schematic representation of the regulatory responses mediated by microRNAs in the plants, when they are tolerating different stresses. The several steps involved in the synthesis of microRNAs from their corresponding miRNA genes as well as the miRNA-mediated gene regulation that occurs in the plants when they are implemented to different stresses are shown in the figure (such as salinity, drought, and temperature).

The pattern of miRNA expression concerning abiotic stress varies among different plant species, and the stress severity and the kind of miRNA being examined have been noted. This has been shown in a vast array of species of plants, including (but not limited to) rice poplar, *Arabidopsis*, maize, and tomato (Wani et al., 2020). Multiple studies support this notion. The miR408 has revealed its role in the responses of *Anthurium andraeanum* to a broad range of stimuli, including salt, cold, drought, and oxidative stresses (Jiang et al., 2018). This study resulted in the confirmation of the stress-sensitive nature of the miRNAs studied and their potential participation as major modulators of stress among a variety of plant species under different stress conditions.

11.4 FUNCTIONAL ANALYSIS AND COMPUTATIONAL TOOLS OF MIRNAS

The first step in understanding miRNA-mediated gene regulation and its other processes is the identification of the MIR genes, miRNAs with their targeted genes. Direct cloning and sequencing processes of small RNA populations were used in the first studies to identify plant miRNAs (Park et al., 2002). The identification of new and also conserved plant miRNAs within the tissue has been greatly facilitated by developing high-throughput sequencing, known as the next-generation sequencing (NGS) technology. In plants, NGS has allowed for unprecedented coverage and depth in the discovery of miRNAs throughout the whole genome or transcriptome (Djami-Tchatchou & Dubery, 2015). Another method for discovery of the miRNAs in various agricultural plants is the use of computational tools, such as (BLAST) and other homology-based techniques (Xie et al., 2011).

Various plant species, notably agricultural plants, provide a massive quantity of data to modern high-throughput systems. All of these results like miRBase, the NCBI-Gene Expression, and the Plant MicroRNA Database are available in public databases. The homology-based, in silico investigation of putative miRNAs and their gene products yields plausible findings because of the conserved nature of the miRNA of plants (Chaudhary et al., 2021). In addition, recent advances in computational biology methods provide substantial assistance in processing massive amounts of raw data and giving it biological context. Reads are assembled into contigs or singletons once quality filters are applied. Nucleotide BLASTn is then used to compare the unique contigs to the unique plant miRNA, yielding an aligned sequence. The two basic requirements the candidate miRNA has to meet are it must have at least 18 nucleotides in length and assembled sequences that are similar to previously characterized miRNAs (Panda et al., 2014). The aligned sequence is then BLASTxed against a plant protein database (Uniport). To get rid of all code, we need to delete the line. To see the secondary structure, you must first remove the protein-coding sequence and then run it via the MFOLD program. MiRNAs that did not meet the MFOLD criteria were discarded, but the candidate pre-miRNAs that remained might be promising and considered new miRNAs. The psRNATarget service may be used to determine which genes are regulated through miRNA complementary binding between different target gene sequences and miRNA. Finally, target genes' functioning may be verified by assigning gene ontology (GO) words using QuickGO (Chaudhary et al., 2021).

There is a pressing need to speed up the functional analysis of miRNAs so that they may be used effectively in plant improvement studies in light of the recent explosion of data describing annotated, sequenced miRNAs in agricultural plants. Methods often used to analyze the function of miRNAs in plant metabolism include overexpression or the repression of miRNA encoding genes and cleavage of target genes. Target mimicry and short tandem target mimic (STTM) are all potent post-transcriptional small RNA inactivating tools. These methods for characterizing the roles of miRNAs are quite efficient and helpful. The miRNAs may also be repressed with the help of amiRNAs and atasiRNAs/syn-tasiRNAs, which are artificial/synthetic trans-acting siRNAs (Wani et al., 2020).

It is challenging to determine the role of miRNA families since each family is regulated by numerous loci (Gupta, 2015). As a result, silencing the expression of all the members of the miRNA family is the best technique to investigate their role. Understanding how miRNA family members are regulated differentially across tissues, and treatment is crucial for investigating their function under different stress in plants (Mangrauthia et al., 2017). When scientists want to silence a whole family of miRNAs, they employ targeted mimic technology. A recent technological breakthrough in this area is the short tandem target mimic. The method uses a pair of complementary miRNAs which can target sites that are separated by the short DNA artificial spacer and have a middle tri-nucleotide (CTA) bulge. Both of these binding sites are complementary to the target miRNAs. The STTM can cleave its target miRNAs because of this confrontation. The miRNA expression suppression using STTM technology have achieved in both the dicots and monocots. Recently, Professor J-K Zhu's group generated a huge array of lines of transgenic rice inhibiting 35 miRNA different families. Many essential agronomic parameters, that is, plant height, number of tillers, and grain number, were shown by phenotypic analysis to alter significantly throughout up to five generations in these lines (P. Zhang et al., 2017). Inhibiting miRNAs in this way is a great way to learn more about the miRNAs involved with stress in plants and how to employ them in breeding and genetic engineering efforts to boost agricultural yields.

11.5 MIRNA-BASED STRATEGIES FOR PLANT IMPROVEMENT

A powerful genetic engineering tool for inserting, removing, or replacing DNA at specified endogenous loci is genome editing employing sequence-specific nucleases (SSNs) (Voytas, 2013). Zinc finger nuclease (ZFN), transcription activator-like effector nuclease (TALEN), clustered regularly interspaced short palindromic repeats and associated protein 9 (CRISPR-Cas9), and clustered regularly interspaced short palindromic repeats and associated protein 1 are a few of the new tools that have been discovered as a result of recent developments in SSN platforms (CRISPR-Cpf1) (Zhao et al., 2017).

CRISPR-Cas9-based tools have taken genomics research to the next level, revolutionizing biology in the process. The effectiveness of the bacterial immune system in combating and invading viruses is the inspiration for this innovation. CRISPR, which can be used to modify the genome, was discovered in 2012 (Wani et al., 2020). The CRISPR-associated (Cas) nuclease9 from the Streptococcus pyogenes is the widely used enzyme in the CRISPR-based genome editing technique for cleaving targeted DNA detected by the single guide RNA (sgRNA). The Cas9-induced double-stranded DNA breaks are repaired by the homology-directed repair (HDR) and non-homologous end-joining (NHEJ) processes, respectively. While HDR is used to insert a whole gene or a small number of nucleotide sequences, NHEJ is utilized for the production of insertion/deletion (Indel) alterations to develop knockout mutants. The CRISPR/Cpf1 system, in which the Cpf1 enzyme is used instead of the Cas9 nuclease to make breaks in the DNA of the targets, represents a significant step forward in this technology. Due to its diminutive size, the enzyme from Acidaminococcus and Lachnospiraceae facilitates efficient intracellular delivery. In addition, it gives you greater leeway in choosing which citations you want to modify (Aglawe et al., 2018).

CRISPR/Cas9 technology presents a greater hurdle when used for the deletion or knockdown of non-coding RNA (ncRNA) genes than for protein-coding genes. The lack of available room for target design in non-coding genes is solely to blame (Wani et al., 2020). Not so long ago, Chang and his colleagues used the CRISPR/cas9 technology to control the expression of miRNA. They observed that the CRISPR/cas9 system may strongly downregulate mature miRNAs expression after transfecting human colon cancer cells with CRISPR/cas9 vectors and targeting miR-200c, miR-141, and miR-17 loci (Chang et al., 2016). Zhou and his colleagues provided more evidence that the CRISPR-Cas9 system is an efficient method for creating miRNA knockout mutants in plants. The researchers used the CRISPR-Cas9 technology to target the miRNA gene families (miR820a/b/c and miR815a/b/c) and including single miRNA genes (OsMIR528 and OsMIR408) and discovered the rates of targeted mutagenesis among the regenerated T0 lines varied in between 48% to 89%. Based on these findings, OsMIR528 may be a useful tool in managing rice's response to salt stress. In addition to aiding in the functional characterization of miRNAs linked with abiotic stress in plants, CRISPR-driven modulation of miRNAs expression can lead to enhance stress-tolerance in plants by appropriately modifying candidate miRNAs via the development of miRNA-edited lines in crop plants (Zhou et al., 2017).

11.5.1 Case Studies

11.5.1.1 Rice

The most widely cultivated cereal crop throughout the world is rice. About three billion people in Asia are dependent on rice as a source of dietary protein and energy. Its quality and production are adversely affected by abiotic stress. High temperature impacts the milling quality properties like immature kernels, dimensions of kernels, fissuring, amylopectin chain length, protein content, amylose content, and chalkiness. So abiotic stress at any growth stage in rice leads to economic loss due to decreased market value (Fahad et al., 2019).

Rice crops show more sensitivity to salt stress. MicroRNAs (miRNAs) are a type of small noncoding RNAs. They are widely reported as gene regulators, suppressing gene expression by the degradation of mRNA or through the inhibition of the translation process. Among different plant species, microRNA168 (miR168) is the conserved miRNA. It targets the gene ARGONAUTE1 (AGO1) and regulates the expression of all miRNA indirectly. Wan et al. (2022) used STTM (short tandem target mimic) method to inactivate miR168. STTM is an advanced method in which mature miRNAs are inactivated to study their functions. STTM168 seedlings were obtained after genetic transformation. Three independent lines were selected randomly followed by genotyping and reproduction. These lines were named, STTM168–1, STTM168–2, and STTM168–3 respectively. Further, phenotypic analysis was performed. The STTM168 plants have decreased levels of expression of miR168. The control and STTM18 seedlings exhibited almost no difference in root length and shoot height under normal conditions. But under salt stress, the root number and length and shoot height were significantly increased however Na + content was lowered in STTM168 seedlings. Hence, STTM168 exhibited more tolerance to salinity stress other than the control. (Wan et al., 2022)

Osa-miR820, a rice-specific miRNA belongs to the class II family of transposon-derived small RNA. Its size varies from 21-nt to 24-nt. A *de novo* methylase, OsDRM2 (*domains rearranged methyl transferase*) is negatively regulated by Osa-miR820. As a result, the methylation of a transposon (CACTA) is prevented in the rice genome. Sharma et al. (2021) overexpressed 21-nt Osa-miR820 in rice plants (OX-820) by using an artificial miR-based approach. It resulted in a 25–30% increase in spikelet per panicle, enhanced vigour, and an increase in grain filling under both normal conditions and salt stress. These lines showed a higher accumulation of proline and better water use efficiency under salt stress.

The miR5505 is a conserved miRNA that has been found through high-throughput sequencing. Various studies revealed that it responds to drought stress in Dongxiang wild rice (DXWR). Fan et al. (2022) overexpressed it to study its effect on salt and drought resistance in rice. After selecting 2 lines out of 18 transgenic lines, they evaluated their salt and drought resistance ability. Both the lines exhibited more resistance towards salinity and drought stress in contrast with wild type (WT). So the overexpression of miR5505 leads to the regulation of drought and salt responses (Fan et al., 2022).

Nitrogen is the most important nutrient in rice cultivation, but its excessive use increases agricultural costs and causes environmental degradation. So the nutrient use efficiency of the crop must be improved to achieve sustainable agriculture. *OsMYB305* is an MYB transcription factor that encodes a transcriptional activator. When the root of a rice plant is deficient in nitrogen, its expression is induced, and when it is overexpressed, the shoot's dry weight, the total concentration of N, and the total number of tillers significantly increased. Wang et al. (2020) found out that during Nitrogen-deficient circumstances, the overexpression of *OsMYB305* leads to suppressing cellulose biosynthesis. As a result, free carbohydrates are available for the uptake and assimilation of nitrates, ultimately improving rice growth. So *OsMYB305* is the potential candidate to manipulate for increased Nitrogen uptake in the cultivation of rice.

Zhang et al. (2018) used the short tandem target mimic (STTM) system to knock down microRNA166 in *Oryza sativa*, which resulted in drought resistance in the respective plant. They have selected STTM166 transgenic plants. They noticed that these plants had rolled leaves which are among the essential traits of rice plants during drought stress. The rolled leaves in STTM166 were likely because of the abnormal sclerenchymatous cells and smaller bulliform cells. Moreover, the xylem vessel diameter was also reduced, which caused a change in the stem xylem and reduced hydraulic conductivity. Moreover, transpiration rates were also decreased due to the reduction in stomatal conductance. Further analysis revealed that the main target of miR166 is rice, *HOMEODOMAIN CONTAINING PROTEIN4* (*OsHB4*), which belongs to the HD-ZIP III gene family.

The miR166 can be a favourable candidate to manipulate to confer drought resistance in rice plants. It will result in some developmental changes, such as the reduced diameter of the xylem, and rolled leaves making plants able to tolerate water deficiency at crucial stages (Zhang et al., 2018).

In an experiment, Um et al. (2022) discovered that osa-miR171, which belongs to the osa-miR171 gene family that regulates *SCARECROW-LIKE6-I* (*SCL6-I*), and *SCL6-II* transcript levels in rice plants (*Oryza sativa*) to confer drought resistance. *SCL6* genes are associated with leaf morphology and shoot branching. The osa-miR171f is expressed mainly under drought conditions, and in transgenic plants its overexpression results in drought tolerance. Um et al. (2022) generated transgenic lines osa-miR171f-OE (osa-miR171f overexpressed) and osa-miR171f-K/O (osa-miR171f knockout) through CRISPR/Cas9 (clustered regularly interspaced short palindromic repeats) technique. The transcriptome analysis shows that the genes for flavonoid biosynthesis are regulated by osa-miR171f, and this regulation leads to drought resistance. So osa-miR171f can be manipulated to maintain stress-tolerant rice plants.

11.5.1.2   Maize

Maize (*Zea mays*) is an essential food crop that is not only needed by a huge growing population but is also an important feed for the livestock industry. It is one of the most-grown crops worldwide. For the last 20 years, maize production per unit area has been reduced adversely throughout the world due to abiotic stresses and thus increasing stress intensity day by day (Vaughan et al., 2018).

In maize, MicroRNAs play the role of a regulator against abiotic stresses. Li et al. (2020) has discovered the novel functions of miR166 in maize crop. They used STTM (short tandem target mimics) technology to produce maize STTM166 transgenic plants. It resulted in different changes in morphology such as increased resistance to abiotic stresses, rolled leaves, change in epidermis structures, vascular patterns, and tassel shapes, inferior yield-related traits, enhanced abscisic acid (ABA) level, and also decreasing indole acetic acid (IAA) level in transgenic plants. The profiling of miR166-regulated genes was done by using qRT-PCR analysis and RNA-seq. Out of the total 178 differently expressed genes, 60 were downregulated, while 118 were upregulated.

The research revealed that miR166 regulates the interaction between ABA and Auxin in monocots, while in dicots the case may be different. The change in vascular structure and cell membrane along with the rolled leaves and increased ABA content are important components in enhancing resistance to abiotic stresses. So this research indicates that miRNA 166 can be manipulated to produce stress-resistant varieties in maize (Li et al., 2020).

11.5.1.3 Soybean

Soybean is an essential source of oil production and protein globally. Its contribution to the total edible oil production in the world is about 50%. Like all other legume crops, it also has a symbiotic relationship with nitrogen-fixing bacteria. These bacteria enhance soil fertility. Its yield and production are also affected by abiotic stresses like heavy metals, salt, water submergence, and drought. Moreover, it is very difficult for the soybean crop to adapt to non-traditional growing areas (Deshmukh et al., 2014).

The miR172a is a miRNA in soybean that cleaves AP2/EREBP-type a transcription factor gene SSAC1 to enhance salt tolerance. This cleavage relieves the protein inhibition of SSAC1 on the gene THI1 used in thiamine biosynthesis. THI1 gene encodes for a positive regulator for salinity resistance. NCED3 and RD22 are miR172a-regulated downstream genes. They are also suspected to participate in salinity tolerance. Pan et al. (2016) have discovered a pathway that involves miR172a-SSAC1-THI1-thiamine and has a subtle role in salinity tolerance in soybean. Scientists can further research and manipulate these genes to adapt soybean crops to saline environments so that their yield can be enhanced under salinity stress (Pan et al., 2016).

11.5.1.4 Banana

Banana is a subsistence as well as a commercially important crop with high economic value. Temperature is the crucial factor in the growth rate and the emergence of new leaves in banana plants. So abiotic stress is fatal to crop yield, causing a potential loss of 65–87%. There are various stress coping mechanisms in plants like tolerance (allowing plants to withstand the stress), acclimation (change in physiology due to stress), and avoidance (no exposure to stress conditions) (Ravi et al., 2016). The miRNA are control factors for stress tolerance, growth, and development, but as compared to other crops, they are unexplored in bananas.

Patel et al. (2019) found *Musa*-miR397 as a potential miRNA in increasing plant biomass as well as in contributing to plant tolerance against abiotic stress. During copper deficiency, it is upregulated about eight to ten folds in leaves and roots, and it correlates with the expression of copper deficiency marker genes in roots such as *Musa*-FRO2 and *Musa*-COPT. It is also upregulated about two folds during heat treatments, MV and ABA, while under salt stress, it was downregulated. Moreover, plant growth is enhanced two to three folds due to the overexpression of *Musa*-miR397 without compromising tolerance during salinity stress and CU deficiency. When the RNA of wild and transgenic plants was sequenced, it was found that about 71 genes that were related to the development and growth, contributing to enhancement in biomass, were modified. So *Musa*-miR397 is a desirable candidate to manipulate and be the reason for abiotic stress-tolerant plants (Patel et al., 2019).

11.6 CHALLENGES AND FUTURE PROSPECTS

Plant miRNAs are crucial factors in gene silencing and gene regulation. They either degrade mRNA or cause translational repression of the target mRNA (Messenger RNA). This regulation of gene expression is also important in adaptation to different environmental stresses. By producing new miRNAs or utilizing the already present ones, plants can increase their tolerance under abiotic stresses. As a result, growth and the development of plants will be enhanced (Zhang et al., 2022).

But the focus of current studies related to miRNA is only on model plants like *Arabidopsis* or rice, while miRNA-related research in other important crops is scarce. Especially in the case of species-specific miRNA, a lot more research can be conducted leading to the discovery of useful results (Zhang et al., 2022). Moreover, the miRNA-mediated pathways are very complex, and they have not been fully discovered yet. There are huge numbers of unknown miRNA and their target genes that play an influential role in stress adaptation in different plants. Besides, miRNA may have unfavourable cleavage sites that may result in unexpected outcomes. It is crucial to fully discover and understand the miRNA mediated-regulation of gene expression so that effective stress-tolerant plants can be developed (Chaudhary et al., 2021; Krishnatreya et al., 2021). To go one step further, the proteome (including regulatory cis-elements), the degradome, the transcriptome, and the microme (microRNome or microtranscriptome) should be linked in a comprehensible way so that miRNA gene regulation pathways can be generated and modified without adverse physiological consequences. With the advancement in bioinformatics, system biology, and sequencing technologies, it has become more feasible nowadays (López et al., 2012).

The miRNA research domain is compelling and useful, but still, the researchers should also shift their focus from the traditional plant models to less explored plants, for example, cassava and mangroves. The research should be done to discover miRNA pathways to induce drowning and salinity tolerance in mangroves and drought tolerance in cassava (López et al., 2012). To fully comprehend the role of miRNA in the gene regulation pathways, the researchers should create the target transcripts with altered nucleotide sequences so that miRNA cannot bind with them. As a result, the actual function of the respective miRNA can be discovered. Moreover, the expression of genes related to the concerned miRNA should be altered at different levels to manipulate stress resistance mechanisms to develop stress-tolerant plants (Gupta et al., 2014).

The miRNA synthesis is regulated step by step from decoding miRNA genes to producing a mature miRNA in plants. Mature miRNA identifies the target mRNA by a complementary sequence that it carries along. The downregulation is performed through DNA methylation, repressing the translational process, or mostly through site-specific cleavage. Despite unleashing the interesting phenomenon of miRNA biogenesis, various unanswered questions must have been solved to avoid unexpected results. For example, we do not know what is the actual trigger for miRNA synthesis. What is their mode of action? And how do they adjust at different stages of development? The answer to these questions is yet to be found to manipulate these intricate molecules in plants for stress resistance (Kaur et al., 2020).

REFERENCES

Aglawe, S. B., Barbadikar, K. M., Mangrauthia, S. K., & Madhav, M. S. (2018). New breeding technique "genome editing" for crop improvement: Applications, potentials and challenges. *3 Biotech, 8*(8), 1–20.

Anwar, A., & Kim, J. K. (2020). Transgenic breeding approaches for improving abiotic stress tolerance: Recent progress and future perspectives. *International Journal of Molecular Sciences, 21*(8), 2695.

Bartel, D. P. (2004). MicroRNAs: Genomics, biogenesis, mechanism, and function. *Cell, 116*(2), 281–297.

Bhatnagar-Mathur, P., Vadez, V., & Sharma, K. K. (2008). Transgenic approaches for abiotic stress tolerance in plants: Retrospect and prospects. *Plant Cell Reports, 27*(3), 411–424.

Bologna, N. G., & Voinnet, O. (2014). The diversity, biogenesis, and activities of endogenous silencing small RNAs in Arabidopsis. *Annual Review of Plant Biology, 65*(1), 473–503.

Chahal, G. K., Ghai, M., & Kaur, J. (2022). Stress mechanisms and adaptations in plants. *Journal of Pharmacognosy and Phytochemistry, 11*(2), 109–114.

Chang, H., Yi, B., Ma, R., Zhang, X., Zhao, H., & Xi, Y. (2016). CRISPR/cas9, a novel genomic tool to knock down microRNA in vitro and in vivo. *Scientific Reports, 6*(1), 1–12.

Chaudhary, S., Grover, A., & Sharma, P. C. (2021). MicroRNAs: Potential targets for developing stress-tolerant crops. *Life, 11*(4), 289.

Deshmukh, R., Sonah, H., Patil, G., Chen, W., Prince, S., Mutava, R., Vuong, T., Valliyodan, B., & Nguyen, H. T. (2014). Integrating omic approaches for abiotic stress tolerance in soybean. *Frontiers in Plant Science, 5*, 244.

Djami-Tchatchou, A. T., & Dubery, I. A. (2015). Lipopolysaccharide perception leads to dynamic alterations in the microtranscriptome of Arabidopsis thaliana cells and leaf tissues. *BMC Plant Biology, 15*(1), 1–19.

Ebert, M. S., & Sharp, P. A. (2012). Roles for microRNAs in conferring robustness to biological processes. *Cell, 149*(3), 515–524.

Fahad, S., Nie, L., Chen, Y., Wu, C., Xiong, D., Saud, S., Liu, H., Cui, K., & Huang, J. (2015). Crop plant hormones and environmental stress. *Sustainable Agriculture Reviews, 15.* https://doi.org/10.1007/978-3-319-09132-7_10

Fahad, S., Noor, M., Adnan, M., Khan, M. A., Rahman, I. U., Alam, M., Khan, M. A., Ullah, H., Mian, I. A., Hassan, S., Saud, S., Bakhat, H. F., Hammad, H. M., Nasim, W., & Ahmad S. (2019). Abiotic stress and rice grain quality. In *Advances in rice research for abiotic stress tolerance* (pp. 571–583). Woodhead Publishing.

Fan, Y., Zhang, F., & Xie, J. (2022). *Overexpression of miR5505 enhanced drought and salt resistance in rice (Orayza sativa).* bioRxiv.

Farooq, M., Hussain, M., Usman, M., Farooq, S., Alghamdi, S. S., & Siddique, K. H. (2018). Impact of abiotic stresses on grain composition and quality in food legumes. *Journal of Agricultural and Food Chemistry, 66*(34), 8887–8897.

Gao, S., Yang, L., Zeng, H. Q., Zhou, Z. S., Yang, Z. M., Li, H., Sun, D., Xie, F., & Zhang, B. (2016). A cotton miRNA is involved in regulation of plant response to salt stress. *Scientific Reports, 6*(1), 1–14.

Gupta, O. P., Sharma, P., Gupta, R. K., & Sharma, I. (2014). MicroRNA mediated regulation of metal toxicity in plants: Present status and future perspectives. *Plant Molecular Biology, 84*(1), 1–18.

Gupta, P. K. (2015). MicroRNAs and target mimics for crop improvement. *Current Science, 108*(9), 1624–1633.

Haroon, M., Wang, X., Afzal, R., Zafar, M. M., Idrees, F., Batool, M., Khan, A. S., & Imran, M. (2022). Novel plant breeding techniques shake hands with cereals to increase production. *Plants, 11*(8), 1052.

Iki, T. (2017). Messages on small RNA duplexes in plants. *Journal of Plant Research, 130*(1), 7–16.

Jamla, M., Patil, S., Joshi, S., Khare, T., & Kumar, V. (2022). MicroRNAs and their exploration for developing heavy metal-tolerant plants. *Journal of Plant Growth Regulation, 41*(7), 2579–2595.

Jiang, L., Tian, X., Fu, Y., Liao, X., Wang, G., & Chen, F. (2018). Comparative profiling of microRNAs and their effects on abiotic stress in wild-type and dark green leaf color mutant plants of Anthurium andraeanum 'Sonate'. *Plant Physiology and Biochemistry, 132*, 258–270.

Jones-Rhoades, M. W., Bartel, D. P., & Bartel, B. (2006). MicroRNAs and their regulatory roles in plants. *Annual Review of Plant Biology, 57*(1), 19–53.

Kaur, R., Bhunia, R. K., & Rajam, M. V. (2020). MicroRNAs as potential targets for improving rice yield via plant architecture modulation: Recent studies and future perspectives. *Journal of Biosciences, 45*(1), 1–17.

Khraiwesh, B., Zhu, J.-K., & Zhu, J. (2012). Role of miRNAs and siRNAs in biotic and abiotic stress responses of plants. *Biochimica et Biophysica Acta (BBA)-Gene Regulatory Mechanisms, 1819*(2), 137–148.

Krishnatreya, D. B., Agarwala, N., Gill, S. S., & Bandyopadhyay, T. (2021). Understanding the role of miRNAs for improvement of tea quality and stress tolerance. *Journal of Biotechnology, 328*, 34–46.

Li, N., Yang, T., Guo, Z., Wang, Q., Chai, M., Wu, M., Li, X., Li, W., Li, G., Tang, J., Tang, G., & Zhang, Z. (2020). Maize microRNA166 inactivation confers plant development and abiotic stress resistance. *International Journal of Molecular Sciences, 21*(24), 9506.

Li, Y., Li, C., Bai, L., He, C., & Yu, X. (2016). MicroRNA and target gene responses to salt stress in grafted cucumber seedlings. *Acta Physiologiae Plantarum, 38*(2), 1–12.

Liu, H.-H., Tian, X., Li, Y.-J., Wu, C.-A., & Zheng, C.-C. (2008). Microarray-based analysis of stress-regulated microRNAs in Arabidopsis thaliana. *Rna, 14*(5), 836–843.

López, C., & Pérez-Quintero, A. L. L. (2012). The micromics revolution: MicroRNA-mediated approaches to develop stress-resistant crops. In *Improving crop resistance to abiotic stress* (pp. 559–590). Wiley.

Mangrauthia, S. K., Maliha, A., Prathi, N. B., & Marathi, B. (2017). MicroRNAs: Potential target for genome editing in plants for traits improvement. *Indian Journal of Plant Physiology, 22*(4), 530–548.

Minhas, P. S., Rane, J., & Pasala, R. K. (2017). Abiotic stresses in agriculture: An overview. In *Abiotic stress management for resilient agriculture* (pp. 3–8). Springer.

Pan, W. J., Tao, J. J., Cheng, T., Bian, X. H., Wei, W., Zhang, W. K., Ma, B., Chen, S. Y., & Zhang, J. S. (2016). Soybean miR172a improves salt tolerance and can function as a long-distance signal. *Molecular Plant, 9*(9), 1337–1340.

Panda, D., Dehury, B., Sahu, J., Barooah, M., Sen, P., & Modi, M. K. (2014). Computational identification and characterization of conserved miRNAs and their target genes in garlic (Allium sativum L.) expressed sequence tags. *Gene*, *537*(2), 333–342.

Park, W., Li, J., Song, R., Messing, J., & Chen, X. (2002). Carpel factory, a dicer homolog, and HEN1, a novel protein, act in microRNA metabolism in Arabidopsis thaliana. *Current Biology*, *12*(17), 1484–1495.

Patel, P., Yadav, K., Srivastava, A. K., Suprasanna, P., & Ganapathi, T. R. (2019). Overexpression of native Musa-miR397 enhances plant biomass without compromising abiotic stress tolerance in banana. *Scientific Reports*, *9*(1), 1–15.

Patil, S., Joshi, S., Jamla, M., Zhou, X., Taherzadeh, M. J., Suprasanna, P., & Kumar, V. (2021). MicroRNA-mediated bioengineering for climate-resilience in crops. *Bioengineered*, *12*(2), 10430–10456.

Rai, M. K., Kalia, R. K., Singh, R., Gangola, M. P., & Dhawan, A. K. (2011). Developing stress tolerant plants through in vitro selection—an overview of the recent progress. *Environmental and Experimental Botany*, *71*(1), 89–98.

Ramachandran, V., & Chen, X. (2008). Small RNA metabolism in Arabidopsis. *Trends in Plant Science*, *13*(7), 368–374.

Ravi, I., & Vaganan, M. M. (2016). Abiotic stress tolerance in banana. In *Abiotic stress physiology of horticultural crops* (pp. 207–222). Springer.

Raza, A., Tabassum, J., Zahid, Z., Charagh, S., Bashir, S., Barmukh, R., Khan, R. S. A., Barbosa, F. Jr., Zhang, C., Chen, H., Zhuang, W., & Varshney, R. K. (2022). Advances in "omics" approaches for improving toxic metals/metalloids tolerance in plants. *Frontiers in Plant Science*, *12*, 794373.

Sarkar, S., Khatun, M., Era, F. M., Islam, A. M., Anwar, M. P., Danish, S., Datta, R., & Islam, A. K. M. A. (2021). Abiotic stresses: Alteration of composition and grain quality in food legumes. *Agronomy*, *11*(11), 2238.

Seleiman, M. F., Al-Suhaibani, N., Ali, N., Akmal, M., Alotaibi, M., Refay, Y., Dindaroglu, T., Abdul-Wajid, H. H., & Battaglia, M. L. (2021). Drought stress impacts on plants and different approaches to alleviate its adverse effects. *Plants*, *10*(2), 259.

Shabbir, R., Singhal, R. K., Mishra, U. N., Chauhan, J., Javed, T., Hussain, S., Anuragi, H., Lal, D., & Chen, P. (2022). Combined abiotic stresses: Challenges and potential for crop improvement. *Agronomy*, *12*(11), 2795.

Sharma, N., Kumar, S., & Sanan-Mishra, N. (2021). *Osa-miR820 regulatory node primes rice plants to tolerate salt stress in an agronomically advantageous manner*. bioRxiv.

Shriram, V., Kumar, V., Devarumath, R. M., Khare, T. S., & Wani, S. H. (2016). MicroRNAs as potential targets for abiotic stress tolerance in plants. *Frontiers in Plant Science*, *7*, 817.

Sun, G. (2012). MicroRNAs and their diverse functions in plants. *Plant Molecular Biology*, *80*(1), 17–36.

Tang, J., & Chu, C. (2017). MicroRNAs in crop improvement: Fine-tuners for complex traits. *Nature Plants*, *3*(7), 1–11.

Um, T., Choi, J., Park, T., Chung, P. J., Jung, S. E., Shim, J. S., Kim, Y. S., Choi, I. Y., Park, S. C., Oh, S. J., Seo, J. S., & Kim, J. K. (2022). Rice microRNA171f/SCL6 module enhances drought tolerance by regulation of flavonoid biosynthesis genes. *Plant Direct*, *6*(1), e374.

Vaughan, M. M., Block, A., Christensen, S. A., Allen, L. H., & Schmelz, E. A. (2018). The effects of climate change associated abiotic stresses on maize phytochemical defenses. *Phytochemistry Reviews*, *17*(1), 37–49.

Voytas, D. F. (2013). Plant genome engineering with sequence-specific nucleases. *Annual Review of Plant Biology*, *64*, 327–350.

Wan, J., Meng, S., Wang, Q., Zhao, J., Qiu, X., Wang, L., Li, J., Lin, Y., Mu, L., Dang, K., Xie, Q., Tang, J., Ding, D., & Zhang, Z. (2022). Suppression of microRNA168 enhances salt tolerance in rice (Oryza sativa L.). *BMC Plant Biology*, *22*(1), 563.

Wang, D., Xu, T., Yin, Z., Wu, W., Geng, H., Li, L., Yang, M., Cai, H., & Lian, X. (2020). Overexpression of OsMYB305 in rice enhances the nitrogen uptake under low-nitrogen condition. *Frontiers in Plant Science*, *11*, 369.

Wani, S. H., Kumar, V., Khare, T., Tripathi, P., Shah, T., Ramakrishna, C., Aglawe, S., & Mangrauthia, S. K. (2020). miRNA applications for engineering abiotic stress tolerance in plants. *Biologia*, *75*(7), 1063–1081.

Xie, F., Frazier, T. P., & Zhang, B. (2011). Identification, characterization and expression analysis of MicroRNAs and their targets in the potato (Solanum tuberosum). *Gene*, *473*(1), 8–22.

Yousuf, P. Y., Shabir, P. A., & Hakeem, K. R. (2021). miRNAomic approach to plant nitrogen starvation. *International Journal of Genomics*, *2021*.

Zhang, B. (2015). MicroRNA: A new target for improving plant tolerance to abiotic stress. *Journal of Experimental Botany*, *66*(7), 1749–1761.

Zhang, B., & Unver, T. (2018). A critical and speculative review on microRNA technology in crop improvement: Current challenges and future directions. *Plant Science*, *274*, 193–200.

Zhang, B., & Wang, Q. (2015). MicroRNA-based biotechnology for plant improvement. *Journal of Cellular Physiology*, *230*(1), 1–15.

Zhang, F., Yang, J., Zhang, N., Wu, J., & Si, H. (2022). Roles of microRNAs in abiotic stress response and characteristics regulation of plant. *Frontiers in Plant Science*, *13*, 919243.

Zhang, J., Zhang, H., Srivastava, A. K., Pan, Y., Bai, J., Fang, J., Shi, H., & Zhu, J. K. (2018). Knockdown of rice microRNA166 confers drought resistance by causing leaf rolling and altering stem xylem development. *Plant Physiology*, *176*(3), 2082–2094.

Zhang, P., Meng, X., Chen, H., Liu, Y., Xue, J., Zhou, Y., & Chen, M. (2017). PlantCircNet: A database for plant circRNA–miRNA–mRNA regulatory networks. *Database: the journal of biological databases and curation*, 2017, bax089. https://doi.org/10.1093/database/bax089.

Zhao, Y., Wen, H., Teotia, S., Du, Y., Zhang, J., Li, J., Sun, H., Tang, G., Peng, T., & Zhao, Q. (2017). Suppression of microRNA159 impacts multiple agronomic traits in rice (Oryza sativa L.). *BMC Plant Biology*, *17*(1), 1–13.

Zhou, J., Deng, K., Cheng, Y., Zhong, Z., Tian, L., Tang, X., Tang, A., Zheng, X., Zhang, T., Qi, Y., & Zhang, Y. (2017). CRISPR-Cas9 based genome editing reveals new insights into microRNA function and regulation in rice. *Frontiers in Plant Science*, *8*, 1598.

Index

Note: Page numbers in *italics* refer to figures and those in **bold** refer to tables.

A

ABA INSENSITIVE 5 (ABI5), 43
abscisic acid, 7, 26, 46–47, 67, 82, 100, 120, 122, 150, 155
ALTERED MERISTEM PROGRAM1 (AMP1), 27
APETALA2, 8, 67, 70
apoptosis, 20–21
Argonaute, 3, 6, 9, 70, 77, 80, *93*, 97, 109, 119, 149, 153
Artificial miRNA, 11, 129–132, 140–141
Atppc4, 69
ATP sulfurylase/serine acetyltransferase (APS/SAT), 81

C

Caenorhabditis elegans, 2, 20, 23, 89
Cardamine hirsute, 40
CDGS (chromatin dependent gene silencing), 20
chlorosis, 58
CHR2, 27
constitutive photomorphogenic1 (COP1), 26
cup-shaped cotyledon, 30, 40, 44–45, 70, 77, 101
cyclophilin, 27

D

DAWDLE (DDL), 77, 90, 97, 149
DCL1, 5, 9, 23–26, 30, 65, 77, 89–91, 97–98, 149
Dicer-like proteins, 6, 55, 127
DNA methylation, 3, 9, 119, 156

E

EasiRNA, 3
Ethylene, 40, 46–47, 78, 82
ETHYLENE-INSENSITIVE3 (EIN3), 40

G

GLl5 (Glossyl5), 41

H

HD-ZIP (homeodomain-leucine zipper protein), 69
HD-ZIP III, 7, **28**, 37–41, 68, 154
HST(HASTY), 10, 77, 91, 149
HYL1, 2, 10, 24–27, 30, 46, 77, 90–92, 97, 149

I

inverted repeats, 21

J

jasmonic acid, 7, 40, 46, 100, 106, 109, 120, 122

K

KATANIN1, 27
KETCH1, 26
Kyoto Encyclopedia of Genes and Genomes (KEGG), 68

L

laccase, 42, 61, 68, 70, 80, 103
Leaf Curling Responsiveness (LCR), **28**, 37, 135, 138
Leaf Senescence, 40, 47, 78
LEAFY, 7
Lin-4, 3, 20, 89
Lipoxygenase (LOX1) gene, 40, 78, 109

M

MiRNA Active 1 (EMA1), 27
miRNA SPONGE, 129
MITE (miniature inverted-repeat transposable element), 21
mitogen-activated protein kinase (*MAPK*), 26, 68, 78, 83, 180
monothiol glutaredoxin-S12 (*GRXS*12), 82
MYB (*MYELOBLASTOSIS*), 24, **28–29**, 38, 43–44, 58, 67–69, **70**, 78, 82, 100, *102*, **123**, 130, 154

N

NAC1, 7–8, 14, 44, 71, 73, 133
necrosis, 58
next-generation sequencing (NGS), 52, 92, 94, 111–112, 119–120, 150
NODULATION SIGNALING PATHWAY, 109

P

pectin methyl esterase inhibitors (*PMEI*s), 81
PEPC (phosphoenolpyruvate carboxylase), 69
phasiRNA, 3
Phosphatidylinositol, 68
Physcomitrella patens, 3–4, 9, 22
PHYTOCHROMEINTERACTING FACTOR, 47
piRNA (Piwi-interacting RNAs), 20, 27, 29
PPT proteins, 9
PTGS (Post-transcriptional gene silencing), 20

Q

quelling, 21

R

RISC (RNA-inducing silencing complex), 2–3, 6, 8, 9, *25*, 27, 45, 55, 77, 89, 91, 97–98, 128, 149
RNA-induced silencing complex (RISC), 3, 6, 8–9, *25*, 27, 45, 55, 77, 89, 91–92, 97–98, 128, 149, 156, 159
RNA interference, 21, 36, 150

S

shoot apical meristem (SAM), 29, 37, 47
SiRNA, 1, 3, 9–11, 20–21, 23, 30, 39, 42, 45–46, 89, 102,
    105, 111–112, 139, 141, 152
SMEK1, 26
SnRK2 (SNF1-related protein kinase 2), 47
SQN, 27
STABILIZD1 (STA1), 26
SWI/SNF complex, 27

T

target mimicry, 59, 129, 131, 152
TRANSPORT INHIBITOR RESPONSE1 (TIR1), 7–8,
    29, 30, 67, 70, 77–78, 81, 102, 111, 120–121, **122–123**,
    131–132

U

UBC24, 46, 58
ubiquitin-conjugating enzyme (UBC), 46,
    58, 77

V

vernalization, 21

X

XAP5 CIRCADIAN TIMEKEEPER (XCT), 126

Z

zinc-finger nucleases (ZFN), 112

For Product Safety Concerns and Information please contact our EU
representative GPSR@taylorandfrancis.com
Taylor & Francis Verlag GmbH, Kaufingerstraße 24, 80331 München, Germany